D0205690

NUMERICAL METHODS
for CHEMICAL ENGINEERS
Using Excel®, VBA, and MATLAB®

NUMERICAL METHODS
for CHEMICAL ENGINEERS
Using Excel®, VBA, and MATLAB®

VICTOR J. LAW

CRC Press
Taylor & Francis Group
Boca Raton London New York

CRC Press is an imprint of the
Taylor & Francis Group, an **informa** business

CRC Press
Taylor & Francis Group
6000 Broken Sound Parkway NW, Suite 300
Boca Raton, FL 33487-2742

© 2013 by Taylor & Francis Group, LLC
CRC Press is an imprint of Taylor & Francis Group, an Informa business

No claim to original U.S. Government works

Printed on acid-free paper
Version Date: 20130125

International Standard Book Number-13: 978-1-4665-7534-9 (Hardback)

Dedication

To my wonderful family:

Penny (wife)

Preston (elder son)

Sandy (younger son)

Vicki (daughter)

Jo (daughter-in-law)

CeCe (granddaughter)

Blake (grandson)

They make even writing a book worthwhile.

Contents

Preface

This book has been written using notes developed for the course Numerical Methods for Chemical Engineers at Tulane University. The author has written two previous textbooks: one on FORTRAN® programming and one using the language Pascal. On a personal note, when I completed the Pascal book, I asked my wife to *break my fingers* if I ever decided to write another book! Well, that was a long time ago, and having been granted a sabbatical leave to write this book, my wife decided that she would *look the other way*.

While there are many textbooks whose title would indicate that they are suitable for the course Numerical Methods for Chemical Engineers, every one that has been tried has been a failure in one way or another. Either they were too elementary and the applications and problems were not ideal or they did not offer instruction in Excel® and Visual Basic® for Applications (VBA). This led to the development, over a 6-year period, of detailed notes to be used in place of a textbook. These notes have been enhanced and put into textbook form to produce the present book.

The primary reason for using Excel is that it is generally available software, and it comes with every computer system (both PC and Mac) with Microsoft Office® installed. VBA is a programming environment that comes with Excel and greatly enhances the capabilities of basic Excel spreadsheets. It is available on systems running Microsoft operating systems and Mac OS. Beware, however, that VBA is available only on the latest (2011) version of Microsoft Office for the Mac.

Other programming software systems that are often used in chemical and biomolecular engineering numerical methods courses are the following:

- MATLAB®
- Mathematica®
- MathCad®
- C/C++
- FORTRAN
- PolyMath

C/C++ and FORTRAN are compiler-based programming languages. Courses that deal with them must devote large amounts of time to learning the language itself rather than emphasizing problem solving.

The first three examples are programming environments with relatively easy-to-use interfaces. Mathematica and MathCad offer powerful built-in methods for solving many common problem types, and both are particularly suited to symbolic problem solving (such as performing analytical differentiation or integration and solving differential equations). MATLAB is by far the most popular of the "proprietary" packages, and at least two textbooks have been written that combine chemical engineering problem solving with the MATLAB system. A significant difficulty with using MATLAB is that it requires rather expensive licenses. In all likelihood,

MATLAB will not be available to practicing engineers in industry. This adds to the attractiveness of using Excel with VBA.

PolyMath is a specialized software package and has its roots in academia. It is especially suitable for a number of chemical engineering applications, and at least one textbook has been written using PolyMath as the base programming tool.

Obviously, there is no panacea when choosing which programming system to use, and any choice will have both backers and detractors. As a compromise, MATLAB is *introduced* in the last chapter of this text. This introduction is sufficient for students to grasp the basics of MATLAB and how it differs from using Excel and VBA. Also, MATLAB programming is easily mastered by those who know VBA.

The vast majority of problems presented in this text, including in-class examples, homework problems, and exam problems, are related to chemical and biomolecular engineering. Application areas include (but are not limited to)

- Material and energy balances
- Thermodynamics
- Fluid flow
- Heat transfer
- Mass transfer
- Reaction kinetics (including biokinetics)
- Reactor design and reaction engineering
- Process design
- Process control

In the course taught by the author, exams (including most of the final exam) are of the "take-home" variety. It is not practical to give a timed, in-class exam when numerical methods and using a computer are involved. In order to encourage individual work, each student is given a unique set of input data so that no two students are expected to get the same "answers."

In the text, when mentioning a topic for which there is neither time nor space to elaborate, the statement "Google it" appears. This is not a plug for any specific search engine but an easy way for the author to suggest getting more information if the reader's interest is sparked. Another feature is the use of "Did You Know" boxes. These are used to remind about features of Excel that are assumed known.

MATLAB® is a registered trademark of The MathWorks, Inc. For product information, please contact:

The MathWorks, Inc.
3 Apple Hill Drive
Natick, MA 01760-2098 USA
Tel: 508-647-7000
Fax: 508-647-7001
E-mail: info@mathworks.com
Web: www.mathworks.com

Author

In the Fall of 2012, Victor J. Law, PhD, FAIC hE, FICH EME, CE, started his 50th year on the faculty of Tulane University. Dr. Law graduated with a BS (ChE) degree from Tulane in 1960, an MS (ChE) degree in 1962, and a PhD degree in 1963. He joined the faculty of the Tulane School of Engineering (Department of Chemical Engineering) on July 1, 1963. Dr. Law's PhD thesis was in the area of automatic process control. Early on, he taught graduate and upper-level undergraduate courses in process control, transport phenomena, applied mathematics, and applied statistics. He began using computers while still an undergraduate, and his research centered on the use of computers for process control and process simulation. Dr. Law worked for several summers at the Monsanto research campus in St. Louis, Missouri, and was peripherally involved in the development of one of the first "process simulators" called FLOWTRAN. While working at Monsanto, Dr. Law became interested in numerical optimization methods. He worked with Monsanto colleague Dr. Robert H. Fariss on general-purpose software for nonlinear equations, nonlinear regression, nonlinear programming, and constrained nonlinear regression. In 1967, Dr. Law was promoted to associate professor (with tenure) and in 1970 to professor.

In 1973, Dr. Law initiated a program that was to become the Department of Computer Science at Tulane. He was the head of that department from 1979 to 1982. During his tenure in computer science, Dr. Law wrote two textbooks on introductory computer programming (one on FORTRAN77 and another on Pascal). In 1988, Dr. Law returned to the chemical engineering department in order to resume his research career. Since returning to chemical engineering, he has taught classes in process control, transport phenomena, process design, engineering statistics, and numerical methods for chemical engineers. His research has included projects in coastal erosion, methane emissions from rice paddies, thermochemical processes for hydrogen production from water, and butanol production from biomass.

Dr. Law is a fellow of the American Institute of Chemical Engineers; a fellow of the Institution of Chemical Engineers; a chartered engineer in the United Kingdom and Europe, No. 20514794; and a registered professional engineer in the State of Louisiana, No. 10961. He is a member of Tau Beta Pi, Sigma Xi, and Omega Chi Epsilon.

Notes to the Instructor

HISTORY OF AND THE REASON FOR EXCEL®/VBA

The material in this book has been developed over a 6-year period while teaching a class entitled Numerical Methods for Chemical Engineers. The author has taught this class (not continually) for the past 15 years. Early on, the computing platform was FORTRAN® running on a mainframe. At that time, students (as freshmen) took a required course in FORTRAN programming. By the time they took this class (as juniors), they needed considerable refreshing in FORTRAN. The course concentrated on linear algebra and the solution of ordinary differential equations. Many of the problems were generic rather than chemical engineering oriented.

When PCs became prevalent, a switch was made to the MATLAB® platform. Considerable time was spent getting students familiar with MATLAB, but the range of problems was greater because of MATLAB's function availability. However, students returning from summer internships complained that MATLAB was not available at their employer's sites. They wanted a tool that they could use in any setting. The result was to settle on Excel and VBA. There is no panacea; other students who went to graduate school came back for reunions and complained that MATLAB, MathCad, and Mathematica were the popular computing tools at the institutions they attended. It was then decided to add some MATLAB training to the Numerical Methods for Chemical Engineers class. In recent years, almost all example and homework problems have been related to chemical and biomolecular engineering.

Mac Users Beware: The most recent version (2011) of MS Office for the Mac *does* include VBA. Students have reported that the Mac version presents no significant differences from the PC version. In Chapter 3, a *free* software package called Matrix.xla is introduced. It offers a host of matrix-based functions not available directly in Excel. While this package can be downloaded to a Mac, the Matrix.xla functions are not available at the Excel level. They can, however, be utilized in VBA mode. If nothing else, these functions allow students to view very well written code in VBA. At some point, it is hoped that the publishers of Matrix.xla will support its features on the Mac.

MATERIAL AVAILABLE FOR INSTRUCTORS

Files with Excel/VBA or MATLAB programs for all of the examples in the book, any concluding comments for those programs, and solutions to all of the end-of-chapter exercises are available on a DVD from the publisher with a qualifying course adoption. Additionally, solutions to all of the end-of-chapter exercises are provided. When possible, concluding comments are included along with the programs.

HOW THE AUTHOR TEACHES THE CLASS

All classes are held in a departmental computer lab. Usual class size is about 20, and each student has his or her own computer. At the beginning of coverage of a new chapter, a short lecture (at the chalk board—no PowerPoint) presents the highlights of the material. Chapter examples are then shown on a projector connected to a PC. Sometimes the lecture/example sequence is repeated when appropriate. Finally, students are given an in-class exercise to perform—usually one of the end-of-chapter exercises. They are given time to attempt the solution on their own; after a while, they are given "helpful hints" as to how to proceed. If they do not complete the exercise within the allotted class time, they are encouraged to finish on their own. A homework assignment is given that is again one of the end-of-chapter exercises. Students are usually given about 1 week to complete a homework assignment. They are expected to turn in Excel/VBA or MATLAB files along with a Word file if needed. The homework assignments are sent via email to a teaching assistant (TA), who grades the work. The instructor usually spends some time with the TA regarding the assignment and how to grade it.

Exams are all of the "take-home" variety. It is not reasonable to give a programming assignment within the time limits of a typical class. In order to discourage *collaboration*, each student is given his or her own set of data for the exam. In addition, an honor code statement is made at the beginning of the exam document. Students are warned that plagiarism on programs is usually very easy to detect and that the spirit of the honor code will be upheld. The instructor grades all exams. Typically, two exams are given during the semester and often consist of about six problems.

The final exam is usually handed out during the second to last week of the class. The due date of the exam is the day that an in-class final exam would have taken place (this is pre-scheduled by the university). So, students usually have about 10–14 days to work on the final exam. Enterprising students can complete the exam well before other final exams begin (students are encouraged to do this). The final exam usually consists of about 10–12 problems. Again, each student is given his or her own set of data required for the exam problems. Since MATLAB is the last subject covered, there are several MATLAB problems on the final exam.

1 Roots of a Single Nonlinear Equation

1.1 INTRODUCTION

Many engineering problems require the solution of a single nonlinear equation. Such an equation can always be cast into the form

$$f(x) = 0 \tag{1.1}$$

The objective of this chapter is to study methods and learn of Excel® tools for finding the root(s) of a nonlinear equation, that is, for finding x such that $f(x) = 0$.

Simple algebra provides the root for a *linear* equation. However, for more complex (nonlinear or transcendental) equations, it is often the case that no analytical solution is available, or is difficult to obtain, so that numerical methods must be used.

An example of a nonlinear equation is the van der Waals equation of state, which is given by

$$\left(P + \frac{a}{V^2}\right)(V - b) = RT \tag{1.2}$$

where

$$a = \frac{27}{64}\left(\frac{R^2 T_c^2}{P_c}\right)$$

$$b = \frac{RT_c}{8P_c}$$

with subscript c referring to the "critical" values of temperature and pressure for the gas. Note that, if a and b are zero, this reduces to the ideal gas equation of state.

A typical problem is to find the molar volume, V, given the temperature and pressure (and the type of gas). While it is possible to find analytic solutions to the van der Waals equation of state for V (this is simply a cubic equation), a numerical solution is often preferred. The equation of state in the form $f(x) = 0$ is obtained by simple rearrangement:

$$f(V) = \left(P + \frac{a}{V^2} \right)(V - b) - RT = 0 \qquad (1.3)$$

Note that this particular form $f(x) = 0$ is not unique, and other algebraic rearrangements are possible.

1.2   ALGORITHMS FOR SOLVING $f(x) = 0$

Clearly there is a need to find good methods for determining roots. Four such methods are now presented (though *many* more have been developed): fixed-point iteration (direct substitution), bisection, Newton's method, and the secant method. These methods are *iterative*. That is, given a *guess* of a root, or the interval in which a root lies, the algorithm refines that guess repeatedly, obtaining (hopefully) better and better guesses, until a value "close enough" to the true root is found. Also, graphing the equation can add insight into the roots of interest and can provide good initial estimates of the roots for the iterative algorithms. It is recommended to always prepare a graph of the function to give insight into possible solutions.

1.2.1   PLOTTING THE EQUATION

Excel has very good plotting capabilities. Unfortunately, it is not possible in Excel to simply give a command such as plot($f(x)$). It is necessary to produce a list or table of x and $f(x)$ values and to graph the resulting data. This is best illustrated by an example.

Example 1.1: Plotting the Equation

A table of data and an Excel graph of the van der Waals equation for ammonia at 250°C and 10 atm are shown in Figure 1.1.

From the graph, it is easy to see that there is one real root between 0 and 5 (the other roots are complex conjugates). Remember that the van der Walls equation of state is an attempt to model the *nonideality* of the gas. Since the given temperature and pressure are not severe, it is expected that the calculated molar volume from the equation of state should not be greatly different from that predicted by the ideal gas law. From the ideal gas law, the molar volume is 4.29 L/gmol, which is very close to the root shown in the graph.

1.2.2   FIXED-POINT ITERATION (DIRECT SUBSTITUTION)

To apply this method, the equation must be cast into the form

$$x = g(x)$$

or, more generally,

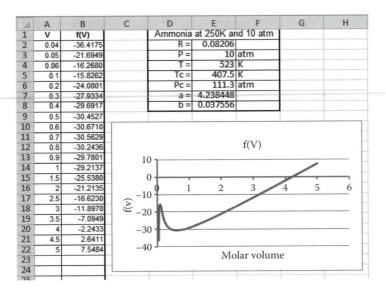

	A	B	C	D	E	F	G	H
1	V	f(V)			Ammonia at 250K and 10 atm			
2	0.04	-36.4175		R =	0.08206			
3	0.05	-21.6949		P =	10	atm		
4	0.06	-16.2680		T =	523	K		
5	0.1	-15.8262		Tc =	407.5	K		
6	0.2	-24.0801		Pc =	111.3	atm		
7	0.3	-27.9334		a =	4.238448			
8	0.4	-29.6917		b =	0.037556			
9	0.5	-30.4527						
10	0.6	-30.6710						
11	0.7	-30.5629						
12	0.8	-30.2436						
13	0.9	-29.7801						
14	1	-29.2137						
15	1.5	-25.5380						
16	2	-21.2135						
17	2.5	-16.6230						
18	3	-11.8978						
19	3.5	-7.0949						
20	4	-2.2433						
21	4.5	2.6411						
22	5	7.5484						
23								
24								

FIGURE 1.1   Roots of the van der Walls equation for ammonia at 250°C and 10 atm.

$$x^{k+1} = g(x^k) \tag{1.4}$$

where k is an iteration counter.

Example 1.2: Direct Substitution

Note that this can always be accomplished by adding x to each side of $f(x) = 0$, if necessary. The van der Walls equation can be cast into the following form:

$$V = b + \frac{RT}{\left(P + \dfrac{a}{V^2}\right)} \tag{1.5}$$

The pertinent data for ammonia are shown in Figure 1.1. If a value for $V = 4$ is guessed (based on the graph of Figure 1.1) and is used on the right-hand side, a new (and hopefully better) value is calculated from Equation 1.5. The iterations produce the following sequence:

```
4.00000  4.21854  4.22946  4.22996  4.22998.
```

The solution is 4.22998 L/gmol. If more significant digits are required, then more iterations can be carried out.

It should be noted that direct substitution can be a *divergent* process. That is, the successively calculated values actually get worse rather than closer to the correct value. Without proof, the following statements apply:

Let g' be the first derivative of the function g in Equation 1.4. Then,

- If $|g'| < 1$, the error will decrease with each iteration.
- If $|g'| > 1$, the error grows at each iteration.
- If $g' > 0$, the error will have the same sign at each iteration.
- If $g' < 0$, the error will alternate signs at each iteration.

Clearly, the equation should be arranged so that the magnitude of g' is less than 1. This might take some experimentation. Often, a form of g is tried, and if the process does not converge, then other forms are attempted.

1.2.3   BISECTION

If it is (somehow) known that a root lies in the interval $[a, b]$, then by simply halving the interval in which the root lies, the interval can be reduced to an acceptable level. This idea is at the heart of the bisection method as shown in Figure 1.2.

The restriction is that $f(a)$ and $f(b)$ *must* have opposite signs—one of them must be positive, the other negative (it does not matter which). Then, because f is assumed to be continuous, it must be a zero somewhere in $[a, b]$. Let c be the midpoint of $[a, b]$. Either c is the root, or the root lies in $[a, c]$ or in $[a, b]$. If $f(c)$ is close enough to zero (see below regarding tolerance), then the root has been found. Otherwise, one pair of $[f(a),f(c)]$or $[f(c),f(b)]$ has opposite signs. Keep the half-interval with opposite signs and discard the other. Repeat the process until either (1) f, evaluated at the midpoint of the interval, is sufficiently small or (2) the interval has been shrunk to a suitably small value.

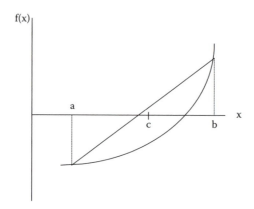

FIGURE 1.2   Bisection: $c = (a + b)/2$.

The term "sufficiently small" is usually tested using a "tolerance," which represents a number small enough to be considered zero based on the application. This can be stated more formally as follows:

If $|f(x)|$<tolerance then (x) is sufficiently small.

Example 1.3: Bisection Applied to the van der Waals EOS

Recall the van der Waals EOS in the form

$$f(V) = \left(P + \frac{a}{V^2}\right)(V - b) - RT = 0 \tag{1.6}$$

The table below shows the progression of applying the bisection method. It uses the Excel IF function to choose whether f(a) or f(b) is replaced by f(c) (and correspondingly whether a or b is to be replaced by c). The syntax for the IF function is as follows (see the explanation of the Excel spreadsheet below for a full explanation of how the IF function is used):

IF(test, True Value, False Value)

In this example (again for ammonia at 10 atm and 250°C), the function in the cell below the label V1 (corresponding to the point a in the nomenclature of the figure) is = IF(F2<0, E2, A2), where cell F3 holds f(Vnew) [corresponding to f(c)], E2 contains Vnew (corresponding to c), and A2 holds V1. So, if f(Vnew) is negative (test is TRUE), V1 (corresponding to a) gets the Vnew value; otherwise, it retains the old V1. Likewise, the formula in the cell below the V2 label is = IF(F2<0,B2,E2), which replaces V2 (corresponding to b) with the value Vnew if the test (F2 < 0) is FALSE. After the first row of formulas has been entered, these cells are copied down to repeat the iterations. Iterations should be repeated until f(Vnew) is sufficiently small (in absolute value). The initial guesses for V are 3 and 5, respectively, which give opposite signs for the function.

	A	B	C	D	E	F
1	V1 (a)	V2 (b)	f(V1)	f(V2)	Vnew (c)	f(Vnew)
2	3	5	-11.8978	7.548387	4	-2.24327
3	4	5	-2.24327	7.548387	4.5	2.641081
4	4	4.5	-2.24327	2.641081	4.25	0.195534
5	4	4.25	-2.24327	0.195534	4.125	-1.02479
6	4.125	4.25	-1.02479	0.195534	4.1875	-0.41485
7	4.1875	4.25	-0.41485	0.195534	4.21875	-0.10971
8	4.21875	4.25	-0.10971	0.195534	4.234375	0.042899
9	4.21875	4.234375	-0.10971	0.042899	4.226563	-0.03341
10	4.226563	4.234375	-0.03341	0.042899	4.230469	0.004744
11	4.226563	4.230469	-0.03341	0.004744	4.228516	-0.01433
12	4.228516	4.230469	-0.01433	0.004744	4.229492	-0.00479
13	4.229492	4.230469	-0.00479	0.004744	4.22998	-2.5E-05

> **Did You Know?:** That there are two ways to copy cell contents down. One way is to select the cells to be copied and then grab the small box in the lower right corner of the selection. Pulling down will copy the cells. The second method is to select the cells to be copied and while holding down the mouse button pull down as far as desired and finally hitting CTRL/d (hold down the CTRL key and hit the d key).

1.2.4 NEWTON'S METHOD

The next algorithm to be considered is the Newton's method for finding roots. Newton's method does not always converge. But, when it does converge, it usually does so very rapidly (at least once it is "close enough" to the root). Newton's method also has the advantage of not requiring a bracketing interval with positive and negative values. So Newton's method allows the solution of equations such as

$$x^2 - 2x + 1 = 0 \tag{1.7}$$

whereas bisection does not. This parabola just *touches* the zero axis and has no negative values.

The basic idea of the Newton algorithm is this: given an initial guess, call it x^1 to a root of $f(x) = 0$, a refined guess, x^2, is computed based on the x-intercept of the line tangent to $f(x)$ at x^1. That is, consider the equation of the line tangent to $f(x)$ at x^1 (this is just the Taylor series expansion of the function ignoring all but linear terms):

$$f(x) = f(x^1) + f'(x^1)(x - x^1) \tag{1.8}$$

This is the point-slope equation of a line, where x^1 is the base point and $f'(x^1)$ is the slope [derivative of $f(x)$ evaluated at x^1]. Solving Equation 1.8 for x at which $f(x) = 0$ gives

$$0 = f(x^1) + f'(x^1)(x^2 - x^1)$$

$$x^2 = x^1 - \frac{f(x^1)}{f'(x^1)}$$

or more generally,

$$x^{k+1} = x^k - \frac{f(x^k)}{f'(x^k)} \tag{1.9}$$

The value of x^{k+1} is the new guess at the root. The process is repeated, computing successively x^2, x^3, x^4,... until an x^K is found at which

$$|f(x^K)| < tol \tag{1.10}$$

where *tol* is a prescribed tolerance.

Example 1.4: Newton's Method Applied to the van der Waals Equation

Newton's method is now applied for ammonia at 10 atm and 250°C with an initial guess for $V = 1$ L/gmol. To begin using Newton's method, the derivative of Equation 1.2 is required and is as follows:

$$f'(V) = \left(P + \frac{a}{V^2}\right) + (V - b)\left(\frac{-a}{V^3}\right) \tag{1.11}$$

Here is an Excel spreadsheet that solves this problem:

V	f(V)	f'(V)
1	−29.21366	10.15918
3.87559	−3.45397	10.00273
4.22090	−0.08875	10.00212
4.22977	−0.00209	10.00210
4.22998	−0.00005	10.00210

The root is found in about 3 iterations. Note the rapid rate of convergence.

An examination of Equation 1.9 for the new iterate x^k reveals a potential for failure of Newton's method, namely,

$$f'(x^k) \approx 0 \tag{1.12}$$

This may lead to a wildly divergent iterative process. There are other possible reasons why this method might not converge. In general, Newton's method is prone to failure, but when it does work, it converges rapidly. Good initial guesses are the key to success.

1.2.5 SECANT METHOD

In Chapter 4, methods for approximating the derivative of a function using finite differences are presented. The secant method uses the idea of finite differences to approximate the derivative in the Newton method formula. Starting with *two* initial guesses x^0 and x^1, which **need not** bracket the root of interest, the approximation to $f'(x)$ can be written as follows:

$$f'(x^1) \cong \frac{f(x^1) - f(x^0)}{x^1 - x^0} \tag{1.13}$$

Or, in general, after k steps or iterations,

$$f'(x^k) \cong \frac{f(x^k) - f(x^{k-1})}{x^k - x^{k-1}} \tag{1.14}$$

Substituting this approximation into the Newton formula (Equation 1.9), the following iteration formula results for the secant method:

$$x^{k+1} = x^k - \frac{f(x^k)}{f(x^k) - f(x^{k-1})}(x^k - x^{k-1}) \tag{1.15}$$

Example 1.5: The Secant Method for van der Waals Equation

The following shows the van der Waals example solved using the secant method. The two initial guesses for V are 1.50 and 1.51—the second was chosen arbitrarily close to the first one. Note that the rate of convergence is about the same as that of Newton's method. Beware that the secant method is subject to the same potential shortcomings of Newton's method.

Secant method				
V0	V1	f(V0)	f(V1)	V2
1.50000	1.51000	−25.53805	−25.45583	4.60602
1.51000	4.60602	−25.45583	3.67996	4.21498
4.60602	4.21498	3.67996	−0.14652	4.22995
4.21498	4.22995	−0.14652	−0.00028	4.22998
4.22995	4.22998	−0.00028	0.00000	4.22998
4.22998	4.22998	0.00000	0.00000	4.22998

1.3   USING EXCEL® TO SOLVE NONLINEAR EQUATIONS (GOAL SEEK)

Any of the methods discussed previously can be implemented quite easily using Excel. However, Excel has *built into it* two tools for solving nonlinear equations, Goal Seek and Solver. Solver is discussed at length in Chapter 9. The *Goal Seeking* feature can solve many single nonlinear equations (good initial guesses are important). Goal Seek uses whatever value is placed in the "By Changing Cell" location as an initial guess. According to Microsoft® documentation on Goal Seek,

> The Goal Seek command uses a simple linear search beginning with guesses on the positive or negative side of the value in the source cell (**By Changing Cell**). Excel uses the initial guesses and recalculates the formula. Whichever guess brings the formula result closer to the targeted result (**To Value**) is the direction (positive or negative) in which Goal Seek heads. If neither direction appears to approach the target value, Goal Seek makes additional guesses that are further away from the source cell. After the direction is determined, Goal Seek uses an iterative process in which the source cell is incremented or decremented at varying rates until the target value is reached. (http://esupport.lenovo.com/mss/mss.pl?doctype=kb&docid=MTAwNzgy)

Therefore, the algorithm used by Goal Seek is somewhat similar to the secant method with some enhancements. If Goal Seek fails, it is usually due to a poor initial guess, so changing the guess might lead to success. To see how Goal Seek works, consider the following simple example:

$$x - x^{1/3} - 2 = 0$$

In the following spreadsheet, cell B1 contains the *initial guess* for the independent variable, x. When a correct value is found for x, then the function $f(x) = 0$ [the formula for $f(x)$ is in cell B2 and is = B1-B1^(1/3) - 2].

Initial spreadsheet:

	A	B
1	x =	0
2	f(x) =	-2

The following is the "Goal Seek" window obtained from the Data/What If Analysis/Goal Seek Menu.

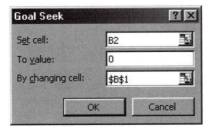

After hitting the OK button on the Goal Seek window, the spreadsheet changes to the following:

	A	B
1	x =	3.521423
2	f(x) =	4.38E-05

The function $f(x)$ is "close" to zero and the solution is shown for x.

Did You Know?: That the F4 key can be used to toggle between the four cell referencing methods. When a $ sign appears before a row or column indicator, the reference is absolute (if the cell contents are copied, the reference with the $ sign is unchanged). If there is no $ sign and a cell's contents are copied to another cell, the reference is relative and changes accordingly. The four cell references are (for cell B3, for example) B3, $B3, B$3, and $B$3.

Example 1.6: Automating Goal Seek

This example gives a very first look at Visual Basic® for Applications (VBA), which is the programming language associated with Excel (as well as all other Microsoft Office applications). The example also introduces the use of Keystroke Macros, which allow the recording of a sequence of keystroke and mouse commands that are recorded for repeated use.

The following spreadsheet applies to the same problem as in previous examples of this chapter. It is desired to compute the molar volume of ammonia at 250°C, but now for a *range* of pressure from 10 to 20 atm (in increments of 1 atm). To do this, Goal Seek can be manually applied for each pressure, but a much more efficient way is shown that involves first recording a Keystroke Macro and then *Editing* the resulting VBA program that was automatically produced by the Macro recorder. When invoking the edited VBA program, the molar volume is found at each pressure automatically. In the spreadsheet, the molar volume from the ideal gas law appears in the second column and is copied to the third column to give a good initial guess of the nonideal volume.

	A	B	C	D
1	R =	0.08206		
2	T =	523	K	
3	Tc =	407.5	K	
4	Pc =	111.3	atm	
5	a =	4.238448		
6	b =	0.037556		
7	P	Videal	V	f(V)
8	10	4.2917	4.2917	0.6034
9	11	3.9016	3.9016	0.6628
10	12	3.5764	3.5764	0.7220
11	13	3.3013	3.3013	0.7810
12	14	3.0655	3.0655	0.8399
13	15	2.8612	2.8612	0.8986
14	16	2.6823	2.6823	0.9571
15	17	2.5246	2.5246	1.0155
16	18	2.3843	2.3843	1.0736
17	19	2.2588	2.2588	1.1317
18	20	2.1459	2.1459	1.1895

The first step is to record a *Keystroke Macro* to find the molar volume (V) at 5 atm. Begin by going to the Developer tab and choosing (in the upper left set of menus) Use Relative Reference. This is necessary so that the recorded Macro works from any proper initial cell. Place the cursor on the f(V) value when P = 10 (the current value is 0.6034). Next choose Record Macro and the following window appears. The name of the Macro (FindVolume) and the Shortcut Key (v) were chosen to be appropriate for the problem.

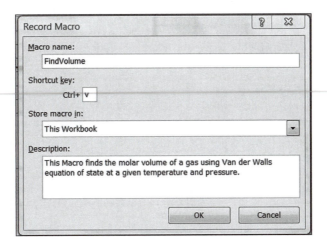

After clicking the OK button, invoke the `Data/What-If-Analysis/Goal Seek` menu and select the appropriate cells so that the volume for the first pressure is calculated. Go back to the `Developer` menu and select `Stop Recording`. The recorded Macro could now be used manually at each pressure by placing the cursor over the next f(V) and hitting Ctrl/v (hold down Ctrl and hit v). However, the Macro can be Edited to *make it work repeatedly* until the data are exhausted.

By choosing `Developer/Macros/FindVolume/Edit`, the following VBA code appears:

```
Sub FindVolume()
'
' FindVolume Macro
' This Macro finds the molar volume of a gas using Van der Walls
' equation of state at a given temperature and pressure.
'
' Keyboard Shortcut: Ctrl+v
'
    ActiveCell.GoalSeek Goal:=0, ChangingCell:= _
            ActiveCell.Offset(0, -1).Range("A1")
    ActiveCell.Offset(1, 0).Range("A1").Select

End Sub
```

Note that this code was generated *automatically* during the recording of the Keystroke Macro. The code has been edited somewhat to fit properly on the printed page without altering its accuracy. The lines beginning with apostrophe (') are comments and can be ignored. The first executable line begins with `ActiveCell.GoalSeek`, which invokes the Goal Seek algorithm. The remainder of the line specifies a goal value of 0 and that the "changing cell" is on the same row, one column to the left (0,-1). The underscore at the very end of the first line is a "continuation" marker, which states that information on the next line is part of the current line (but there was no room for it). The second line that involves the word `Offset` is a command to move the cursor down one row but in the same column (1,0).

This code can be altered so that, when invoked, it repeats over and over until the next item in the f(x) column is blank (which also represents a value of 0). The edited code appears below, where the additional code is shaded for emphasis:

```
Sub FindVolume()
'
' FindVolume Macro
' This Macro finds the molar volume of a gas using Van der Walls
' equation of state at a given temperature and pressure.
'
' Keyboard Shortcut: Ctrl+v
'
While ActiveCell.Value <> 0
    ActiveCell.GoalSeek Goal:=0, ChangingCell:= _
                ActiveCell.Offset(0, -1).Range("A1")
    ActiveCell.Offset(1, 0).Range("A1").Select
Wend
End Sub
```

Details concerning the VBA language are covered in Chapter 2. The statement `While ActiveCell.Value <> 0` is a looping command that says, in effect, perform all statements below this until the `Wend` is encountered as long as the value in the `ActiveCell` (the one where the cursor is) is not zero (or not blank). When this revised Macro is invoked by typing Ctrl/v, the spreadsheet changes to that shown below. As expected, the difference between ideal and non-ideal volume becomes larger as pressure increases.

P	Videal	V	f(V)
10	4.2917	4.2300	0.0001
11	3.9016	3.8398	0.0001
12	3.5764	3.5146	0.0001
13	3.3013	3.2394	0.0002
14	3.0655	3.0036	0.0003
15	2.8612	2.7991	0.0004
16	2.6823	2.6203	0.0005
17	2.5246	2.4624	0.0006
18	2.3843	2.3221	0.0007
19	2.2588	2.1966	0.0009
20	2.1459	2.0835	0.0000

The graph shown in Figure 1.3 illustrates the difference between the molar volumes for ammonia at 250°C computed by the ideal gas law and by the van der Waals equation of state.

Example 1.7: Fraction Vaporized of a Hydrocarbon Mixture

Figure 1.4 gives data for a mixture of four hydrocarbons. Included in the data are constants (A, B, and C) for the Antoine equation, a correlation that allows the calculation of the vapor pressure of pure components as follows:

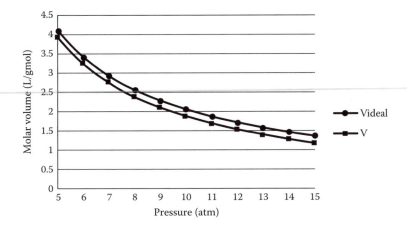

FIGURE 1.3   Comparison of ideal gas and van der Waals molar volume.

	Ethylene	Ethane	Propane	n-Butane
A	3.86690	3.93264	3.97721	3.84431
B	584.146	659.739	819.296	909.65
C	−18.307	−16.719	−24.417	−36.146
z	0.10	0.30	0.40	0.20

FIGURE 1.4   Feed mole fraction and antoine coefficients. z is the mole fraction of each component in the feed. Antoine constants are from the NIST Chemistry Database.

$$\log_{10} P_i^* = A_i - \frac{B_i}{C_i + T} \qquad (1.16)$$

where
   P_i^* = vapor pressure of component i (atm)
   T = temperature (K)
   A_i, B_i, C_i = Antoine coefficients

It is desired to find the fraction of the mixture in the vapor phase when the mixture is flashed at 60°C over a pressure range of 18, 19, ..., 40 atm. A plot of the fraction vaporized versus pressure is also desired.

Consider the schematic of a flash tank as shown in Figure 1.5.

F is the molar feed rate, and z is the mole fraction of each component in the feed; V is the molar vapor rate, and y is the mole fraction of each component in the vapor; L is the molar liquid rate, and x is the mole fraction of each component in the liquid.

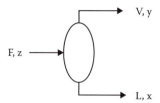

FIGURE 1.5   Schematic diagram of a flash tank.

Assuming an ideal mixture (where Raoult's law applies), the following equilib-rium expression applies:

$$k_i = \frac{y_i}{x_i} = \frac{P_i^*}{P}; i = 1, 2, \ldots, n_c \tag{1.17}$$

where k_i is called the equilibrium constant for component i, P is the total pressure, and n_c is the number of components.

The material balance equations for the flash tank can then be written as follows:

$$z_i F = x_i L + y_i V; \quad i = 1, 2, \cdots, n_c$$

$$F = L + V$$

$$\sum_{i=1}^{n_c} x_i = \sum_{i=1}^{n_c} y_i = 1 \tag{1.18}$$

Let $\alpha = V/F$, the fraction of the feed flashed to the vapor phase. These equa-tions can be manipulated (see Henley and Rosen, p. 341) into the following single nonlinear equation in α:

$$f(\alpha) = \sum_{i=1}^{n_c} (x_i - y_i) = \sum_{i=1}^{n_c} \frac{z_i (1 - k_i)}{1 + \alpha(k_i - 1)} \tag{1.19}$$

Once α has been determined, the liquid and vapor mole fractions can be found from

$$x_i = \frac{z_i}{1 + \alpha(k_i - 1)}$$

$$y_i = k_i x_i \tag{1.20}$$

The following Excel spreadsheet shows the setup for $T = 60°C$ and $P = 18, 19, \ldots,$ 22 atm. Initial guesses for Alpha are entered as 0.5, which is a reasonable first estimate since the result must be between 0 and 1. The first of Equation 1.19 is used for f(Alpha).

	A	B	C	D	E	F	G	H	I	J	K	L	M	N	O
1	Temp (C)=	60													
2		Ethylene	Ethane	Propane	n-Butane										
3	A	3.86690	3.93264	3.97721	3.84431										
4	B	584.146	659.739	819.296	909.65										
5	C	−18.307	−16.719	−24.417	−36.146										
6	z	0.10	0.30	0.40	0.20										
7															
8	P (atm)	k1	k2	k3	k4	x1	x2	x3	x4	y1	y2	y3	y4	f(Alpha)	Alpha
9	18	5.694	3.903	1.167	0.335	0.030	0.122	0.369	0.300	0.170	0.478	0.431	0.100	−3.577E-01	0.5000
10	19	5.394	3.698	1.105	0.317	0.031	0.128	0.380	0.304	0.169	0.472	0.420	0.096	−3.146E-01	0.5000
11	20	5.124	3.513	1.050	0.301	0.033	0.133	0.390	0.307	0.167	0.467	0.410	0.093	−2.735E-01	0.5000
12	21	4.880	3.346	1.000	0.287	0.034	0.138	0.400	0.311	0.166	0.462	0.400	0.089	−2.342E-01	0.5000
13	22	4.658	3.194	0.954	0.274	0.035	0.143	0.409	0.314	0.165	0.457	0.391	0.086	−1.965E-01	0.5000

Goal Seek can be used to find α at all of the pressures individually, or a Keystroke Macro can be recorded and edited as in Example 1.6. The result of having done this is shown in the next spreadsheet, again for pressures of 18, 19, ..., 22 atm. The graph in Figure 1.6 summarizes the final result for all pressures in the range 18–40 atm.

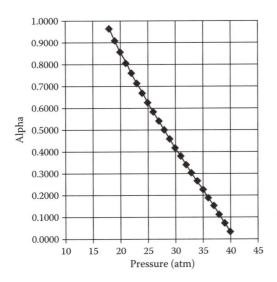

FIGURE 1.6 Results for Example 1.7. Fraction vaporized for a hydrocarbon mixture.

	A	B	C	D	E	F	G	H	I	J	K	L	M	N	O
1	Temp (C)=	60													
2		Ethylene	Ethane	Propane	n-Butane										
3	A	3.86690	3.93264	3.97721	3.84431										
4	B	584.146	659.739	819.296	909.65										
5	C	−18.307	−19.719	−24.417	−36.146										
6	z	0.10	0.30	0.40	0.20										
7															
8	P (atm)	k1	k2	k3	k4	x1	x2	x3	x4	y1	y2	y3	y4	f(Alpha)	Alpha
9	18	5.694	3.903	1.167	0.335	0.018	0.079	0.345	0.558	0.103	0.308	0.402	0.187	−4.136E-05	0.9647
10	19	5.394	3.698	1.105	0.317	0.020	0.087	0.365	0.528	0.108	0.321	0.403	0.167	−4.149E-04	0.9094
11	20	5.124	3.513	1.050	0.301	0.022	0.095	0.384	0.500	0.113	0.334	0.403	0.151	−3.460E-04	0.8584
12	21	4.880	3.346	1.000	0.287	0.024	0.104	0.400	0.472	0.118	0.347	0.400	0.135	−7.712E-04	0.8075
13	22	4.658	3.194	0.954	0.274	0.026	0.112	0.414	0.447	0.123	0.359	0.395	0.122	−3.020E-04	0.7604

1.4   A NOTE ON IN-CELL ITERATION

Suppose that in a spreadsheet the following formula is entered into cell B1:

$$= B1^{\wedge}(1/3)+2$$

Since cell reference B1 is in the formula of the same cell, this is called a "circular reference." Normally, Excel will complain about this. However, if the *Enable Iterative Calculations* (see File/Options/Formulas) box is checked, Excel will immediately do the following:

1. It will automatically use an *initial guess of zero* (there is no control over this).
2. It will iterate up to 100 times or until successive values are within 0.001 (these values can be changed).
3. The final result will be the answer (hopefully).

In-cell iteration is not to be encouraged when robustness is a goal. It often fails to find a solution.

EXERCISES

Exercise 1.1: For the following functions, graph the function and then use Goal Seek to find the root(s).
  a. $f(x) = x - x^{1/3} - 2$
  b. $f(x) = x \tan x - 1$
  c. $f(x) = x^4 - e^x + 1$
  d. $f(x) = x^2 e^x - 1$

Exercise 1.2: Find the roots of the functions given in Exercise 1.1 using the bisection method. Use the graph of each function to choose points that bracket the root of interest.

Exercise 1.3: Set up a spreadsheet that implements the secant method and then solve each of the problems from Exercise 1.1. Use the graph of each function to select an initial guess. Recall the iteration formula for the secant method:

$$x^{k+1} = x^k - \frac{f(x^k)}{f(x^k) - f(x^{k-1})}(x^k - x^{k-1})$$

Hint: Set up the first row of the spreadsheet for your problem with headings such as the following:

xk-1	xk	f (xk-1)	f (xk)	xk+1

Put the formula for the function under the headings f(xk-1) and f(xk). In the cell under xk+1, put the secant method iteration formula. In the second row, replace the previous xk-1 with xk and then xk with xk+1. Now copy the two formulas down one row. At this point, one iteration of the secant method is displayed. To see more iterations, just copy the second row down for as many iterations as desired. If too many iterations are copied and the function difference (the denominator of the iteration formula) becomes exactly zero, a "divide by zero" error will appear.

Exercise 1.4: Use Goal Seek to find root(s) of the following functions. Plot the functions first to obtain an approximation of a desired root.
a. $f(x) = x^3 - 17x + 12 = 0$
b. $J_1(x) = 0$ J_1 is the Bessel function of the first kind of order 1. It can be computed in Excel using the BESSELJ() function. This function has an infinite number of roots; find the root between 2 and 5.
c. Solve for the molar volume of a gas at 400 K and 1200 kPa using the van der Waals equation of state. The critical temperature and pressure are 500 K and 80 atm, respectively. Use the ideal gas solution for your initial guess.
d. Solve the Colebrook equation for the Darcy friction factor, f, for a Reynolds number (N_{Re}) of 10^5 and a roughness factor, ε/D, of 10^{-4} (this equation holds for Reynolds numbers > 4000):

$$\sqrt{\frac{1}{f}} + 0.86 \ln\left(\frac{\varepsilon/D}{3.7} + \frac{2.51}{N_{Re}\sqrt{f}}\right) = 0$$

e. Repeat part d using the same roughness factor, but for a range of Reynolds numbers from 5000 to 30,000 (pick a reasonable increment). Plot the results (friction factor versus Reynolds number). Automate Goal Seek for this.

Exercise 1.5: Use the secant method as described in Exercise 1.3 to find the root(s) of the functions given in Exercise 1.4. Carefully choose the two initial guesses so that the function values have opposite signs. The roots found may or may not correspond to those found using Goal Seek in Exercise 1.3—it depends on the initial guesses.

Exercise 1.6: Solve the problems of Exercise 1.1 using the Newton method.

Exercise 1.7: Repeat Exercise 1.4 parts a and b using the Newton method. The derivative of $J_1(x)$ is given by

$$J_1'(x) = J_0(x) - \frac{1}{x}J_1(x)$$

Exercise 1.8: Repeat Exercise 1.4 using the bisection method.

Exercise 1.9: An additional method for solving single nonlinear equation is the "Regula–Falsi" or method of "false position." Like the bisection method, the false position method starts with two points a_0 and b_0 such that $f(a_0)$ and $f(b_0)$ are of opposite signs, which implies that the function f has a root in the interval $[a_0, b_0]$. The Regula–Falsi method proceeds by producing a sequence of shrinking intervals $[a_k, b_k]$ that always contain a root of f.

At iteration number k, the value

$$c_k = \frac{f(b_k)a_k - f(a_k)b_k}{f(b_k) - f(a_k)}$$

is computed. As explained below, c_k is the root of the secant line through $(a_k, f(a_k))$ and $(b_k, f(b_k))$. If $f(a_k)$ and $f(c_k)$ have the same sign, then set $a_{k+1} = c_k$ and $b_{k+1} = b_k$; otherwise set $a_{k+1} = a_k$ and $b_{k+1} = c_k$. This process is repeated until the root is approximated sufficiently well.

The above formula is also used in the secant method, but the secant method always retains the last two computed points, while the false position method retains two points that bracket a root. On the other hand, the only difference between the false position method and the bisection method is that the latter uses $c_k = (a_k + b_k)/2$.

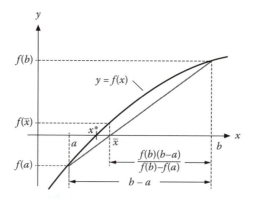

FIGURE 1.7   Regula–Falsi method.

A schematic description of the Regula–Falsi method is shown in Figure 1.7. Repeat either Exercise 1.1 or 1.4 using the method of false position.

Exercise 1.10: Refer to the diagram and data of Example 1.7. The dew point of a vapor is the temperature at a given pressure at which the first drop of liquid is formed. Thus, at the dew point, the ratio of vapor to feed (V/F) is essentially one. Conversely, the bubble point of a liquid is the temperature at a given pressure at which the first bubble of vapor is formed and at this condition $V/F = 0$.

Assume, as in Example 1.7, an ideal mixture (where Raoult's law applies). Also, assume the same mixture and data as shown in Figure 1.4.

a. At the dew point, $V/F = 1$ [there is an infinitesimal amount of liquid, but it *does* exist and has a mole fraction for each component (x_i); however, since there is only an infinitesimal amount of liquid, $V = F$ and $y_i = z_i$]. Applying the fact that the sum of $x_i = 1$, the following equation results:

$$1 - \sum_{j=1}^{n_c} \left(\frac{z_j}{k_j} \right) = 0$$

Since k_j are functions of temperature, this nonlinear equation can be solved for the temperature (dew point).

b. At the bubble point, $V/F = 0$ ($L/F = 1$) and $x_i = z_i$, and if that fact that the sum of $y_i = 1$ is applied, the following equation results:

$$1 - \sum_{j=1}^{n_c} z_j k_j = 0$$

This nonlinear equation can be solved for the temperature (bubble point).

The specific assignment is as follows.

1. Prepare an Excel spreadsheet that calculates for a mixture the bubble point at pressures of 15, 16, ..., 25 atm and produces a plot of the bubble point versus pressure with suitable annotations and title. Use Goal Seek to solve the single nonlinear equation at each pressure. Data for the system are given in Figure 1.4.

   Prepare a *Keystroke Macro* to find the bubble point at each pressure. That is, record the macro and edit it using a While statement so that one keystroke finds all bubble points.

2. Prepare an Excel spreadsheet that is similar to the one for the bubble point, but is used to compute the dew point of the same mixture at the same pressures. Put these calculations on the same Worksheet as those for the bubble point.

3. Transfer the pressure, bubble point, and dew point data to a separate portion of the Worksheet and graph both the bubble and dew points versus pressure. This produces a *phase diagram* for the hydrocarbon mixture. It should look *similar* to the following:

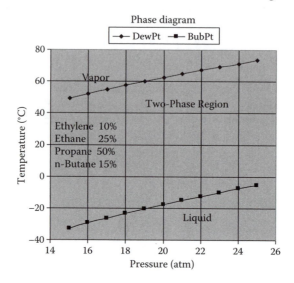

Note that all of the text to fully describe the graph (as in the example plot) must be added manually.

REFERENCES

Henley, E.J., and E.M. Rosen, *Material and Energy Balance Computation*, Wiley, New York (1969).
NIST Data Gateway, Chemistry WebBook. http://webbook.nist.gov/chemistry/.

2 Visual Basic® for Applications Programming

2.1 INTRODUCTION

Visual Basic® for Applications (VBA) is a computer programming language. This short introduction is not intended as a complete course on computer programming. However, programming involves things that engineers are good at: calculating the results of formulas and equations, making logical decisions, and designing algorithms (steps to be used to solve a particular problem). VBA is only one of a host of programming languages, some of which are FORTRAN, C, or C++. The reason for using VBA instead of any other language is that it allows the use of Excel® worksheets for both input and output of data. Excel also makes it easy to put the output data into graphical form. Therefore, VBA provides the Excel system with highly flexible programming capabilities. Furthermore, VBA is *part* of Excel—no licensing is required.

A computer program consists of *code* in the syntax of a specific programming language that implements an *algorithm*. An algorithm can be thought of as a recipe, much like the instruction one sees in a cookbook. If the algorithm can be expressed (e.g., through words or graphically or by any other means), then the algorithm can be *translated* into any programming language with relative ease. For simple algorithms, one can simply begin by writing code in the target language such as VBA; but for complex algorithms, it can be foolhardy and frustrating to try to jump directly to code writing without some means that encourages *algorithm design*. The earliest method for algorithm design was a graphical tool known as a *flowchart*. The method shown here is related to a flowchart but is known as a *structure chart* (Bowles 1979; Law 1983, 1985) since it enforces a high level of structure to the algorithm.

2.2 ALGORITHM DESIGN

Consider, for example, the task of changing a flat tire. The steps involved can be expressed in English as follows (these steps are not unique):

1. Remove the spare and be sure it is not itself flat.
2. If the spare is not flat then proceed; otherwise call AAA!
3. Set the emergency brake and chock the wheels.
4. Remove the jack and set it up in the appropriate place.
5. Loosen the lugs slightly.
6. Jack up the vehicle until the spare no longer touches the ground.

7. Remove the lugs.
8. Remove the flat tire and put on the spare.
9. Put the lugs back on and tighten slightly.
10. Lower the jack all the way.
11. Tighten the lugs snugly.
12. Done!

Another good idea when several sequential steps are involved is to split them into categories (as when preparing an outline for a paper or report). Here is a typical way in which this might be done for the flat tire problem:

1. Get ready.
    a. Remove the spare and be sure it is not flat.
    b. Set the brake and chock the wheels.
    c. Remove the jack and set it up.
2. Remove the flat tire.
    a. Loosen the lugs.
    b. Jack up the car.
    c. Remove the lugs.
    d. Remove the tire.
3. Put on the spare.
    a. Place spare on wheel.
    b. Slightly tighten the lugs.
    c. Lower the jack.
    d. Tighten the lugs.
4. Finish up.
    a. Place the flat tire in the trunk.
    b. Take the spare to be repaired.

Writing out instructions in this manner is fine as long as the steps are consecutive (sequential), do not require a decision (although step 2 involves a simple go/no-go decision), and do not require repetition (doing something over again several times).

Shown in Figure 2.1 is the algorithm expressed as a structure chart.

Comparing the outline form of the algorithm and the structure chart indicates that the sequence of operations is "left to right, depth first." That is, the top (or root) node simply gives a title for the algorithm. Control passes down to "Get ready" then "Remove and check spare,"..., "Set up the jack," then "Remove flat," and so forth. Studies have shown that most algorithm designers work better using a two-dimensional tree rather than a linear outline (or using code directly).

Most algorithms can be expressed using the following logical structures:

- Sequence (as in the change flat example of Figure 2.1)
- Decision making (ask a question and take different actions depending on the answer)
- Repetition (performing similar or identical tasks over and over again)

Shown in Figure 2.2 is a structure chart for the tire problem that includes all three of these types of operations.

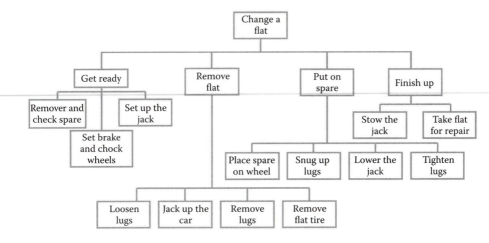

FIGURE 2.1 Structure chart algorithm design for changing a flat tire.

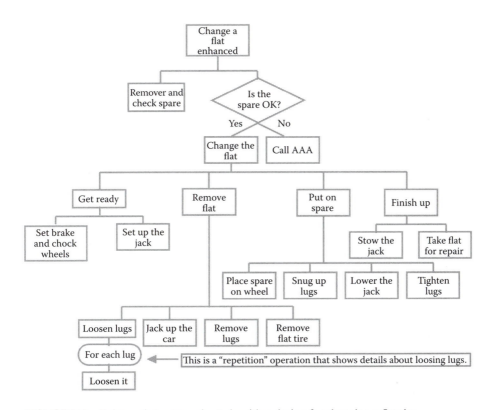

FIGURE 2.2 Enhanced structure chart algorithm design for changing a flat tire.

A "diamond" shape indicates a *decision* and the lines emanating from it are labeled Yes/No or True/False. The "oval" shape indicates *repetition*. A similar repetition operation could be added each time the lugs are manipulated, but only one repetition is shown in order to conserve space.

2.3 VBA CODING

Now that the basic idea about designing an algorithm has been introduced, it is appropriate to look at how to translate the structure chart into VBA code that can be executed from within Excel. VBA is a programming language similar to (but in some instances different from) Visual Basic. The differences in the two languages can be subtle (and frustrating). A simple way to learn VBA is to construct example programs and study them. The process of creating a simple example will be "walked through," and it will be seen how that, at any time, *help* can be obtained from within the VBA Editor. The biggest problem when getting help is to "weed out" the myriad of topics that are likely to come up. That is, there is no lack of help, but it takes a bit of practice to know where to look for what.

Here is a list of VBA topics that are covered (at least partially) in this first example:

- Data types (especially Integer, Double)
- Declarations (especially Dim, ReDim)
- Variables and their declaration
- Strings
- Constants
- Operators
- Expressions
- Statements
- Procedures
- Control flow
  - Conditional
  - Repetition

2.4 EXAMPLE VBA PROJECT

By "project" is meant a VBA program interacting with an Excel spreadsheet. This first example is *not* one that would actually require a VBA program since it is too simple. However, it shows the basics of

- Retrieving data from an Excel spreadsheet
- Performing operations on the input data
- Outputting data to the Excel spreadsheet

To get started, open a blank Excel spreadsheet. The first thing a programmer should do is give the "specifications" for the program. The specifications take the form of a User's Manual. It is good programming practice to write the specifications *first* (before even thinking about the algorithm or the code).

Example Program 2.1: Averaging Numbers

SPECIFICATION

This program takes numeric data from an Excel spreadsheet and calculates the average of the numbers. The input data look similar to the following:

	A
1	Numbers to be Averaged
2	1
3	2
4	3
5	

The user must supply a list of numbers to be averaged in a vertical column ending with a "blank" cell. It calculates the average of the numbers and outputs the average on the spreadsheet row after the blank cell. For the sample data shown, the final appearance of the spreadsheet is as follows:

	A	B
1	Numbers to be Averaged	
2	1	
3	2	
4	3	
5		
6	Average =	2

Note: Before the program is executed, the cursor (selected cell) must be the one with the text "Numbers to be Averaged" before running the program.

ALGORITHM DESIGN

The next step is to "design" the program logic (algorithm) and then (finally) to actually write the code. A structure chart for the program logic is given in Figure 2.3. There are many ways in which the desired operations can be carried out; the steps shown are merely one way in which to accomplish the desired steps.

While designing the algorithm, several variable names have been chosen (invented). These names are chosen by the programmer and should, in some way, be descriptive of the data that they represent. A description of the chosen variable names is as follows:

```
ActRow        An integer to keep tract of the row number
              where data are stored in the flowsheet
Sum           A floating point (double precision) variable
              to hold the sum of the input numbers
NumNumbers    An integer variable representing how many
              numbers
              there are
InputNumber   A floating point number holding one input data
              value
Average       A floating point value representing the average
              of the input data values
```

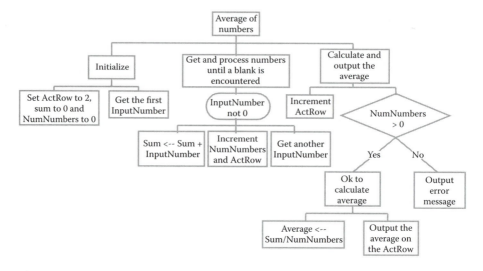

FIGURE 2.3   Algorithm for averaging numbers.

CODING IN VBA

Now that the algorithm has been sketched out in some detail, it is time to implement it in VBA code. Note that the "oval" shape indicates repetition: continue to get input numbers from the spreadsheet until a 0 or blank is encountered (note that a blank is interpreted as zero). The backward arrow (<––) indicates assignment: the entity on the right is computed and placed in the entity on the left. Again, a diamond shape represents a decision block.

To create a VBA program from within the Excel environment, go to Developer/ Macros. A screen like the following appears.

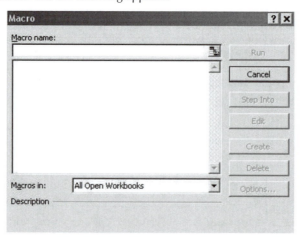

Type a descriptive name, such as CalcAverage into the Macro name field. The Create indicator will activate; hit Create. This opens the VBA Editor and that screen will look similar to the following. Note that the program or macro is automatically called a Sub (for Subroutine) (similar to a procedure or function in C). For now, all of the information to the left of the VBA Editor can be ignored.

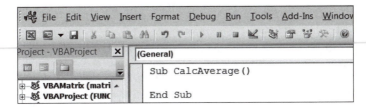

The programmer is presented with a "template" for writing the code. The following is a listing of the code written for this problem. It follows carefully the algorithm design. Any text after an apostrophe (') is a *comment*.

```vba
Option Explicit
Option Base 1
Sub CalcAverage()                 'this subroutine has no parameters

Dim InputNumber As Double     'a number obtained from the spreadsheet
Dim NumNumbers As Integer     'the number of numbers in the list
Dim Sum As Double             'the sum of numbers in the list
Dim Average As Double         'the average of the numbers
Dim ActRow As Integer         'a pointer to the row in the spreadsheet
                              'where data are being input or output
'Make the current Worksheet the selected one:
    Worksheets(ActiveSheet.Name).Activate
'Note: this is a "good housekeeping" command for more complex
'           situations where there could be several Worksheets

'Initialization
ActRow = 2                        'numbers start on row 2.
Sum = 0
NumNumbers = 0

'get the first number; Cells(i, j) gets the contents of that cell
InputNumber = ActiveSheet.Cells(ActRow, 1)

'keep getting numbers until a zero (blank) is encountered
While InputNumber <> 0
  Sum = Sum + InputNumber          'add latest number to the sum
  NumNumbers = NumNumbers + 1      'increment how many numbers
  ActRow = ActRow + 1             'increment the active row
  InputNumber = ActiveSheet.Cells(ActRow, 1) 'get the next number
Wend                               'this indicates the end of the While loop

ActRow = ActRow + 1
If NumNumbers > 0 Then             'begin an If-Then-Else statement
  Average = Sum / NumNumbers       'calculate the average
  ActiveSheet.Cells(ActRow, 1) = "Average = " 'text for the first column
  ActiveSheet.Cells(ActRow, 2) = Average     'average goes to the 2nd col.
Else                               'If no numbers give a message
  ActiveSheet.Cells(ActRow, 1) = "No input numbers to average"
End If                             'this ends the If-Then-Else

Exit Sub                           'another "housekeeping" statement

End Sub
```

Here are some things to notice about this program:

- The first two items are *compiler options*. The `Option Explicit` makes it imperative that the data type of all variables be declared (using a `Dim` statement). The `Option Base 1` makes any arrays (there are none in this first program) start with a subscript of 1.
- All variables *must* be declared in a `Dim` statement.
- Variable names can be any length with a mix of upper and lower case; all names are case sensitive. The first character of a variable name must be alphabetic.
- Any text beginning with an apostrophe is a comment.
- The *built-in* function `Cells(i, j)` refers to the i,jth cell in the spreadsheet, where i is the row number and j is the column number. That is, the spreadsheet is treated as a matrix.
- Assignment statements use an equal sign (=).
- The `While` loop continues as long as the logical (Boolean) statement is `True`; the last statement in the loop range is the `Wend`.
- The `If-Then-Else-End If` allows one kind of action if the logical statement is `True` and another if the logical statement is `False`.
- Character strings are delimited by quote marks as in "No input numbers to average." Remember, anything after an apostrophe is a *comment* and is not part of the code.

To execute the program, go to File/Close and Return to Microsoft Excel®. Enter the data to be averaged (the numbers *must* be in the first column) and then go to Developer/Macros/CalcAverage/Run. The program executes and displays the average of any numbers entered. If no numbers were entered, the message "No input numbers to average" is displayed. While at the Developer/Macros/CalcAverage/Run menu, Options can be chosen, which gives an opportunity to assign a Ctrl key (also called a "hot key") to the Macro (VBA Program). After this, Ctrl+(selected key) can be used to run the program. Note that the selected key is case sensitive. The lowercase letter a was used here.

Another way to run the program is to create a button on the spreadsheet to do so. Select Developer/Controls/Insert and a pop-up box appears. Select the "button-button" (first row, first column) and the cursor turns into a cross-hair. Use this cursor to draw a button on the spreadsheet (be sure to drag the cross-hairs cursor across an area—do not just click on a cell). A window appears and the button macro can be named (Here, RUN was used). Then hit Record (no name need be given here). Hit OK at the next window. Simply type Ctrl+a (or execute the program the "long way" via the Developer/Macros/CalcAverage/Run). When the program executes, hit the Stop Recording button. If the name on the button does not change, right click on it and select Change Text and put RUN in manually. Here is what the new spreadsheet looks like (note that different data have been entered):

	A	B	C	D	E
1	Numbers to be Averaged		Type CTRL-a to run the program.		
2		20			
3		33			
4		55			
5		66			
6		89			
7		32		RUN	
8		45			
9		35			
10					
11	Average =	46.875			

When testing a program, data should be entered that exercises all logic of the program. In this instance, data should be entered that produces the error message as in the following:

	A	B	C	D	E
1	Numbers to be Averaged		Type CTRL-a to run the program.		
2					
3	No input numbers to average		RUN		
4					
5					
6					
7					

Since no numbers were entered, the error message was output. Note that without this safeguard, a divide by zero would be attempted, which causes a system error.

2.5   GETTING HELP AND DOCUMENTATION ON VBA

While in the VBA Editor, much can be learned about the language by clicking the question mark (?). Follow the choices until getting the VBA Language Reference, and there will be a number of choices, some of which will not mean much until later. Another way to learn about VBA is by recording a Keystroke Macro and then looking at the code (recall the exercise when a While Loop was inserted into a recorded Macro involving Goal Seek).

Click on any of the items listed to get detailed information. Ones of probable particular interest are the following:

- Constants
- Data types
- Groups
- Operators
- Statements

It might be helpful to peruse the available Functions as well as other Help topics. A word of warning: there is much more information available than many people will want to investigate.

2.6   VBA STATEMENTS AND FEATURES

In Example 2.1 that calculates the average of a list of numbers, many statements and features of VBA have been introduced informally. The following VBA components are now discussed in some detail:

- Assignment statement
- Expressions
- Object-oriented programming (OOP) and the properties of objects
- Built-in functions (Abs, Exp, etc.)

- Program control
  - Branching (If-Then-Else)
  - Looping (For...Next, While-Wend)
- Data types (Integer, Long, etc.)
- Subroutines and functions
- Objects and methods
- Getting data from a worksheet
- Putting data onto a worksheet
- Alternative I/O methods
- Arrays in VBA (static and dynamic arrays)

2.6.1 ASSIGNMENT STATEMENT

The general form of the assignment statement is

```
Variable = Expression
```

Variable names must begin with an alphabetic character and can contain digits (0...9) and the underscore character (_). Variable names can be of any length, and good practice is to mix uppercase and lowercase letters to make the names self-documenting. For example, SumOfNumbers takes on obvious significance, whereas sumofnumbers is not so obvious. Variable names cannot be the same as a VBA "keyword." There are hundreds of these keywords, and this can be problematic.

2.6.2 EXPRESSIONS

There are many different types of expressions in VBA. Among these are numerical, logical (Boolean), and character string expressions.

2.6.2.1 Numerical Expressions

The most common type is a numerical expression involving the algebraic operators as follows:

^	Exponentiation
*	Multiplication
/	Division
+	Addition
−	Subtraction

Parentheses can be used (and should be used liberally) to reflect the exact hierarchy of operations. The operational hierarchy is the same as for standard algebraic expressions as follows:

1. Parentheses
2. Exponentiation
3. Multiplication/division
4. Addition/subtraction

When operations are at the same hierarchical level, they are performed from left to right.

2.6.2.2 Logical (Boolean) Expressions

The result of a logical expression is either True or False (these are two VBA "keywords"). Boolean expressions are composed of comparative sub-expressions and/or logical operators. Comparative operators are <, < =, >, > =, =, < > and the logical operators are And, Or, and Not (these are VBA keywords). Here are a few examples:

`a < b And c = a`	In the operator hierarchy, comparative operators come before logical operators, so this is the same as the next one. Putting parentheses around the comparatives makes the meaning clear and this practice is encouraged.
`(a < b) And (c = a)`	The expression is True only if both comparisons are `True`.
`(a < b) Or (c = a)`	The expression is True if either or both comparisons are `True`.
`Not (a = b)` Same as `a <> b`.	

2.7 OBJECTS AND OOP

OOP may be seen as a collection of cooperating *objects*, as opposed to a traditional view in which a program may be seen as a list of instructions to the computer. In OOP, each object is capable of receiving messages, processing data, and sending messages to other objects. Each object can be viewed as an independent little machine with a distinct role or responsibility.

A thorough treatment of OOP is far beyond what can or need be covered in this text. It is only important to grasp the basic ideas and know where to get help when it is needed.

Some examples of VBA objects are the Workbook object, the Worksheet object, the Chart object, and the Range object. To get an alphabetic list of objects, when in the VBA editor, click the question mark and then select Microsoft Visual Basic Reference/Objects.

Objects have *methods*, which are invoked in the traditional syntax of OOP using a period separator. To get an alphabetic list of methods, when in the VBA editor, click the question mark and then select Microsoft Visual Basic Reference/Methods.

As an example, the following is a commonly used invocation of an `object.method` command:

```
Worksheets(ActiveSheet.Name).Activate
```

The object `Worksheets(ActiveSheet.Name)` takes on the identity of the active worksheet (the one that is currently showing when the Excel window is active).

The `Activate` method tells the current VBA program that any further references to an object (`Range`, for example) refer to cells of the active worksheet. Effectively, this statement makes the current worksheet the "active" one. For Excel files with multiple worksheets, this can be an important distinction. A VBA program with this statement in it can then be invoked from any of the worksheets, and all references to, for example, `Cells(I, J)` refer to objects within the worksheet from which the VBA program is executed. In other words, if the VBA program was created while `Sheet 1` was active and later it is used when `Sheet 2` is active, then `Cells(I, J)` refers to cells on `Sheet 2`. Without the `Activate` command, the cells in Sheet 1 would be referred to by default.

Here is another example:

```
Range("A1:E150").Sort "Last Name", xlAscending
```

This says sort the data contained in the range `A1:E150` in ascending order using as the sort key the values in the column headed by the label `Last Name`. The label `xlAscending` is one of the many VBA built-in constants.

A list of all of the VBA objects and associated "members" can be viewed by clicking on View/Object Browser when in the VBA Editor. The Object Browser looks like this:

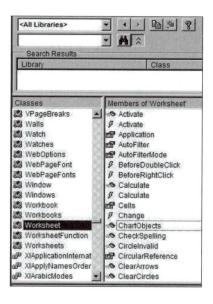

Notice that the `Worksheet` object is selected, so its members appear in the right-hand column (and among these members is `Activate`). There are two `Activate` entries: one is a "subroutine" and one is an "event." *Right clicking* on the one with the "lightning bolt" and selecting "Help" gives the following:

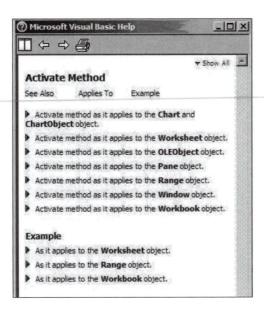

By selecting "Activate method as it applies to the **Worksheet** object," the following window appears:

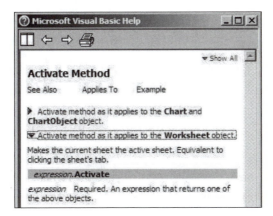

Thus, `Worksheet.Activate` does the same as clicking on the sheet's tab, but it is done under program control instead of with the mouse pointer.

2.8   BUILT-IN FUNCTIONS OF VBA

There are many built-in functions in VBA, but only a small number of them apply to scientific and engineering applications. Table 2.1 shows a list of just a few of them.

TABLE 2.1

Short List of VBA Functions and Their Excel®
Counterparts

Purpose	VBA Function	Excel Function
Absolute value	Abs(x)	ABS(x)
Truncate to integer	Int(x)	INT(x)
Round x to n digits after decimal	Round(x, n)	ROUND(x, n)
Square root	Sqr(x)	SQRT(x)
Exponential	Exp(x)	EXP(x)
Natural log	Log(x)	LN(x)
Base 10 log	–	LOG10(x)
Base b log	–	LOG(x, b)
Value of π	–	PI()
Sine	Sin(x)	SIN(x)
Cosine	Cos(x)	COS(x)
Tangent	Tan(x)	TAN(x)
ArcSine	–	ASIN(x)
ArcCosine	–	ACOS(x)
ArcTangent	Atn(x)	ATAN(x)
ArcTangent (4 quadrant)	–	ATAN2(x, y)
Degrees to radians	–	RADIANS(x)
Radians to degrees	–	DEGREES(x)
Remainder (x modulo y)	x Mod y	MOD(x, y)
Random number	Rnd()	RAND()

To see the entire list, click the question mark for help. Then choose Microsoft Visual Basic Documentation/Visual Basic Language Reference/Functions. The functions are grouped alphabetically.

Also shown in Table 2.1 are the corresponding Excel functions (ones that can be accessed directly from a Worksheet). Note that the VBA functions are capitalized. Perhaps the most confusing one is **Sqr** for taking the square root. The corresponding Excel function is the more usual SQRT.

Excel functions that have no VBA counterpart can be called from VBA. For example, if it is desired to use the Pi() function within a VBA program, the Application method can be used as in the following Function:

```
Function PiCalc()
    PiCalc = Application.Pi()
End Function
```

This code produces the value of π when the Function PiCalc is called.

2.9   PROGRAM CONTROL

Program control statements include those for decision making (branching) and for looping. Many such statements available in VBA are discussed below. Ones less frequently used (or ones that are repetitious) are not covered.

2.9.1   BRANCHING

Two popular branching or decision making statements in VBA are the If-Then-Else and the Select Case statements. Since anything that can be done with the Select Case statement can be implemented just as well using If-Then-Else, Select Case is not presented here.

2.9.1.1   If-Then-Else

The syntax of the If-Then-Else statement is

```
If Logical Expression Then
    statements 1       'this can be any number of statements on separate lines
Else
    statements 2       'this can be any number of statements on separate lines
End If
```

Simpler forms of this statement are possible, but it is easier to always use this form, which is more formally called the Block If statement. The Else clause can be omitted if there is no else "branch" in the program logic. It is good programming practice to indent the statements in each branch. Here is a simple example:

```
IF a < b Then
        c = a + b
Else
        c = a - b
End If
```

The amount of indentation to use is up to the programmer. The simplest thing is to use the Tab key to provide indentation, but at least two space indentation is recommended.

2.9.2   LOOPING

There are four looping statements in VBA (For...Next, Do While...Loop, Do...Loop While, For Each...Next). Only the first two of these are presented since the other two are repetitious (or can be confusing).

2.9.2.1   For...Next

This is a simple "counting" loop and it has the following syntax:

```
For Counter = Start To End [Step Increment]
        Statements
Next [Counter]
```

Counter is a variable name (usually of an integer type). On entering the loop (executing Statements the first time), Counter is initialized to the variable (or expression) Start. After the Statements are executed the first time, Counter is incremented by 1 unless the optional Step Increment is used. For example, using Step 2 means that Counter is incremented by 2 instead of 1. The Increment can be negative so that counting is backward. Looping continues until Statements are executed with Counter having the value of End. Note that if Counter is not an integer type, there can be some questions about its value the last time through the loop. Here is a very typical example:

```
For i = 1 To n
   x(i) = i
Next i
```

In the example, the Counter is the variable i, Start is 1, and End is the value of the variable n. So, i takes on the values 1, 2, ..., n, after which control passes to the statement after the loop.

2.9.2.2 While...Wend Statement

This looping statement executes a series of statements as long as a given condition is **True**. The syntax of this statement is

```
While condition
   statements
Wend
```

If condition (a logical expression) is True, all statements are executed until the Wend statement is encountered. Control then returns to the While statement and condition is again checked. If condition is still True, the process is repeated. If it is not True, execution resumes with the statement following the Wend statement. While...Wend loops can be nested to any level. Each Wend matches the most recent While.

2.10 VBA DATA TYPES

All of the VBA data types are shown in Table 2.2. For floating point numbers, it is best to always use Double, which provides about 15 significant digits. Most modern computers are equipped with floating point processors, so using Double (as opposed to Single) costs essentially nothing. For integers, it is best to use Long (instead of Integer) since integers outside the range +/– 32,767 occur sometimes. Boolean data types can be useful in some applications.

The keyword DIM is used to specify the data type of a variable. When *Option Explicit* is used (as is always recommended), the type of every variable in the program *must* be specified explicitly. Later it will be seen that DIM is also used to specify an Array data structure. Table 2.2 summarizes some of the VBA built-in data types. Only the types most often used are included. A full list of data types can be viewed via the Help system.

TABLE 2.2
VBA Built-In Data Types

Data Type	Storage Required	Range of Values
Boolean (logical)	2 bytes	True or False
Integer	2 bytes	–32,767 to 32,768
Long	4 bytes	–2,147,283,648 to 2,147,283,647
Real (single precision)	4 bytes	–3.402823E+38 to –1.401298E-45 for negatives 1.401298E-45 to 3.402823E+38 for positives
Double (double precision)	8 bytes	–1.79769313486232E+308 to –4.94065645841247E-324 for negatives 4.94065645841247E-324 to 1.79769313486232E+308 for positives
String	1 byte/char.	Delimited by quote marks (").
Variant	16 bytes + 1/char.	Any numeric value up to the range Double or any text.

2.11 SUBS AND FUNCTIONS

A VBA program is automatically a "Sub" or Subroutine. The introductory VBA example program for getting the average of a list of numbers began with

```
Sub CalcAverage()
```

In addition to Subs, VBA also has Function subprograms. Here are the differences between the two:

1. A Sub can receive information (properties) and it can change or set properties.
2. A Function can only receive properties.
3. A Sub is invoked by a "Call" statement.
4. A Function is invoked simply by using its name.

A Sub has the following syntax:

```
Sub name ([arglist])
    Statements
End Sub
```

Things within square brackets are *optional*. The definition of arglist is as follows:

arglist: A list of variables representing arguments that are passed to the Sub procedure when it is called. Multiple arguments are separated by commas.

In its simplest form, `arglist` consists of variable names separated by commas. Consider for example

```
Call Alpha(Beta, Gamma)
```

The name of the subroutine is `Alpha`. `Beta` and `Gamma` are the arguments or parameters (often called the actual arguments) of the subroutine in the calling program. When the subroutine is coded, it might appear as follows:

```
Sub Alpha (Delta, Epsilon)
```

Here, `Delta` and `Epsilon` are what we call *dummy* parameters since they take on information provided by the actual arguments when `Alpha` is called.

The actual parameters can be "sent" to the subroutine in one of two modes:

- By *reference*: What is sent to the subroutine is the memory location of the parameter. Therefore, if the parameter is altered by the subroutine, such changes will be known to the caller when control is passed back at the end of the subroutine (often called "returning" from the subroutine). This is the *default mode* for parameter passing.
- By *value*: If an actual argument is a constant, then it is automatically passed by value (it would be chaotic if the constant 2 suddenly represented the number 3, for example). It is rare that a variable name used as an argument is to be passed by value, but if so, simply enclose it in parentheses as in

```
Call Alpha((Beta), Gamma)
```

Here, only the current value of `Beta` is sent to the subroutine. Using the same dummy variables as previously (`Delta, Epsilon`), any changes made to `Delta` by the subroutine are not transmitted back to the calling program.

2.12   INPUT AND OUTPUT

2.12.1   GETTING DATA FROM THE WORKSHEET

In traditional programming (such as with C++), the source of input data is usually either the keyboard or (usually when there are lots of data) a file. When using VBA, which is an integral part of Excel, the natural source of input is from the Worksheet itself.

In Example Program 2.1, the following statement appeared:

```
InputNumber = ActiveSheet.Cells(ActRow, 1)
```

The `Cells` method treats the spreadsheet as a two-dimensional array representing the rows and columns. If the variable `ActRow` is 3, for example, then the variable

`InputNumber` is assigned whatever value is contained in the cell in the third row and first column. It is possible using the `Range` object to write a statement that inputs values from a collection of cells (but this tends to make things more complex than they need be).

2.12.2 PUTTING DATA ONTO THE WORKSHEET

The procedure for "writing" output data to the active worksheet simply reverses the assignment used for input. Another example taken from the sample program is as follows:

```
ActiveSheet.Cells(ActRow + 1, 2) = Average
```

The value currently stored in the variable `Average` is written into the cell referenced on the left of the assignment.

2.13 ARRAY DATA STRUCTURES

An Array is the equivalent of a subscripted variable. There are one-, two-, or even higher-dimensional arrays. As with scalar variables, arrays are declared in a `Dim` statement as in

```
Dim x(4) as Double
```

In this example, the array x consists of 5 elements: $x(0)$, $x(1)$, $x(2)$, $x(3)$, and $x(4)$. By default, the first subscript is 0. However, starting at zero can lead to confusion, and it is recommended to always include

```
Option Base 1
```

Then subscripts begin with 1, which many find more natural. For those comfortable with subscripts starting at zero, this option need not be used. For all programs in this book use, `Option Base 1` is used.

For a two-dimensional array (matrix), an example declaration is as follows:

```
Dim y (3, 4)
```

The matrix y has 3 rows and 4 columns (assuming subscripts start at 1).

Example Program 2.2: Using an Array

The following is a modified version of the program of Example Program 2.1. Here an Array is used to store the numbers to the averaged. A "For" loop is then used to compute the `Sum` of the numbers. An additional "check" is included to be sure that the number of data items does not exceed 20, which is an arbitrary hard-coded dimension for the array `InputNumbers`.

```
Option Explicit
Option Base 1    'This option makes arrays start with subscript 1

Sub CalcAverage2()              'this version uses an array to store the numbers
Dim InputNumbers(20) As Double  'array of numbers obtained from the spreadsheet
Dim ANumber As Double           'Use this for getting a number from the sheet
Dim NumNumbers As Integer       'the number of numbers in the list
Dim Sum As Double               'the sum of numbers in the list
Dim Average As Double           'the average of the numbers
Dim ActRow As Integer           'a pointer to the row in the spreadsheet
                                'for getting a number or outputting data
Dim i As Integer                'a counter
'Make the current Worksheet the selected one
    Worksheets(ActiveSheet.Name).Activate
'Initialize
ActRow = 2 'number start on row 2 (first cell has text)
Sum = 0
NumNumbers = 0

'get the first number; Cells(i, j) gets the contents of that cell
ANumber = ActiveSheet.Cells(ActRow, 1)
'keep getting numbers until a zero (blank) is encountered
While (ANumber <> 0) And (NumNumbers < 20)  'Note the "check" on NumNumbers!
  NumNumbers = NumNumbers + 1                'increment how many numbers
  InputNumbers(NumNumbers) = ANumber         'store the number in the array
  ActRow = ActRow + 1                        'increment the active row
  ANumber = ActiveSheet.Cells(ActRow, 1) 'get the next number
Wend                                         'end of the While loop
Sum = 0
For i = 1 To NumNumbers           'this is a "counted" loop
  Sum = Sum + InputNumbers(i)     'add each number to the Sum
Next i

ActRow = ActRow + 1
If NumNumbers > 0 Then                  'begin an If-Then-Else statement
  Average = Sum / NumNumbers            'calculate the average
  ActiveSheet.Cells(ActRow, 1) = "Average = " 'text to the first column
  ActiveSheet.Cells(ActRow, 2) = Average      'average to the second column
Else                                    'if no numbers give a message
  ActiveSheet.Cells(ActRow, 1) = "No input numbers to average"
End If                                   'this ends the If-Then-Else
Exit Sub                                 'a "housekeeping" statement
End Sub
```

2.13.1 ARRAY ARGUMENTS

Each of the next two versions of the example program uses a subroutine to do some of the work. The subroutine is named Summer and is declared as follows:

```
Sub Summer(N, Nums, Avg)
```

The "dummy" parameters N, Nums, and Avg do not appear in a Dim statement. They "inherit" the data type and structure of the actual argument when the subroutine is called. The calling statement in the next two examples is

```
Call Summer(NumNumbers, InputNumbers(), Average)
```

The empty parentheses after InputNumbers identify it as an array argument, and the dummy Nums becomes an array structure of whatever length InputNumbers happens to be. These parentheses are not essential but they help to document the array argument.

Example Program 2.3: Using a Subroutine

The same program is changed again, this time using a Sub with arguments to perform the summing and averaging operations:

```
Sub CalcAverage3()              'this version uses a subroutine to calculate
                                'the sum and average
Dim ANumber As Double           'A number obtained from the spreadsheet
Dim InputNumbers(20) As Double  'numbers obtained from the spreadsheet
Dim NumNumbers As Integer       'the number of numbers in the list
Dim Average As Double           'the average of the numbers
Dim ActRow As Integer           'a pointer to the row in the spreadsheet
                                'for inputting a number or outputting data
Dim i As Integer                'a counter

'Make the current Worksheet the selected one
    Worksheets(ActiveSheet.Name).Activate
'Initialize things
ActRow = 2 'number start on row 2 (first cell has text)
NumNumbers = 0

'get the first number; Cells(i, j) gets the contents of that cell
ANumber = ActiveSheet.Cells(ActRow, 1)

'keep getting numbers until a zero (blank) is encountered
While (ANumber <> 0) And (NumNumbers < 20)  'note again the NumNumbers check
   NumNumbers = NumNumbers + 1        'increment how many numbers
   InputNumbers(NumNumbers) = ANumber 'store the number in the array
   ActRow = ActRow + 1                'increment the active row
   ANumber = ActiveSheet.Cells(ActRow, 1) 'get the next number
Wend                                  'this indicates the end of the While loop

ActRow = ActRow + 1
If NumNumbers > 0 Then               'begin an If-Then-Else statement
   Call Summer(NumNumbers, InputNumbers(), Average)  'calculate the average
   ActiveSheet.Cells(ActRow, 1) = "Average = " 'text to the first column
   ActiveSheet.Cells(ActRow, 2) = Average       'average to the second column
Else                                 'if no numbers give a message
   ActiveSheet.Cells(ActRow, 1) = "No input numbers to average"
End If                               'this ends the If-Then-Else

Exit Sub                             'a "housekeeping" statement
End Sub
```

```
Sub Summer(N, Nums, Avg)   'Note that these are "dummy" parameters so
                           'the names used here have no relationship to
                           'those in the calling program
                           'Also, these arguments are not "Dim"ed'
                           '- their type is "inherited" from the calling
                           'program

Dim i As Integer           'These are "local" variables that are
Dim Sum As Double          'not "known" to the calling program.

Sum = 0
For i = 1 To N
   Sum = Sum + Nums(i)
Next i
Avg = Sum / N              'it is KNOWN that N > 0, so no need to check

Exit Sub
End Sub
```

2.13.2   Dynamic Arrays in VBA

It is often the case that it is not known ahead of time how large an array's dimensions need to be. It is wasteful of memory (and poor programming practice) to simply make the dimensions arbitrarily large. To circumvent this problem, VBA provides the `ReDim` statement so that the dimensions of an array do not have to be "hard coded" in the program.

Example Program 2.4: Use of `ReDim`

To illustrate the use of `ReDim`, one more version of the example program is shown. The input data format is changed so that the very first number in the list is not one of the numbers to be averaged, but it indicates how many numbers there are (and thus the required size of the array that holds the numbers). In this version, the dimension of `InputNumbers` is left *blank* (note the empty parentheses), and then the following statement makes the dimension what it needs to be:

```
ReDim InputNumbers(NumNumbers)
```

Another approach that does not require the number of data items known in advance is to again use a `While` loop to detect a zero or sentinel value and put a `ReDim` statement within the `While` loop. When designing a program, it is wise to think of different ways in which to accomplish the same task.

```
Sub CalcAverage4()
'This version uses a subroutine to calculate the sum and average
'And the first cell contains how many numbers are to be averaged
'And it uses ReDim for a "dynamic" array
'And it uses a For Loop instead of a While Loop

Dim ANumber As Double       'A number obtained from the spreadsheet
Dim InputNumbers() As Double 'array of numbers (note the Empty parentheses)
Dim NumNumbers As Integer    'the number of numbers in the list
Dim Average As Double        'the average of the numbers
Dim ActRow As Integer        'a pointer to the row in the spreadsheet
                             'where we are getting a number or outputting data
Dim i As Integer             'a counter
'Make the current Worksheet the selected one
    Worksheets(ActiveSheet.Name).Activate
'Initialize things
ActRow = 2 'number start on row 2 (first cell has text)
NumNumbers = ActiveSheet.Cells(ActRow, 1)    'This must be how many numbers
                                             'follow and are to be averaged
ActRow = ActRow + 1                          'increment the active row
ReDim InputNumbers(NumNumbers)             |  'array is now NumNumbers long
'Get all NumNumbers of numbers
For i = 1 To NumNumbers 'this is a "conditional" loop
  InputNumbers(i) = ActiveSheet.Cells(ActRow, 1)
  ActRow = ActRow + 1                        'increment the active row
Next i                                       'this indicates the end of the For loop

ActRow = ActRow + 1
If NumNumbers > 0 Then          'begin an If-Then-Else statement
  Call Summer(NumNumbers, InputNumbers(), Average)    'calculate the average
  ActiveSheet.Cells(ActRow, 1) = "Average = " 'text to the first column
  ActiveSheet.Cells(ActRow, 2) = Average      'average to the second column
Else                                'if no numbers give a message
  ActiveSheet.Cells(ActRow, 1) = "No input numbers to average"
End If                              'this ends the If-Then-Else

Exit Sub                           'a "housekeeping" statement

End Sub
```

2.14 ALTERNATIVE I/O METHODS

Using the `Cells` method is straightforward and the one that is most often used for input and output. There are other methods worthy of mention, however.

2.14.1 USING RANGE.SELECT

The `Range.Select` combined with the `ActiveCell` object can accomplish essentially the same thing that `Cells` does. Here is an example:

```
Range("b10").Select
v = ActiveCell.Value
```

This inputs into the variable v whatever value is in cell B10. The `.Value` method is actually the default for `ActiveCell` and can be omitted. This technique of input is more unwieldy than the `Cells` method.

2.14.2 USING MESSAGE BOX

The built-in function `MsgBox` can be used to send a message to the user without it appearing in a worksheet cell. Here is an example:

```
MsgBox "This program finds the average of a list of numbers"
```

This pops up a box containing the quoted message overlaid onto the active worksheet.

2.14.3 USING INPUT BOX

The built-in function `InputBox` can be used to both send a message to the user and to obtain data from the user without it appearing in a worksheet cell. Consider the example:

```
a = InputBox("Please enter the first value to be added")
```

A box pops up on the worksheet with the quoted prompt and a place for the user to type in the value requested (in this case, for the variable a).

Example Program 2.5: Using `MsgBox` and `InputBox` for Input/Output

Another version of the example program for finding the average of a list of numbers is shown below. In this case, `MsgBox` and `InputBox` are used for input and output. Note that three consecutive quote marks are required for one quote mark to appear.

```
Option Explicit
Option Base 1
Sub CalcAverage()                  'this subroutine has no parameters

Dim InputNumber As Double    'a number obtained from the spreadsheet
Dim NumNumbers As Integer    'the number of numbers in the list
Dim Sum As Double            'the sum of numbers in the list
Dim Average As Double        'the average of the numbers

'This version uses MsgBox and InputBox for input/output.

'Make the current Worksheet the selected one:
    Worksheets(ActiveSheet.Name).Activate
'Initialize things
Sum = 0
NumNumbers = 0

'Output the purpose of the program

MsgBox """Find average of a list of numbers, hit OK to continue"""

'This version uses InputBox to input numbers

'keep getting numbers until a zero (blank) is encountered
'This version uses a different form of looping statement
InputNumber = InputBox("Enter a number, zero to end ") 'get  first number
While InputNumber <> 0
   Sum = Sum + InputNumber          'add latest number to the sum
   NumNumbers = NumNumbers + 1      'increment how many numbers
   InputNumber = InputBox("Enter a number, zero to end ") 'get  next number
Wend

If NumNumbers > 0 Then                'begin an If-Then-Else statement
   Average = Sum / NumNumbers         'calculate the average
   MsgBox "Average = " & Average 'put this text in the first column
Else                                  'if no numbers give a message
   MsgBox "No input numbers to average"
End If                                'this ends the If-Then-Else

Exit Sub                             'another "housekeeping" statement
End Sub
```

The & symbol is the "concatenation" operator. It appends the value of the variable Average to the preceeding text.

2.15 USING DEBUGGER

The Debugger allows many operations that assist in finding errors in programs or in the input data. When opening an Excel spreadsheet file that contains a VBA macro then going to Tools/Macros and Edit, the Debug menu is among those that appear. By selecting from the Debug menu Step Into, the Sub statement is highlighted (in yellow). The program is now running but under *your* control. To view the Debug Toolbar, go to View/Toolbars and select Debug. The following is what the screen looks like for the CalcAverage4 VBA example:

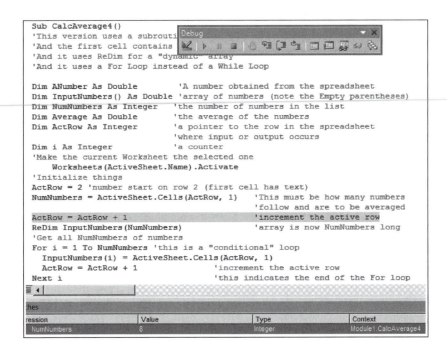

```
Sub CalcAverage4()
'This version uses a subrouti
'And the first cell contains
'And it uses ReDim for a "dynamic array
'And it uses a For Loop instead of a While Loop

Dim ANumber As Double          'A number obtained from the spreadsheet
Dim InputNumbers() As Double 'array of numbers (note the Empty parentheses)
Dim NumNumbers As Integer    'the number of numbers in the list
Dim Average As Double        'the average of the numbers
Dim ActRow As Integer        'a pointer to the row in the spreadsheet
                             'where input or output occurs
Dim i As Integer             'a counter
'Make the current Worksheet the selected one
    Worksheets(ActiveSheet.Name).Activate
'Initialize things
ActRow = 2 'number start on row 2 (first cell has text)
NumNumbers = ActiveSheet.Cells(ActRow, 1)    'This must be how many numbers
                                             'follow and are to be averaged
ActRow = ActRow + 1                          'increment the active row
ReDim InputNumbers(NumNumbers)               'array is now NumNumbers long
'Get all NumNumbers of numbers
For i = 1 To NumNumbers 'this is a "conditional" loop
   InputNumbers(i) = ActiveSheet.Cells(ActRow, 1)
   ActRow = ActRow + 1                 'increment the active row
Next i                                 'this indicates the end of the For loop
```

ression	Value	Type	Context
NumNumbers	8	Integer	Module1.CalcAverage4

To single step through the program, use F8 or the Step tool button, which is the one just to the right of the "Hand" icon in the Debug Toolbar. Here is what the screen looks like after stepping down to the ReDim statement. Here the Debug/Add Watch menu was used to display at the bottom the values of the variables ActRow and NumNumbers.

From the Debug menu, several useful operations can be invoked to aid in tracking down a programming error. For example, a "breakpoint" can be placed on any statement so that when Run (instead of single stepping) is chosen, execution halts when that statement with a breakpoint is reached. At that time, the value of any variable can be viewed. Note that in addition to using the Set Watch feature, values of a scalar variable can be observed simply by placing the cursor over the variable's name.

Using the Debugger can provide a powerful tool in tracking down programming errors and errors in input data. This discussion has only touched on a few features of the Debugger.

Example Program 2.6: Modified False Position

The method of false position is a "hybrid" of the bisection and secant methods. This method assumes two starting points having opposite function signs. A new point is found by drawing a straight line between the bracketing points. One of the old points is *discarded* depending on the sign of the function at the new point. This is depicted in Figure 2.4.

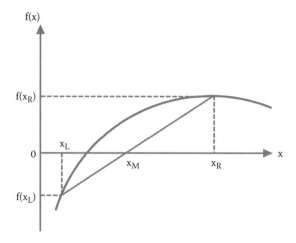

FIGURE 2.4   Modified false position method.

The formula for finding improved values for x using the notation of Figure 2.4 is as follows:

$$x_m = x_L - \frac{(x_R - x_L)f(x_L)}{f(x_R) - f(x_L)}. \qquad (2.1)$$

An inefficiency of the method of false position is that one of the end points tends to become *stagnant*. That is, the end point remains unchanged for several iterations. A modification to overcome stagnation is to detect when it occurs and to take action to avoid it. An effective modification is to simply *halve* the stagnated $f(x)$ value in the update formula. While the modified false position method can be implemented directly in Excel, the logic required is sufficiently complex to make using VBA a better choice. Program specification, algorithm design, coding, and testing are now given where the problem is to find the root between 0.4 and 1.2 of the function

$$f(x) = \tan x - x - 0.5 \qquad (2.2)$$

SPECIFICATION

It is necessary to place on the spreadsheet user interface the two initial values of x that bracket the root. If the initial values do not bracket the root, the program should terminate with an error message. It is desired that at each iteration the values of x_L, x_R, $f(x_L)$, $f(x_R)$, x_m, and $f(x_m)$ be displayed along with an iteration counter. Shown below is one way the spreadsheet might appear before program execution:

	A	B	C	D	E	F	G
1	Iteration	xL	xR	f(xL)	f(xR)	xM	f(xM)
2	1	0.1	1.4				

The program then computes and displays the remaining values for Iteration 1. The update formula, Equation 2.1, is used to compute x_m. Next, either x_L or x_R is replaced by x_m (based on the sign of $f(x_m)$), and a new iteration is performed. When the absolute value of the difference between x_L and x_R is less than a tolerance (1.E-8), the following message is shown:

```
Solution is within tolerance
```

To prevent continued iterations when convergence is very slow, a maximum number of iterations (100) is invoked, and the following error message is displayed:

```
Maximum iteration reached; Solution might not be valid
```

If the initial values of x_L and x_R do not bracket a root, the following error message is displayed:

```
Root is not bracketed, please try again
```

ALGORITHM DESIGN

An "outline" summary of the logic for this program is as follows:

1. Get input data and fill out the first line of the output.
2. Compute x_m and $f(x_m)$.
3. If $f(x_m)$ has the same sign as $f(x_R)$, then replace x_R with x_m and set a repetition counter for x_R to zero while incrementing a repetition counter for x_L.
4. Otherwise, $f(x_m)$ has the same sign as $f(x_L)$ so replace x_L with x_m and set a repetition counter for x_L to zero while incrementing a repetition counter for x_R.
5. Output a line on the spreadsheet.

This process is repeated until the absolute value of the difference between x_L and x_R is less than the tolerance or the maximum number of iterations has occurred (Figure 2.5).

Coding could be attempted using the brief outline summary of the algorithm, but by using a structure chart, even more detail can be displayed and envisioned unambiguously prior to coding. Such a structure chart appears below, accompanied by a list of variable names used within the chart.

DICTIONARY OF VARIABLES

```
Name        Definition
Tol         Tolerance to detect small differences between x
            values
iMax        Maximum iteration before terminating with error
            message
xL          One of the x values
xR          The other of the x values
fxL         f(xL)
fxR         f(xR)
```

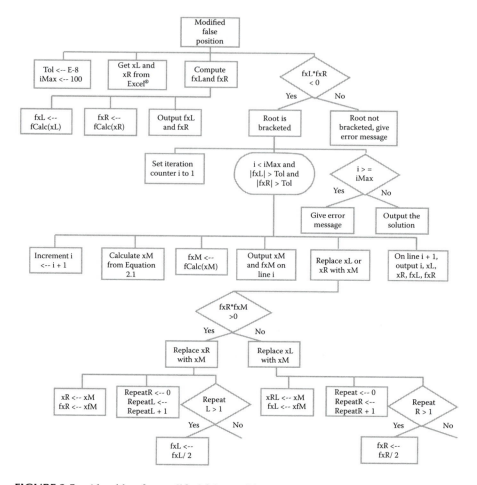

FIGURE 2.5 Algorithm for modified false position.

DICTIONARY OF VARIABLES (cont.)

```
i         Iteration counter and row counter for output
fCalc     Function subprogram that computes the function to
          zero
xM        Updated x value from Equation 2.1
fxM       f(xM)
RepeatR   Repetition indicator for xR
RepeatL   Repetition indicator for xL
```

VBA CODE

Shown below is the VBA code that matches the logic of the structure chart. At the end of the code is the Function subprogram fCalc that provides the value of the function whose root is desired.

```
Option Base 1
Option Explicit

Sub ModFalsePosition()

Dim xL As Double
Dim xR As Double
Dim fxL As Double
Dim fxR As Double
Dim xM As Double
Dim fxM As Double
Dim i As Integer
Dim iMax As Integer
Dim Tol As Double
Dim RepeatL As Integer
Dim RepeatR As Integer

'Make the current Worksheet the selected one:

Worksheets(ActiveSheet.Name).Activate

iMax = 100              'Maximum number of iterations
Tol = 0.000001          'Tolerance for zero

xL = Cells(2, 2)
xR = Cells(2, 3)
fxL = Fcalc(xL)
fxR = Fcalc(xR)
Cells(2, 4) = fxL
Cells(2, 5) = fxR
If (fxL * fxR < 0) Then
  i = 1
  fxM = fxR
  While (i <= iMax And (Abs(fxL) > Tol) And (Abs(fxR) > Tol))
    i = i + 1
    xM = xL - fxL * (xR - xL) / (fxR - fxL)
    fxM = Fcalc(xM)
    Cells(i, 6) = xM
    Cells(i, 7) = fxM
If (fxR * fxM > 0) Then   'they have the same sign
  xR = xM
  fxR = fxM
  RepeatR = 0              'xR is not repeated
  RepeatL = RepeatL + 1 'xL is repeated
  If (RepeatL > 1) Then
    fxL = fxL / 2
  End If
Else                      'fxM and fxL have the same sign
  xL = xM
  fxL = fxM
  RepeatL = 0             'xL is not repeated
  RepeatR = RepeatR + 1 'xR is repeated
  If (RepeatR > 1) Then
    fxR = fxR / 2 '
  End If
End If
Cells(i + 1, 1) = i
```

```
        Cells(i + 1, 2) = xL
        Cells(i + 1, 3) = xR
        Cells(i + 1, 4) = fxL
        Cells(i + 1, 5) = fxR
        Cells(i + 1, 8) = RepeatL
        Cells(i + 1, 9) = RepeatR
     Wend
     If (i >= iMax) Then
        Cells(i + 2, 1) = "Maximum iteration reached;"
        Cells(i + 3, 1) = "Solution might not be valid"
     Else
        Cells(i + 2, 1) = "Solution is within tolerance"
        Cells(i + 3, 1) = "x ="
        Cells(i + 3, 2) = xM
     End If
  Else
     Cells(3, 1) = "Root is not bracketed, please try again"
  End If
  End Sub
```

```
Function Fcalc(x)
  Fcalc = Tan(x) - x - 0.5
End Function
```

TESTING

The spreadsheet after executing the VBA program appears as follows:

	A	B	C	D	E	F	G	H	I
1	Iteration	xL	xR	f(xL)	f(xR)	xM	f(xM)	RepeatL	RepeatR
2	1	0.1	1.2	-0.49967	0.872152	0.50066	-0.4535		
3	2	0.50066	1.2	-0.4535	0.872152	0.739901	-0.32699	0	1
4	3	0.739901	1.2	-0.32699	0.436076	0.937064	-0.07624	0	2
5	4	0.937064	1.2	-0.07624	0.218038	1.005181	0.07012	0	3
6	5	0.937064	1.005181	-0.07624	0.07012	0.972546	-0.00535	1	0
7	6	0.972546	1.005181	-0.00535	0.07012	0.974859	-0.00034	0	1
8	7	0.974859	1.005181	-0.00034	0.03506	0.975154	0.000298	0	2
9	8	0.974859	0.975154	-0.00034	0.000298	0.975017	-1E-07	1	0
10	9	0.975017	0.975154	-1E-07	0.000298			0	1
11	Solution is within tolerance								
12	x =	0.975017							

It can be detected from the output that xR tends to be stagnant. Each time RepeatR is greater than 1, the fxR value is halved in Equation 2.1. This has a marked effect on the rate of convergence. Exercise 2.1 at the end of this chapter involves programming the unmodified false position method. Upon executing that program for the same test problem as shown in this example, the stagnation of xR will be observed.

EXERCISES

Exercise 2.1: Further test the VBA program of Example 2.6 as follows:
   a. Put in two initial points that *do not* bracket the roots. This should produce the appropriate error message.

b. Test the program using different initial points. Some suggested ones are (0.1, 1.5) and (0.01, 1.55).

c. Test the program for the van der Waals problem as given in Example 1.1 of Chapter 1. Choose two initial points from the graph of the function shown in Figure 1.1. Shown below is one way to alter the Function Subprogram Fcalc for this problem:

```
Function Fcalc (V)
Dim R, T, P,  Tc, PC ,a, b As Double

  R = 0.08206
  T = 250
  P = 10
  Tc = 407.5
  Pc = 111.3
  a = 4.238448175
  b = 0.037555537

  Fcalc = (P + a / V ^ 2) * (V - b) - R * T

End Function
```

Exercise 2.2: In order to appreciate the increased efficiency of the modified false position method, redesign and reprogram Example 2.6 so that the original false position method is implemented.

a. Compare the output of the program for the same test function as Example 2.6 and using the same initial points with that of Example 2.6.

b. Repeat part a using (0.01, 1.55) as the initial point. Does the algorithm converge?

c. Repeat this exercise using the van der Waals function as described in Exercise 2.1c.

Exercise 2.3: Write a VBA program (Macro) to automate the process of finding x_L and x_R that give function values with opposite signs. Use the function of Example 2.6 for testing the program. This can be done in a wide variety of ways, but here is a suggested method:

a. Prepare a spreadsheet that looks like the following:

xL	xR	fxL	fxR
0.001	0.005		

The program "reads" the values of xL and xR and computes the associated function (that should be zero at a solution). The program displays these two function values. If the functions have opposite signs, the program terminates; otherwise

b. Use the InputBox feature to prompt for a new xR value. This new value replaces the original one in the spreadsheet and again displays the two function values. This is all displayed on the next line in the spreadsheet. The process continues until the program terminates with function values having opposite signs. Note that xR values less than xL values are possible.

Exercise 2.4: Design and Code a VBA program to perform the following tasks:

a. Generate a list of 20 numbers between 1 and 100 using the Excel function RANDBETWEEN. Select these numbers and copy them (by value) to another column (the reason for this is that RANDBETWEEN updates the list each time an operation is performed in the spreadsheet).

b. Input the list of numbers generated in part a. Any method can be used to indicate the "end of data," such as a blank cell as used in Example 1.1.

c. Compute the following "statistics" for the input numbers:
    1. Average
    2. Standard deviation
    3. Largest in absolute value
    4. Smallest in absolute value

Necessary formulas are as follows:

$$\text{average} = \bar{x} = \frac{1}{n}\sum_{i=1}^{n} x_i$$

$$\text{standard deviation} = s = \sqrt{\frac{1}{n-1}\sum_{i=1}^{n}\left(x_i - \bar{x}\right)^2}$$

Check your results by using the Excel functions AVERAGE, STDEV, MAX, and MIN.

Exercise 2.5: A very inefficient way to compute the value of π is as follows: Consider a circle of radius ½ circumscribed by a unit square (each side is 1 unit). Recall that the equation of a circle is

$$x^2 + y^2 = r^2$$

If N random values are generated for x and y between 0 and ½, the number of times the random coordinates are within the circle ($x^2 + y^2 \le r^2$) can be "tallied." Call this N_{in}. Then the area of the circle can be estimated as

$$A = N_{in}/N$$

It is also known that

$$A = \pi r^2$$

So, π can be estimated as

$$\pi = (N_{in}/N)/r^2 = (N_{in}/N)/0.25 = 4(N_{in}/N)$$

Write a VBA Macro to "read" N from the spreadsheet and output the estimated value of π. Each time the Macro is run, a different estimate of π appears since each time, different random numbers are generated. To "smooth" out the randomness of the estimate, for each N, repeat the execution of the Macro 10 times (the program should be altered to do this automatically) and then output only the average of the 10 estimates.

Repeat this exercise for $N = 10$, 100, 1000, and 10,000. Make a table of N versus the absolute value of the error in the π estimates and show these on a graph using a logarithmic scale for N.

Even though this is a very inefficient way to compute π, it makes a good problem for doing VBA programming.

By the way, the VBA function to generate a random number between 0 and 1 is RND.

Exercise 2.6: Sorting is a frequently used operation in many data processing applications. Excel has a powerful sorting tool, available from the Data/Sort menu. If it is desired to sort data under control of a VBA program, such code might need to be written. The so-called bubble sort algorithm is not a very efficient one, but it is easy to program and it easily will suffice for small data sets. The essence of the bubble sort method is as follows: Repeatedly step through the list to be sorted, comparing two items at a time and swapping them if they are in the wrong order. The process is repeated until no swaps are needed, which indicates that the list is sorted. The algorithm gets its name from the way smaller elements (when sorting in a descending order) "bubble" to the top of the list. The following is a *pseudo-code* description of the process. Pseudo-code looks similar to VBA code, but no particular attention is paid to strict coding rules. The pseudo-code assumes that the array name is A and its length is N.

```
swapped = true            'swapped is a logical (boolean) variable
while swapped
  swapped = false         'temporarily assume the list is sorted
  for i =1 to N - 1
    if A(i) > A(i+1) then
      swap A(i) and A(i+1)
      swapped = true       'a swap indicates an unsorted list
    end if
  end for
end while
```

This logic could just as easily have been illustrated with a structure chart.

Generate a list of 20 numbers between 1 and 100 using the Excel function RANDBETWEEN. Select these numbers and copy them (by value) to another column (the reason for this is that RANDBETWEEN updates the list each time an operation is performed in the spreadsheet). Write a VBA program to input the copied numbers and sort them into descending order using bubble sort. Output the sorted numbers to the spreadsheet. Be sure to give an error message if no numbers are entered.

REFERENCES

Bowles, K., *Microcomputer Problem Solving Using Pascal*, Springer-Verlag, New York (1979).

Law, V.J., *ANSI Fortran 77: An Introduction to Structured Software Design*, Wm. C. Brown, Dubuque, Iowa (1983).

Law, V.J., *Standard Pascal: An Introduction to Structured Software Design*, Wm. C. Brown, Dubuque, Iowa (1985).

3 Linear Algebra and Systems of Linear Equations

3.1 INTRODUCTION

Linear algebra is a topic that many students of numerical methods will have been exposed to in mathematics classes. In this chapter, a brief review of linear algebra is given along with numerical methods for solving problems that are common in engineering and scientific applications. Linear algebra involves the manipulation of linear relationships and usually involves the use of vectors and matrices. The most common problem class of linear algebra is the solution of a set of linear algebraic equations.

3.2 NOTATION

- Scalars are indicated by a lowercase letter.
- Vectors are also identified by a lowercase letter. The context distinguishes between a scalar and a vector.
- Matrices are designated by a capital letter.
- By default, vectors are column vectors. To show a row vector, the transpose operator (a superscript T) is used (as in x^T).
- When necessary, the dimensions of matrices and vectors are shown by scalar subscripts. For example, $A_{m \times n}$ indicates a matrix with m rows and n columns. Further, $y_{n \times 1}$ designates a column vector of n elements.

Definition: The equation

$$ax + by + cz + dw = h \tag{3.1}$$

where a, b, c, d, and h are known numbers, while x, y, z, and w are unknown numbers (variables), is called a *linear equation*. If $h = 0$, the linear equation is said to be *homogenous*. A *linear system* is a set of linear equations, and a *homogenous linear system* is a set of homogenous linear equations.

For example,

$$\begin{aligned} 2x_1 - 3x_2 &= 1 \\ x_1 + 3x_2 &= -2 \end{aligned} \tag{3.2}$$

is a linear system. But

$$2x_1 - 3x_2^2 = -1$$
$$x_1 + x_2 = 1 \tag{3.3}$$

is a nonlinear system (because of x_2^2).

The system

$$2x_1 - 3x_2 - 3x_3 = 0$$
$$x_1 + 3x_2 = 0 \tag{3.4}$$
$$x_1 - x_2 + x_3 = 0$$

is a homogenous linear system.

Vectors and matrices offer a convenient, compact way of representing, manipulating, and solving linear systems. These are introduced in the next few sections.

3.3 VECTORS

A vector is an ordered set of numbers arranged as a column (the default). An m-element vector takes the form

$$x = \begin{bmatrix} x_1 \\ x_2 \\ \vdots \\ x_m \end{bmatrix} \tag{3.5}$$

The use of square brackets to enclose the elements is common notation.

Note that lowercase letters (without subscripts) are used to represent the entire vector. Individual elements of a vector are subscripted according to their placement within the vector. For example, x_3 indicates the third element in the column vector x.

Note on vectors of physics and mechanics: Engineers are familiar with vectors as they appear in problems of physics and mechanics. For example, Newton's law of motion takes the form

$$\vec{F} = m\vec{a} \tag{3.6}$$

where $\vec{F}$ is the applied force vector, m is the mass of the object, and $\vec{a}$ is the acceleration vector. The overbar arrow notation is commonly used to distinguish vectors from scalars. The physical vectors are one-, two-, or three-dimensional since they represent quantities in physical space. There is no difference between these physical vectors and those of linear algebra, but the notation is slightly different and linear algebra vectors can be of any size. The vector $\vec{F}$ can be written in component form as

$$\vec{F} = \begin{bmatrix} F_x \\ F_y \\ F_z \end{bmatrix} \tag{3.7}$$

where each component represents the magnitude of the force in each of the three Cartesian coordinates. In summary, physical vectors are a special case (three-dimensional) of the more general mathematical vector concept.

3.4 VECTOR OPERATIONS

3.4.1 VECTOR ADDITION AND SUBTRACTION

Only vectors of the same size (same number of elements) and shape (column or row) can be added or subtracted. The shorthand notation

$$c = a + b$$

is equivalent to element by element addition:

$$c_i = a_i + b_i;\ i = 1, n \tag{3.8}$$

3.4.1.1 Multiplication by a Scalar

If a is a vector of length n and σ is a scalar, then

$$b = \sigma a$$

is equivalent to the following where each element of a is multiplied by σ:

$$b_i = \sigma a_i;\ i = 1, n \tag{3.9}$$

3.4.2 VECTOR TRANSPOSE

The transpose operator indicates an interchange of rows with columns and *vice versa*. If u is a (column) vector, then a row vector is indicated by u^T.

3.4.3   LINEAR COMBINATIONS OF VECTORS

A linear combination of two vectors involves the multiplication of each vector by a scalar and then adding the results. The shorthand notation

$$\alpha u + \beta v = w \tag{3.10}$$

can be summarized as follows:

$$\alpha \begin{bmatrix} u_1 \\ u_2 \\ \vdots \\ u_m \end{bmatrix} + \beta \begin{bmatrix} v_1 \\ v_2 \\ \vdots \\ v_m \end{bmatrix} = \begin{bmatrix} \alpha u_1 + \beta v_1 \\ \alpha u_2 + \beta v_2 \\ \vdots \\ \alpha u_m + \beta v_m \end{bmatrix} = \begin{bmatrix} w_1 \\ w_2 \\ \vdots \\ w_m \end{bmatrix} \tag{3.11}$$

This concept can be extended to any number of scalars and vectors as long as all of the vectors have the same "shape" (row or column vector) and size.

3.4.4   VECTOR INNER PRODUCT

The vector inner product is the same as the familiar "dot" product of physical vectors. The following are equivalent notations, where x is an n-element vector and y is an n-element vector:

$$\sigma = x \cdot y = x^T y = y^T x = \sum_{i=1}^{n} x_i y_i \tag{3.12}$$

The inner product is always the result of multiplying a row vector by a column vector (the reverse is called the "outer" product and is discussed later). The result of the inner product is a scalar (a single number).

Example: Let $x = \begin{bmatrix} 1 \\ 2 \\ 3 \end{bmatrix}$ and $y = \begin{bmatrix} 2 \\ 1 \\ 3 \end{bmatrix}$. Then

$$x \cdot y = x^T y = y^T x = 1 \cdot 2 + 2 \cdot 1 + 3 \cdot 3 = 13 \tag{3.13}$$

3.4.5   VECTOR NORM

The usual (L_2) norm of a vector is the square root of the sum of squares of elements. For a vector, v, of n elements,

$$\|v\| = \sqrt{\sum_{i=1}^{n} v_i^2} = \sqrt{v^T v} \qquad (3.14)$$

A less useful norm, called the L_1 norm, is the sum of absolute values of the vector elements.

3.4.6 ORTHOGONAL VECTORS

The vector inner product has an interesting and useful geometric interpretation. The angle θ between two nonzero vectors u and v is given by the following (which is easily derived from the law of cosines for the two-dimensional case):

$$\cos \theta = \frac{u^T v}{\|u\| \|v\|} \qquad (3.15)$$

If the angle is 90° ($\pi/2$ radians), then the inner product is zero, which in two or three dimensions means that the vectors are perpendicular to each other (also called *orthogonal* to each other). The concept of orthogonality can be extended to any number of dimensions. That is, two vectors u and v are orthogonal if

$$u^T v = 0 \qquad (3.16)$$

3.4.7 ORTHONORMAL VECTORS

Orthonormal vectors are *unit* vectors (having a magnitude of 1) that are orthogonal to each other. Any vector can be converted to a unit vector by dividing by its L_2 norm:

$$\hat{u} = \frac{u}{\|u\|} \qquad (3.17)$$

which is a unit vector in the direction of u. Note that since

$$\|u\| = \sqrt{u^T u} \qquad (3.18)$$

it follows that $u^T u = 1$ if u is a unit vector.

3.5   MATRICES

A matrix is a two-dimensional array of numbers. It can also be viewed as a vector of vectors. Typically, matrix A with m rows and n columns is written as follows:

$$A = \begin{bmatrix} a_{11} & a_{12} & \cdots & a_{1n} \\ a_{21} & a_{22} & \cdots & a_{2n} \\ \vdots & \vdots & \ddots & \vdots \\ a_{m1} & a_{m2} & \cdots & a_{mn} \end{bmatrix} \qquad (3.19)$$

An uppercase letter represents an entire matrix. Among the matrix elements, the first subscript is the row number and the second subscript is the column number.

If u, v, and w are n-element vectors, then a matrix B with three rows and n columns can be constructed from them as follows:

$$B = \begin{bmatrix} u^T \\ v^T \\ w^T \end{bmatrix} \qquad (3.20)$$

Likewise, if d, e, and f are m-dimensional vectors, a matrix C with m rows and three columns can be formed from them as follows:

$$C = [d \, e \, f] \qquad (3.21)$$

3.6   MATRIX OPERATIONS

3.6.1   MATRIX ADDITION AND SUBTRACTION

This is defined only for matrices of exactly the same shape (same number of rows and columns). For addition of two $m \times n$ matrices A and B, the sum C is given by

$$C = A + B$$
$$c_{ij} = a_{ij} + b_{ij}; \quad i = 1, \cdots, m; \quad j = 1, \cdots, n \qquad (3.22)$$

3.6.2   MULTIPLICATION BY A SCALAR

If σ is a scalar, then the operation σA involves multiplying every element of A by σ.

3.6.3 TRANSPOSITION OF MATRICES

When applied to a matrix, the transposition operator converts each row to a column (and conversely, each column into a row). That is, each row becomes a column in the resulting transposed matrix. This can be represented as follows, where A is an $n \times m$ matrix:

$$B = A^T$$
$$b_{ij} = a_{ji}; \quad i = 1, \cdots, m; \quad j = 1, \cdots, n \tag{3.23}$$

Note that if A is $n \times m$, then B is $m \times n$.

3.6.4 SPECIAL MATRICES

Square: $m = n$, same number of rows and columns
Diagonal: Square, only elements on the diagonal are nonzero
Identity: Diagonal, diagonal elements are all 1 (rest are 0)
Upper triangular: Usually square, all elements below the diagonal $= 0$
Lower triangular: Usually square, all elements above the diagonal $= 0$

3.6.5 MATRIX MULTIPLICATION

Any two *conformable* matrices A and B can be multiplied in the order AB. A and B are conformable if the number of columns of A is the same as the number of rows of B. This is summarized as follows, where A is $n \times p$ and B is $p \times m$:

$$C = AB$$
$$c_{ij} = \sum_{k=1}^{p} a_{ik} b_{kj}; \quad i = 1, \cdots, n; \quad j = 1, \cdots, m \tag{3.24}$$

Another view is that the i, j-th element of C is the inner product of the ith row of A with the jth column of B.

Example 3.1: Matrix Multiplication

Given the following matrices and vectors:

$$A = \begin{bmatrix} 4 & 3 & 2 \\ -4 & 2 & -2 \\ 0 & -1 & 0 \\ -1 & 2 & 2 \end{bmatrix} \quad B = \begin{bmatrix} 1 & -3 \\ -2 & 2 \\ 3 & 4 \end{bmatrix} \quad x = \begin{bmatrix} -3 \\ 2 \\ 4 \end{bmatrix} \quad y = \begin{bmatrix} 1 & -2 & 3 \end{bmatrix}$$

the following are the results of various valid matrix multiplications:

$$AB = \begin{bmatrix} 4 & 2 \\ -14 & 8 \\ 2 & -2 \\ 1 & 15 \end{bmatrix} \quad Ax = \begin{bmatrix} 2 \\ 8 \\ -2 \\ 15 \end{bmatrix} \quad Ay^T = \begin{bmatrix} 4 \\ -14 \\ 2 \\ 1 \end{bmatrix}$$

$$yB = \begin{bmatrix} 14 & 5 \end{bmatrix} \quad x^T B = \begin{bmatrix} 5 & 29 \end{bmatrix}$$

3.6.6 MATRIX DETERMINANT

The determinant of a matrix is a scalar that provides significant information about the matrix. Determinants can be computed only for square matrices. The determinant of a 2×2 matrix is easy to compute and is defined as follows:

$$\det(A) = \begin{vmatrix} a_{11} & a_{12} \\ a_{21} & a_{22} \end{vmatrix} = a_{11}a_{22} - a_{12}a_{21} \tag{3.25}$$

Note that this is the product of the diagonal elements minus the product of the off-diagonal elements.

For a 3×3 matrix, there is a trick or shortcut for calculating the determinate. This involves duplicating the first two columns and then adding the products "to the right" and subtracting the products "to the left." The trick does not work for higher dimensional matrices. It is much easier to compute the determinant of a matrix by transforming it to an upper or lower triangular matrix (see below).

An important property of determinates is

$$\det(AB) = \det(A)\det(B) \tag{3.26}$$

That is, the determinate of a product is the product of determinates. This property is useful when finding the determinate of a matrix by numerical manipulations.

3.6.7 MATRIX INVERSE

The inverse is defined only for square matrices (same number of rows and columns). For an $n \times n$ matrix A, the *inverse* of A is designated A^{-1} and has the following properties:

$$A^{-1}A = AA^{-1} = I \quad \text{(the identity matrix)} \tag{3.27}$$

3.6.8 MORE SPECIAL MATRICES

Symmetric matrices are square and have the property $A = A^T$. Another way of saying this is that for all i and j, $a_{ij} = a_{ji}$.

Tridiagonal matrices are square and have nonzero entries on the diagonal (elements with indices $i = j$), just above the diagonal (elements with indices $j = i + 1$), and just below the diagonal (elements with indices $j = i - 1$). Here is an example of a tridiagonal matrix:

$$\begin{bmatrix} 2 & -2 & 0 & 0 \\ -1 & 2 & -1 & 0 \\ 0 & -3 & 2 & -1 \\ 0 & 0 & -1 & 2 \end{bmatrix}$$

Orthogonal matrices have the powerful property that the transpose is the inverse. That is, for an orthogonal matrix Q, $Q^T Q = I$.

3.7 SOLVING SYSTEMS OF LINEAR ALGEBRAIC EQUATIONS

In matrix-vector form, a system of linear algebraic equations takes the form

$$Ax = b \tag{3.28}$$

where A is an $m \times n$ matrix, x is an n-vector, and b is an m-vector.

When $m = n$, the number of equations is the same as the number of unknowns and this is the usual case. If the inverse of A exists, then the solution can be written as

$$x = A^{-1}b \tag{3.29}$$

If A^{-1} does not exist, then, clearly, an associated linear system cannot be solved using the inverse. Another good reason to not use the inverse is that it is inefficient to compute the inverse and then multiply by the right-hand side vector, b. Later, a very general method is covered using the so-called singular value decomposition (SVD) that always produces a useful solution. A method, called Gaussian elimination, leads to a solution if one exists and also gives insight to those cases where a unique solution does not exist.

3.7.1 GAUSSIAN ELIMINATION

Gaussian elimination is most easily described using a specific example. Consider the linear system

$$\begin{aligned} x_1 + x_2 + x_3 &= 0 \\ x_1 - 2x_2 + 2x_3 &= 4 \\ x_1 + 2x_2 - x_3 &= 2 \end{aligned} \tag{3.30}$$

The idea behind Gaussian elimination is to *kill* (eliminate) one of the unknowns in the second and third equations (using the first) and then to eliminate another unknown from the third equation (using the second). This leaves the third equation with only one unknown. For the example, if the first equation is subtracted from the second and third equations, the result is

$$
\begin{aligned}
x_1 + x_2 + x_3 &= 0 \\
-3x_2 + x_3 &= 4 \\
x_2 - 2x_3 &= 2
\end{aligned}
\tag{3.31}
$$

Now, if the second equation is multiplied by 1/3 and the result added to the third, there results

$$
\begin{aligned}
x_1 + x_2 + x_3 &= 0 \\
-3x_2 + x_3 &= 4 \\
-\frac{5}{3}x_3 &= \frac{10}{3}
\end{aligned}
\tag{3.32}
$$

The last equation now contains only x_3; solving gives $x_3 = -2$. Knowing x_3, from the second equation, $x_2 = -2$, and finally from the first equation, knowing x_3 and x_2 leads to $x_1 = 4$. Therefore, the linear system has one solution

$$
x_1 = 4, \, x_2 = -2, \, x_3 = -2
$$

Going from the last equation to the first while solving for the unknowns is called *backsolving* or *backsubstitution*. It is important to see that when multiplying one equation by a scalar and adding the result to another one, *equality is maintained*.

Another way to represent the steps of Gaussian elimination is to use an *augmented matrix*, which is designated as [A|b], where *only the coefficients* of the equations are written (the right-most column contains values for the vector b). For the previous example, the augmented matrix is

$$
\begin{bmatrix}
1 & 1 & 1 & 0 \\
1 & -2 & 2 & 4 \\
1 & 2 & -1 & 2
\end{bmatrix}
$$

The same elementary row operations can be performed on this matrix as with the original equations. Keeping the first row and subtracting it from the second and third ones gives

$$
\begin{bmatrix}
1 & 1 & 1 & 0 \\
0 & -3 & 1 & 4 \\
0 & 1 & -2 & 2
\end{bmatrix}
$$

Then, keep the first and second equations, multiply the second by 1/3, and add the result to the third to get

$$\begin{bmatrix} 1 & 1 & 1 & 0 \\ 0 & -3 & 1 & 4 \\ 0 & 0 & -\dfrac{5}{3} & \dfrac{10}{3} \end{bmatrix}$$

If the variables were inserted to convert the transformed augmented matrix back into equation form, the last equation would involve only x_3 and can be solved for it. Then, proceed to backsubstitute for x_2 and x_1. Here is a summary of the Gaussian elimination procedure.

Gaussian elimination summary: Consider an $n \times n$ linear system.

1. Construct the augmented matrix for the system.
2. Use elementary row operations to transform the augmented matrix into an upper-triangular one.
3. Solve the last equation for the single variable x_n.
4. Complete the backsubstitution for all other variables.

DID YOU KNOW?

Carl Friedrich Gauss, the great German mathematician, did not discover what is called Gaussian elimination.

The method of Gaussian elimination appears in Chapter 8, *Rectangular Arrays*, of the important Chinese mathematical text *Jiuzhang suanshu* or *The Nine Chapters on the Mathematical Art*. Its use is illustrated in eighteen problems, with two to five equations. The first reference to the book by this title is dated to 179 CE, but parts of it were written as early as approximately 150 BCE. It was commented on by Liu Hui in the 3rd century.

The method in Europe stems from the notes of Isaac Newton. In 1670, he wrote that all the algebra books known to him lacked a lesson for solving simultaneous equations, which Newton then supplied. Cambridge University eventually published the notes as *Arithmetica Universalis* in 1707 long after Newton left academic life. The notes were widely imitated, which made (what is now called) Gaussian elimination a standard lesson in algebra textbooks by the end of the 18th century. Carl Friedrich Gauss in 1810 devised a notation for symmetric elimination that was adopted in the 19th century by professional hand computers to solve the normal equations of least-squares problems. The algorithm that is taught in high school was named for Gauss only in the 1950s as a result of confusion over the history of the subject.

Source: http://meyer.math.ncsu.edu/Meyer/PS_Files/GaussianEliminationHistory.pdf.

3.7.2 DETERMINANT REVISITED

Previously, the determinant for a 2×2 matrix was defined, and it was stated that it is easy to calculate the determinant for a 3×3 matrix. Existence, uniqueness, families of solutions, rank, and even the determinant are all determined via the Gaussian elimination process.

Note that each step in the Gaussian elimination process can be expressed as

$$M_i A \rightarrow B_i \tag{3.33}$$

where M_i is an identity matrix altered only with additional elements either below or above the diagonal (but not both). Recall Example 3.1. The steps of zeroing all elements below the first diagonal can be summarized as follows:

$$M_1 A = \begin{bmatrix} 1 & 0 & 0 \\ -1 & 1 & 0 \\ -1 & 0 & 1 \end{bmatrix} \begin{bmatrix} 1 & 1 & 1 \\ 1 & -2 & 2 \\ 1 & 2 & -1 \end{bmatrix} = \begin{bmatrix} 1 & 1 & 1 \\ 0 & -3 & 1 \\ 0 & 1 & -2 \end{bmatrix} = B_1 \tag{3.34}$$

The matrix, M_1, on the left is a lower triangular transformation matrix and its determinant is 1 (this is easy to verify). One algebraic rule for determinants is that *the determinant of a product is the product of determinants*; therefore, the determinant of the transformed matrix (B_1) is the same as that of A, since $\det(M_1) = 1$.

The next step is to eliminate the remaining nonzero element below the diagonal as follows:

$$M_2 M_1 A = M_2 B_1 \begin{bmatrix} 1 & 0 & 0 \\ 0 & 1 & 0 \\ 0 & 1/3 & 1 \end{bmatrix} \begin{bmatrix} 1 & 1 & 1 \\ 0 & -3 & 1 \\ 0 & 1 & -2 \end{bmatrix} = \begin{bmatrix} 1 & 1 & 1 \\ 0 & -3 & 1 \\ 0 & 0 & -5/3 \end{bmatrix} = B_2$$

$$\det(A) = \det(B_2) = (1)(-3)(-5/3) = 5 \tag{3.35}$$

Therefore, when, via Gaussian elimination, matrix A has been transformed into an upper (or lower) triangular matrix, the value of the determinant has not changed. The determinant of a triangular matrix is simply the product of the diagonal elements (also easy to prove). So, $\det(A)$ can be computed simply by multiplying the diagonal elements of the final triangular matrix:

$$\det(A) = \det(M_{n-1} M_{n-2} \cdots M_1 A) = \det(B) = \prod_{i=1}^{n} b_{ii} \tag{3.36}$$

where n is the number of rows and columns of A.

3.7.3 Gauss–Jordan Elimination

A method closely related to Gauss elimination is called the Gauss–Jordan algorithm. As a "bonus" (but it involves more work), the *inverse* of the matrix is also calculated. The basic idea behind the Gauss–Jordan method is to first form an augmented matrix consisting of the original system matrix and the identity matrix as follows:

$$[A \mid I]$$

Next, A is transformed into an identity matrix using elementary row operations (indicated by the matrix T) resulting in

$$T\,[A|I] = [I|T] \tag{3.37}$$

This implies that

$$TA = I \tag{3.38}$$

or, in other words,

$$T = A^{-1} \tag{3.39}$$

Example 3.2: Gauss–Jordan Elimination

If the original square matrix, A, is given by the following expression:

$$A = \begin{bmatrix} 2 & -1 & 0 \\ -1 & 2 & -1 \\ 0 & -1 & 2 \end{bmatrix} \tag{3.40}$$

then, after augmentation by the identity matrix, the following is obtained:

$$[A|I] = \begin{bmatrix} 2 & -1 & 0 & 1 & 0 & 0 \\ -1 & 2 & -1 & 0 & 1 & 0 \\ 0 & -1 & 2 & 0 & 0 & 1 \end{bmatrix} \tag{3.41}$$

By performing elementary row operations on the $[A|I]$ matrix until it is transformed into the identity matrix, the following form results:

$$[I|A^{-1}] = \begin{bmatrix} 1 & 0 & 0 & \dfrac{3}{4} & \dfrac{1}{2} & \dfrac{1}{4} \\ 0 & 1 & 0 & \dfrac{1}{2} & 1 & \dfrac{1}{2} \\ 0 & 0 & 1 & \dfrac{1}{4} & \dfrac{1}{2} & \dfrac{3}{4} \end{bmatrix} \tag{3.42}$$

The matrix augmentation can now be *undone*, which gives the following:

$$I = \begin{bmatrix} 1 & 0 & 0 \\ 0 & 1 & 0 \\ 0 & 0 & 1 \end{bmatrix}$$

$$A^{-1} = \begin{bmatrix} \dfrac{3}{4} & \dfrac{1}{2} & \dfrac{1}{4} \\ \dfrac{1}{2} & 1 & \dfrac{1}{2} \\ \dfrac{1}{4} & \dfrac{1}{2} & \dfrac{3}{4} \end{bmatrix} \qquad (3.43)$$

To solve the system of equations $Ax = b$, use the same set of operations, indicated as the transformation matrix T, as follows:

$$\begin{aligned} Ax &= b \\ Tax &= Tb \\ Ix &= Tb \end{aligned} \qquad (3.44)$$

This is identically the same as

$$x = A^{-1}b$$

Gauss–Jordan is not the most efficient method for solving systems of linear equations, but it is nevertheless popular.

3.7.4 Rank of Matrix

Definition: The rank of a matrix is the number of independent rows or columns in the matrix.

When performing Gaussian elimination, the rank becomes evident as rows are eliminated. The rank can be deduced from the triangular matrix by observing the number of nonzero rows and columns.

Having defined rank and knowing at least one way to compute it, the following statements can be made about the $n \times n$ linear system $Ax = b$:

Consistency: A linear system is consistent if rank(A) = rank($[A \mid b]$). That is, both A and the augmented matrix $[A \mid b]$ have the same rank.

3.7.5 EXISTENCE AND UNIQUENESS OF SOLUTIONS FOR $Ax = b$

Assume that A is an $n \times n$ square matrix. Then the following statements can be made:

For $Ax = b$, there is
1. *No* solution if and only if rank$(A) \neq$ rank$([A| b])$ (i.e., inconsistent)
2. A *unique* solution if and only if rank$(A) =$ rank$([A| b]) = n$
3. An $(n - r)$-parameter *family* of solutions if and only if rank$(A) =$ rank$([A \mid b]) = r < n$

For the homogeneous case, $Ax = 0$
1. Is consistent
2. Admits the trivial solution $x = 0$
3. Admits the unique trivial solution $x = 0$ if and only if rank$(A) = n$
4. Admits an $(n - r)$-parameter family of nontrivial solutions, in addition to the trivial solution, if and only if rank$(A) = r < n$

Stated more succinctly, if A is of full rank and b is independent of any columns of A, then a unique solution exists; otherwise, either no solution exists or an infinite number of solutions exist.

3.8 LINEAR EQUATIONS AND VECTOR/ MATRIX OPERATIONS IN EXCEL®

Table 3.1 shows some of the more widely used vector/matrix operations that are native to Excel®.

MMULT is the only function that operates on both matrices and vectors. This function requires the two arrays to be conformable.

While the MINVERSE and MMULT functions are sufficient to solve well-posed linear systems, there are many more functions that can lead to more robust computations. A completely *free* Excel Add-On called *Matrix.xla* can be downloaded from the website http://digilander.libero.it/foxes/SoftwareDownload.htm. In addition to the software, there is a comprehensive manual (in two volumes). Readers are encouraged to install this software on their own computer. Once Matrix.xla is installed, all Excel applications have an enhanced set of functions available for matrix operations

TABLE 3.1
Excel® Matrix Functions

Function Name	Operation
MDETERM	Returns the determinant of a matrix
MINVERSE	Returns the inverse of a matrix
MMULT	Returns the product of two arrays

and linear algebra. Only a few of these are discussed here; refer to the Matrix.xla manual for many more details.

Within Matrix.xla, there are several functions that can lead to solutions of linear systems of the form $Ax = b$. The simplest of these is called SysLin, which uses the Gauss–Jordan algorithm. The major advantage of using SysLin as opposed to MINVERSE and MMULT is that it is a one-step operation. Also, if the matrix is singular (has no inverse), SysLin provides a suitable error message.

Example 3.3: Use of SysLin for Linear Systems

The following window shows an Excel spreadsheet with the matrix **A** and the vector **b** defined for a linear equation set. Space is available for the solution vector **x**.

	A	B	C	D	E
1		A		b	x
2	1	2	2	1	
3	2	3	1	2	
4	2	1	1	3	

If the three cells allocated for the vector **x** are selected and Formulas/InsertFunction/SysLin is invoked, a window for SysLin appears. In the area for **Mat**, simply drag the cursor across the cells containing the matrix **A**. Then in the area for **v**, drag across the cells containing **b**. At this point, everything appears as follows:

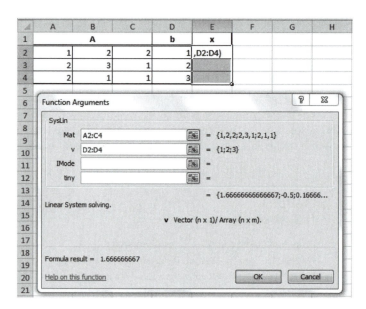

Now (very important), hold down both Ctrl and Shift and hit Enter. The selected area reserved for **x** is filled in as follows:

◢	A	B	C	D	E
1	A			b	x
2	1	2	2	1	1.666667
3	2	3	1	2	-0.5
4	2	1	1	3	0.166667

This example also gives a "feel" for how most of the matrix-related functions work.

3.9   MORE ABOUT Matrix.xla

Instead of having to go through all of the steps shown in the previous example, in Excel, go to AddIns/ToolBar Commands. A small blue icon with the letter M appears. A click on the M produces a menu that looks like the following:

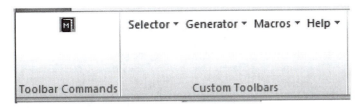

The number of tools available via Matrix.xla is far beyond what can be covered here. Only the most pertinent options are considered.

By clicking the Generator option and then selecting Random gives a convenient way to generate a matrix of random numbers. The following window appears:

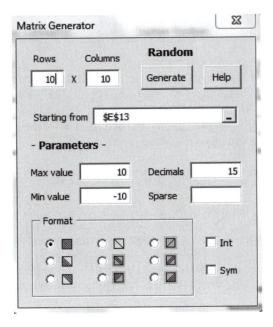

The Rows × Columns selections allow any (reasonable) size matrix of uniformly distributed random numbers to be generated (the matrix does not have to be square). The *Starting From* space indicates where the first matrix element is to be placed (cell E13 in the example). The user can then choose the maximum and minimum values for the random numbers as well as how many significant digits to include (Decimals). A variety of Formats are available, but usually, the solid icon is chosen indicating a full matrix. Other Formats include lower triangular, upper triangular, diagonal, etc. Pressing the Generate button produces the desired random matrix. This is illustrated in the next example.

Perhaps the most useful among the Matrix.xla choices is Matrix Operations under the Macro ribbon. This produces a window like the following:

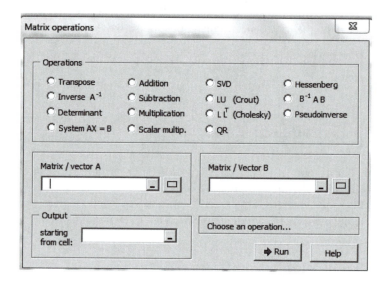

Among the many choices are Transpose, Inverse, Determinant, System AX = B, Multiplication, and Pseudoinverse. The Inverse and Multiplication choices can be used to solve Ax = b instead of using the built-in Excel functions. The System AX = B selection solves a linear system directly and even allows for the right-hand side to be a matrix and, thus, a matrix of solutions.

Example 3.4: Using System AX = B from Matrix.xla

The Excel screen shot shown below indicates how to solve the same problem of the previous example. The button for AX = B is selected. The cell range for the matrix appears in the *Matrix/vector A* selection while the cell range for the right-hand side vector appears in the *Matrix/vector B* selection. The address of the first element of the solution vector is entered into the *Output starting from cell:* selection. When the Run button was clicked, the solution was computed and output to the appropriate cells.

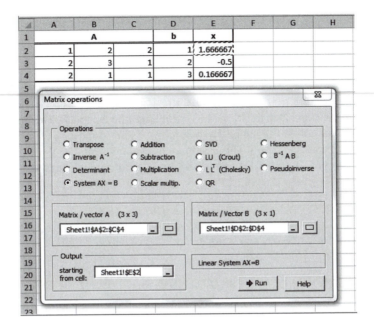

	A	B	C	D	E	F	G	H
1		A		b	x			
2	1	2	2	1	1.666667			
3	2	3	1	2	-0.5			
4	2	1	1	3	0.166667			

3.10 SVD AND PSEUDO-INVERSE OF A MATRIX

The most general way for solving any linear system (consistent, overdetermined, or underdetermined) is to use the pseudo-inverse of the matrix. A consistent system has a unique solution, an overdetermined system is one with more equations than unknowns, and an underdetermined system has an infinite number of solutions. If the pseudo-inverse is denoted by A^+, then the solution of $Ax = b$ can be written as

$$x = A^+ b \qquad (3.45)$$

For a square, nonsingular matrix, A^+ coincides with the inverse, A^{-1}. The pseudo-inverse *always exists*, whether or not the matrix is square or has full rank.

For an $n \times m$ matrix, A, A^+ must satisfy the following four conditions:

$$\begin{aligned} AA^+A &= A \\ A^+AA^+ &= A^+ \\ (AA^+)^T &= (AA^+) \\ (A^+A)^T &= (A^+A) \end{aligned} \qquad (3.46)$$

Note that A^+ is an $m \times n$ matrix.

The conditions that the pseudo-inverse must satisfy are not very helpful for computing A^+. Computation is most easily accomplished by using the *singular value decomposition* (SVD) of A. The SVD can be visualized as a factoring of A into three separate matrices as follows:

$$A = UDV^T \tag{3.47}$$

or, for a linear system, $Ax = b$ and

$$UDV^Tx = b \tag{3.48}$$

where, setting $p = \min(n, m)$, U is an $(n \times p)$ orthogonal matrix ($U^TU = I$), V is an $(m \times p)$ orthogonal matrix, and D is a $(p \times p)$ diagonal matrix (the diagonal elements are called the singular values of A). For simplicity, assume $n > m$, in which case D and V are both square with dimension $(m \times m)$. This assumption does not invalidate the final result.

Multiplying both sides of Equation 3.48 by U^T and remembering that $U^T U = I$, there results

$$U^TUDV^Tx = U^Tb \Rightarrow DV^T x = U^Tb \tag{3.49}$$

The matrix DV^T is square so taking its inverse gives [recall that for any two matrices X and Y, $(XY)^{-1} = Y^{-1}X^{-1}$]

$$DV^T x = U^T b \Rightarrow x = (DV^T)^{-1} U^Tb \Rightarrow x = (V^T)^{-1}D^{-1} U^Tb \tag{3.50}$$

Because $V^T = V^{-1}$, the final result is

$$x = (V^T)^{-1}D^{-1} U^Tb = A^+b \tag{3.51}$$

Therefore, the pseudo-inverse can be computed by the following formula:

$$A^+ = V D^{-1}U^T \tag{3.52}$$

When using the pseudo-inverse, there are three possible situations:

1. $m = n$: The matrix A is square and $A^+ = A^{-1}$.
2. $m > n$: There are more equations than unknowns and the system is overdetermined. In this case, x represents the "least squares" solution. That is, x minimizes $\|Ax - b\|$, which is the sum of squares of "residuals" (the difference between each element of Ax and the corresponding element of b).
3. $m < n$: There are fewer equations than unknowns and the system is underdetermined. Therefore, there is no unique solution; in fact, there are an infinite number of solutions. The most convenient way to express these solutions using the pseudo-inverse is

$$x = A^+b + (I - A^+A)z \tag{3.53}$$

where z is an arbitrary vector. Any z can be specified and the corresponding x will satisfy the m equations. The proof of this equation is lengthy and is offered here without proof. Note that $A^+A \neq I$.

The actual algorithm for computing A^+ is complex and is not covered here. Fortunately, in Matrix.xla, this computation is made available by the function MPseudoinv. Recall that the pseudo-inverse of an $m \times n$ matrix is an $n \times m$ matrix.

Example 3.5: Solution of Linear Systems Using the Pseudo-Inverse

The Excel spreadsheet shown below illustrates solving

- A consistent system
- An overdetermined system (least-squares solution)
- An underdetermined system (infinite number of solutions)

In each case, the matrix and the right-hand side vector are generated using the random matrix feature of Matrix.xla. For the underdetermined system, an arbitrary z vector $[0\ 1\ 1\ 1]^T$ is chosen to show how to produce another of the infinite number of other solutions (see Equation 3.53).

Note the following about the results shown in the spreadsheet:

- For the consistent system, the vector Ax is identically equal to the vector b. This is a unique solution.
- For the overdetermined system, Ax does not equal b. This is because the solution minimizes the sum of squares of differences between Ax and b. The least squares solution is useful in several applications, such as regression, which is covered in Chapter 7.
- For the underdetermined system, the initial (or base) x is the product A^+b and is only one of an infinite number of solutions possible. A second solution is generated using Equation 3.53 with the arbitrary vector $z = [0\ 1\ 1\ 1]^T$. In both cases, Ax is equal to b indicating that indeed the associated x is a solution.

Consistent System

A			b	x	Ax
0.832855	7.902832	−2.18289	−7.67573	−2.73575	−7.67573
3.889161	9.665859	3.044689	3.055222	0.298083	3.055222
−5.28107	8.451514	−5.57466	−2.8325	3.551683	−2.8325
A+					
0.387957	−0.12478	−0.22006			
−0.0273	0.078798	0.053725			
−0.40891	0.23767	0.110541			

Overdetermined System

A			b	x	Ax
6.190156	−0.42224	−5.69814	−9.26793	0.045362	−3.49158
10.65635	−4.55003	1.185262	4.296808	0.484615	−0.97949
2.568451	−6.16077	6.203344	−3.82716	0.626126	1.014982
9.26915	9.534804	10.53612	10.77149	↖	11.63811
				Least-squares solution	

A+			
0.04579	0.04914	−0.00505	0.022212
0.013649	−0.03479	−0.06283	0.048287
−0.0582	−0.00667	0.056641	0.030838

Underdetermined System

A			b	Base x	Ax
-4.90858	-5.57692	0.430979	-4.66382	-6.93659	-0.17947	-6.93659
2.021388	1.200819	7.644887	1.3031	-5.13765	1.595509	-5.13765
-9.45529	0.608641	-6.26193	-9.6764	10.79749	-0.82275	10.79749
					-0.3077	

A+			Base x = A+b
-0.02176	-0.02893	-0.04437	
-0.15092	0.081009	0.089359	
0.030857	0.134928	0.007826	
-0.00819	-0.05395	-0.05943	

I				A+A			
1	0	0	0	0.467865	0.059627	0.04728	0.493129
0	1	0	0	0.059627	0.993319	-0.0053	-0.05526
0	0	1	0	0.04728	-0.0053	0.995799	-0.04381
0	0	0	1	0.493129	-0.05526	-0.04381	0.543017

I – (A+A)				z	Arbitrary Part	Base x	New Solution	New Ax
0.532135	-0.05963	-0.04728	-0.49313	0	-0.60004	-0.17947	-0.77951	-6.93659
-0.05963	0.006681	0.005298	0.055257	1	0.067236	1.595509	1.662745	-5.13765
-0.04728	0.005298	0.004201	0.043815	1	0.053313	-0.82275	-0.76943	10.79749
-0.49313	0.055257	0.043815	0.456983	1	0.556054	-0.3077	0.248349	

New Solution is x = A+ b + (I – A+A)

EXERCISES

Exercise 3.1: Mass balance on a gas absorber

A gas absorber is fed, via stream F_1, 100 mol/min of monoethanolamine (MEA) and CO_2. Stream F_1 is composed of 98% (mol%) MEA and 2% CO_2. Stream F_2 contains CO_2, SO_2, and N_2 (Figure 3.1). Experimental data available for the unit are shown in Table 3.2.

Derive the linear system for the absorber using the supplied information. The three unknowns are the flow rates of P_1, F_2, and P_2. Note that since data are available on only two components, it is necessary to include the overall material balance as one of the equations.

a. Using SYSLIN (or the linear system option from Matrix.xla)
b. By calculating A^{-1} followed by multiplication ($A^{-1}b$)
c. By calculating the pseudo-inverse, A^+, followed by matrix multiplication A^+b

Exercise 3.2: Coffee leaching

A "Mr. Coffee" apparatus for brewing a good "cuppa joe" is a chemical extraction unit. Ingredients include water (W), solubles (S), and grounds (G). A schematic diagram of the "system" is shown in Figure 3.2.

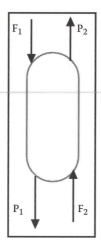

FIGURE 3.1   Gas absorber schematic.

	P_1	F_2	P_2
TABLE 3.2			
Data for Gas Absorber (Mole Fractions)			
SO_2	0.0170	0.0200	0.0022
N_2	0.0000	0.9000	0.9890

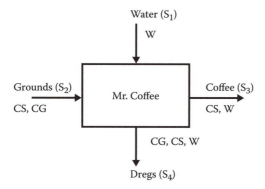

FIGURE 3.2   Automatic coffee maker schematic.

The Grounds input contains components CG and CS. Water input contains only component W. The Coffee stream contains both water (W) and solubles (CS), while the Dregs output has all three components. Other pertinent data are as follows (all percentages are by volume):

- Stream S_1 consists of 1.1 L of pure water.
- Stream S_2 contains 98% solid (CG) and 2% solubles (CS).
- Stream S_3 contains 0.8% CS and 99.2% W.
- Stream S_4 contains 81% CG, 0.5% CS, and 18.5% W.

Write three component balances (these are "volume" balances since percentages are volume based) to give three linear equations in the three unknown flowrates (S_2, S_3, and S_4).

Solve the linear system for the following problem:

a. Using SYSLIN (or the linear system option from Matrix.xla)
b. By calculating A^{-1} followed by multiplication ($A^{-1}b$)
c. By calculating the pseudo-inverse, A^+, followed by matrix multiplication A^+b

Exercise 3.3: Flash tanks in series

Shown in Figure 3.3 is a schematic of a separation system consisting of two flash tanks in series. Experimental data are available on streams F, V_1, V_2, and L_2 as shown in Table 3.3. The feed rate, F, is 1000 kg/min.

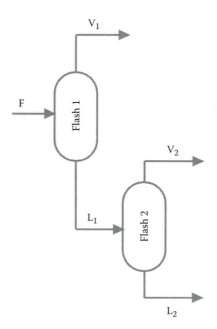

FIGURE 3.3   Two flash tanks in series.

TABLE 3.3

Stream Mass Fractions

| Component | | Mass Fraction | | |
	F	V_1	V_2	L
Methanol	0.3	0.71	0.44	0.08
Ethanol	0.4	0.27	0.55	0.39
Butanol	0.3	0.02	0.01	0.53

Write mass balances on each of the three components using the supplied data. The result is three linear algebraic equations in the three unknown flowrates, V_1, V_2, and L_2. Use any method to find these three unknowns.

Exercise 3.4: For the following exercises, use the random matrix generator available in Matrix.xla as required.

a. Generate a random matrix of size 3 × 3 and a column vector of length 3. Using these data to represent A and b in the linear system $Ax = b$, solve the system

    1. By taking the matrix inverse followed by matrix-vector multiplication

    2. Use the SYSLIN function

    3. Use the function MPseudoinv and matrix multiplication

b. Repeat part a for a 10 × 10 matrix and associated vector.

c. Repeat step 3 of part a for a 4 × 3 matrix.

d. Repeat step 3 of part a for a 3 × 4 matrix.

e. For the 3 × 4 matrix of part d, apply Equation 3.53 to generate a particular solution based on an arbitrary vector, $z = [1\ 1\ 1\ 1]^T$.

Note that for a 3 × 4 matrix, A^+A is not equal to the identity matrix. An infinite number of solutions exist since *any* z gives a proper solution.

Exercise 3.5: A chemical separation system

Benzene, styrene, toluene, and xylene are to be separated with the array of distillation columns shown in Figure 3.4. Experimental data show that the feed rate (stream A) is 100 mole/hr and that the composition of stream A is 10% benzene, 15% styrene, 35% toluene and 40% xylene (compositions are in mole %).

The following information is available for this system:

1. 80% benzene, 60% styrene, 30% toluene and 10% xylene of the feed stream go overhead to stream B. The remainder goes out the bottom to stream C.

2. 70% benzene, 65% styrene, 25% toluene and 5% xylene of stream B go overhead to stream D. The remainder goes out the bottom to stream E.

3. 85% benzene, 65% styrene, 20% toluene and 10% xylene of stream C go overhead to stream F. The remainder goes out the bottom to stream G.

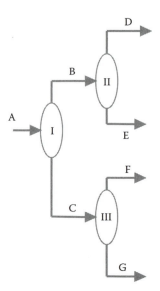

FIGURE 3.4 Distillation column train.

Write four material balances where the unknowns are the flow rates of streams D, E, F, and G. Solve the resulting linear system by any method that has been discussed.

REFERENCE

Gaussian Elimination History; http://meyer.math.ncsu.edu/Meyer/PS_Files/GaussianElimination History.pdf (Oct. 2012).

4 Numerical Differentiation and Integration

4.1 NUMERICAL DIFFERENTIATION

4.1.1 APPROXIMATION OF A DERIVATIVE IN ONE VARIABLE

Recall the definition of the derivative from calculus:

$$\frac{df}{dx} = \lim_{\Delta x \to 0} \left(\frac{f(x + \Delta x) - f(x)}{\Delta x} \right)$$

For a finite Δx, this is an *approximation* of the derivative:

$$\frac{df}{dx} = \left(\frac{f(x + \Delta x) - f(x)}{\Delta x} \right) + E(\Delta x) \tag{4.1}$$

This is the simplest form of "finite difference derivative" and is called the "forward" difference approximation to the derivative. $E(x)$ represents the "error" in the approximation. In order to estimate the size of the error term, consider the Taylor expansion of $f(x + \Delta x)$ in the neighborhood of x:

$$f(x + \Delta x) = f(x) + \Delta x f'(x) + \frac{\Delta x^2}{2!} f''(x) + \frac{\Delta x^3}{3!} f'''(x) + \frac{\Delta x^4}{4!} f^{iv}(x) + \cdots \tag{4.2}$$

Equation 4.1 results from truncating all but the first two terms in the Taylor expansion of Equation 4.2. In order to determine how good the approximation is, consider *temporarily* retaining the term involving f'. This gives

$$f'(x) = \frac{f(x + \Delta x) - f(x)}{\Delta x} - \frac{\Delta x}{2!} f''(x) + \cdots$$

The last term is an estimate of $E(x)$, which is proportional to Δx. A terminology used to describe this is the *Big $\mathcal{O}$ notation* or the order of magnitude notation. The approximation of Equation 4.1 is, therefore, $\mathcal{O}(\Delta x)$, or the error in the approximation is proportional to Δx.

If negative Δx is applied in the Taylor expansion, there results

$$f(x - \Delta x) = f(x) - \Delta x f'(x) + \frac{\Delta x^2}{2!} f''(x) - \frac{\Delta x^3}{3!} f'''(x) + \frac{\Delta x^4}{4!} f^{iv}(x) - \cdots \quad (4.3)$$

Truncating all but the linear terms and solving for $f'(x)$ gives

$$\frac{df}{dx} \cong \left(\frac{f(x) - f(x - \Delta x)}{\Delta x} \right) \quad (4.4)$$

This is called the "backward" difference approximation to the first derivative and is also $\mathcal{O}(\Delta x)$.

Subtracting Equation 4.3 from Equation 4.2 gives

$$f(x + \Delta x) - f(x - \Delta x) = 2 \Delta x f'(x) + 2 \frac{\Delta x^3}{3!} f'''(x) + \cdots \quad (4.5)$$

Solving for f' yields

$$\frac{df}{dx} = \frac{f(x + \Delta x) - f(x - \Delta x)}{2 \Delta x} - \frac{1}{6} \Delta x^2 f''' \quad (4.6)$$

Note that the error is proportional to Δx^2, or it is $\mathcal{O}(\Delta x^2)$. Equations 4.1 and 4.4 are "first-order correct," and Equation 4.6 is "second-order correct." Equation 4.6 (without the error term) is called the central difference approximation to the first derivative.

Adding Equations 4.2 and 4.3 gives the following result:

$$f(x + \Delta x) + f(x - \Delta x) = 2f(x) + 2 \frac{\Delta x^2}{2!} f''(x) + 2 \frac{\Delta x^4}{4!} f^{iv}(x) + \cdots$$

Solving for $f''(x)$ gives

$$f''(x) \cong \frac{f(x + \Delta x) - 2f(x) + f(x - \Delta x)}{\Delta x^2} - \frac{1}{12} \Delta x^2 f^{iv} \quad (4.7)$$

The error term for this approximation to the second derivative is $\mathcal{O}(\Delta x^2)$.

Further manipulations can be performed with the Taylor expansions, such as involving more "points" (e.g., $f(x + 2\Delta x)$, $f(x - 2\Delta x)$, etc.). This leads to (1) better approximations for the first and second derivatives and (2) approximations for higher-order derivatives. Table 4.1 shows the *first-order correct* formulas for the first and second derivatives, while Table 4.2 depicts the *second-order correct* counterparts. In the tables, $\Delta x = x_{i+1} - x_i = x_i - x_{i-1}$.

TABLE 4.1
First-Order Correct Approximations for Derivatives

Forward Difference

$$f'(x_i) = \frac{f(x_{i+1}) - f(x_i)}{\Delta x}$$

$$f''(x_i) = \frac{f(x_i) - 2f(x_{i+1}) + f(x_{i+2})}{\Delta x^2}$$

Backward Difference

$$f'(x_i) = \frac{f(x_i) - f(x_{i-1})}{\Delta x}$$

$$f''(x_i) = \frac{f(x_i) - 2f(x_{i-1}) + f(x_{i+2})}{\Delta x^2}$$

Central Difference

None

TABLE 4.2
Second-Order Correct Approximations for Derivatives

Forward Difference

$$f'(x_i) = \frac{-3f(x_i) + 4f(x_{i+1}) - f(x_{i+2})}{2\Delta x}$$

$$f''(x_i) = \frac{2f(x_i) - 5f(x_{i+1}) + 4f(x_{i+2}) - f(x_{i+3})}{\Delta x^2}$$

Backward Difference

$$f'(x_i) = \frac{3f(x_i) - 4f(x_{i-1}) + f(x_{i-2})}{2\Delta x}$$

$$f''(x_i) = \frac{2f(x_i) - 5f(x_{i-1}) + 4f(x_{i-2}) - f(x_{i-3})}{\Delta x^2}$$

Central Difference

$$f'(x_i) = \frac{f(x_{i+1}) - f(x_{i-1})}{2\Delta x}$$

$$f''(x_i) = \frac{f(x_{i+1}) - 2f(x_i) + f(x_{i-1})}{\Delta x^2}$$

In order to see more clearly where all of these come from, the first one in Table 4.2 is now derived:

$$f(x + \Delta x) = f(x) + \Delta x f'(x) + \frac{\Delta x^2}{2!} f''(x) + \frac{\Delta x^3}{3!} f'''(x) + \frac{\Delta x^4}{4!} f^{iv}(x) + \cdots \quad (4.8)$$

$$f(x + 2\Delta x) = f(x) + 2\Delta x f'(x) + \frac{4\Delta x^2}{2!} f''(x) + \frac{8\Delta x^3}{3!} f'''(x) + \frac{16\Delta x^4}{4!} f^{iv}(x) + \cdots \quad (4.9)$$

Multiplying Equation 4.8 by 4 and subtracting Equation 4.9 from the result gives

$$4f(x + \Delta x) - f(x + 2\Delta x) = 3f(x) + 2\Delta x f'(x) - \frac{4\Delta x^3}{3!} f'''(x) + \cdots$$

Solving for $f'(x)$ yields

$$f'(x) = \frac{-f(x+2\Delta x) + 4f(x+\Delta x) - 3f(x)}{2\Delta x} + \frac{1}{3}\Delta x^2 f''' \tag{4.10}$$

This is the same as the first formula in Table 4.2 and is a second-order correct formula since the error term is $\mathcal{O}(\Delta x^2)$. This is called an *end-point formula* since it can be applied at a boundary where negative perturbations are not allowed.

4.1.2 APPROXIMATION OF PARTIAL DERIVATIVES

Consider a two-dimensional function $f(x, y)$. The difference approximation for the partial derivative

$$f_x = \frac{\partial}{\partial x} f(x, y) \text{ at } x = x_0 \text{ and } y = y_0 \tag{4.11}$$

can be derived by fixing y to y_0 and considering $f(x, y_0)$ as a one-dimensional function. The forward, centered, and backward difference approximations for the above partial derivative may be written as follows:

$$f_x = \frac{f(x_0 + \Delta x, y_0) - f(x_0, y_0)}{\Delta x} + \mathcal{O}(\Delta x)$$

$$f_x = \frac{f(x_0 + \Delta x, y_0) - f(x_0 - \Delta x, y_0)}{2\Delta x} + \mathcal{O}(\Delta x^2) \tag{4.12}$$

$$f_x = \frac{f(x_0, y_0) - f(x_0 - \Delta x, y_0)}{\Delta x} + \mathcal{O}(\Delta x)$$

The centered difference approximations for the second partial derivatives are shown below:

$$f_{xx} = \frac{\partial^2}{\partial x^2} f = \frac{f(x_0 + \Delta x, y_0) - 2f(x_0, y_0) + f(x_0 - \Delta x, y_0)}{\Delta x^2}$$

$$f_{yy} = \frac{\partial^2}{\partial y^2} f = \frac{f(x_0, y_0 + \Delta y) - 2f(x_0, y_0) + f(x_0, y_0 - \Delta y)}{\Delta y^2}$$

$$f_{xy} = \frac{\partial \partial}{\partial \Delta x \partial y} f = \frac{f(x_0 + \Delta x, y_0 + \Delta y) - f(x_0 + \Delta x, y_0 - \Delta y)}{4\Delta x \Delta y} + $$

$$\frac{-f(x_0 - \Delta x, y_0 + \Delta y) + f(x_0 - \Delta x, y_0 - \Delta y)}{4\Delta x \Delta y}$$

$$(4.13)$$

Example 4.1: Solids Volume Fraction in a Fluidized Bed

For a fluidized bed with a gas whose density, ρ_g, is 0.012 kg/m³ and a solid whose density, ρ_s, is 2650 kg/m³, calculate the percentage solids volume as a function of axial position given the following data:

Axial Position (m)	Pressure (kPa)
0	1.8
0.5	1.38
1	1.09
1.5	0.63
2	0.18

Writing a momentum balance for the two-phase flow leads to the following equation for the fraction solids volume, where z = axial position, P = pressure, and g = 9.81 (the gravitational constant):

$$\varepsilon_s = \frac{(-dP/dz) - \rho_g g}{(\rho_s - \rho_g)g} \tag{4.14}$$

This problem requires numerical differentiation. It is recommended to always apply the second-order correct finite difference formulas. The end-point formulas are applied at $z = 0$ and $z = 2$, while centered difference approximations are used at all other points. The results appear in the following spreadsheet, which includes a graph of percent solids volume versus axial position.

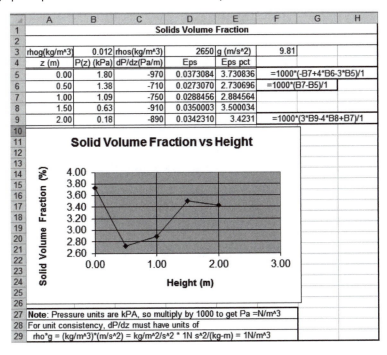

	A	B	C	D	E	F	G	H
1			Solids Volume Fraction					
2								
3	rhog(kg/m^3)	0.012	rhos(kg/m^3)	2650	g (m/s^2)	9.81		
4	z (m)	P(z) (kPa)	dP/dz(Pa/m)	Eps	Eps pct			
5	0.00	1.80	-970	0.0373084	3.730836	=1000*(-B7+4*B6-3*B5)/1		
6	0.50	1.38	-710	0.0273070	2.730696	=1000*(B7-B5)/1		
7	1.00	1.09	-750	0.0288456	2.884564			
8	1.50	0.63	-910	0.0350003	3.500034			
9	2.00	0.18	-890	0.0342310	3.4231	=1000*(3*B9-4*B8+B7)/1		
10								
11			Solid Volume Fraction vs Height					
27	Note: Pressure units are kPA, so multiply by 1000 to get Pa =N/m^3							
28	For unit consistency, dP/dz must have units of							
29	rho*g = (kg/m^3)*(m/s^2) = kg/m^2/s^2 * 1N s^2/(kg-m) = 1N/m^3							

The derivative at $z = 0$ is computed as follows:

$$\frac{dP}{dz} = 1000 \cdot \frac{-3(1.80)+4(1.38)-1.09}{2(0.5)} = 1000 \cdot \frac{-5.4+5.52-1.09}{1} = -970 \qquad (4.15)$$

4.2 NUMERICAL INTEGRATION

4.2.1 TRAPEZOIDAL RULE

The simplest method for numerical integration is the trapezoidal rule, which is based on joining interval end points with a chord to form a trapezoid, whose area is an approximation to the definite integral over the interval. This is depicted in Figure 4.1 with five nonequally spaced subintervals.

The entire integral of $f(x)$ from x_0 to x_5 can be expressed as

$$\int_{x_0}^{x_5} f(x)\,dx = \sum_{i=1}^{5} \frac{f(x_{i-1})+f(x_i)}{2}(x_i - x_{i-1}) + E \qquad (4.16)$$

where the error term, E, can be derived using Taylor expansions for $f(x)$ at the interval end points, $f(x_{i-1})$ and $f(x_i)$. This is a long process whose final result is

$$E \cong -\frac{1}{12}(f(x_{i-1}) - f(x_i))h^2 \overline{f''} \qquad (4.17)$$

where $\overline{f''}$ is the average second derivative of f over the interval. The bottom line is that the error is proportional to h^2, or the method is $\mathcal{O}(h^2)$, where h is a typical interval width.

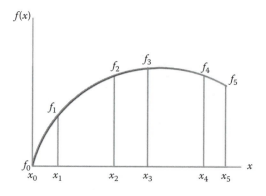

FIGURE 4.1 Schematic of trapezoidal rule.

Unless otherwise stated, all numerical integrations for the remainder of the book will be computed by the trapezoidal rule. It is accurate enough for most engineering applications, does not require equally spaced data, and is easy to implement in Excel® and VBA. For information on more complex and accurate numerical integration methods, Google it!

4.2.2 SIMPSON'S RULE

The trapezoidal rule approximates the function over a subinterval with a straight line. Simpson's rule uses a quadratic function over two successive intervals as shown in Figure 4.2.

By subdividing the interval from a to b into smaller ones, the following formulas apply for N an even number:

$$\int_a^b f(x)\,dx = \frac{h}{3}\left[f(a) + 4\sum_{\substack{i=1 \\ i\,odd}}^{N-1} f(a+ih) + 2\sum_{\substack{i=1 \\ i\,even}}^{N-2} f(a+ih) + f(b) \right] + E$$

$$= \frac{h}{3}[f_0 + 4f_1 + 2f_2 + 4f_3 + 2f_4 + \cdots + + 2f_{N-2} + 4f_{N-1} + f_N] + E$$

(4.18)

The error term, E, can be shown to be proportional to h^4. Major disadvantages are that N must be an *even* number and h must be constant. There exists another version of Simpson's rule when N is odd, but h still must be constant.

There are many more sophisticated methods of quadrature (another name for numerical integration). For reasons already given, the trapezoidal rule is sufficient for most engineering applications. In some specialized cases, Gauss integration (see Section 4.2.3) is the most advantageous method.

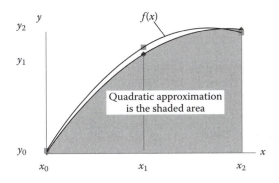

FIGURE 4.2 Schematic of Simpson's rule.

Example 4.2: Using the Trapezoidal Rule

Consider evaluating the integral

$$I = \int_0^2 \pi \left(1 + \left(\frac{x}{2} \right)^2 \right)^2 dx$$

The exact answer is 11.7286. The following Excel spreadsheet shows a solution to this using the trapezoidal rule with different values for h:

Trapezoidal Rule Using $h = 0.5$		
x	$f(x)$	I
0	3.141593	0
0.5	3.546564	1.672039
1	4.908739	3.785865
1.5	7.669904	6.930525
2	12.56637	11.98959

Trapezoidal Rule Using $h = 0.2$		
x	$f(x)$	I
0	3.141593	0
0.2	3.204739	0.634633
0.4	3.397947	1.294902
0.6	3.732526	2.007949
0.8	4.227327	2.803934
1	4.908739	3.717541
1.2	5.81069	4.789484
1.4	6.97465	6.068018
1.6	8.449628	7.610445
1.8	10.29217	9.484625
2	12.56637	11.77048

It can be seen that the answer using $h = 0.5$ is rather poor while that for $h = 0.2$ is close to the correct answer. For most engineering applications, the trapezoidal rule is good enough if a small h is used. In real applications, it is often the case that a good value for h can be estimated. When presented with predetermined data, it is often the case that h is not controllable and might well be variable.

4.2.3 GAUSS QUADRATURE

The mathematician/scientist Gauss developed a particularly unique and interesting method of numerical integration by asking the question, "if I can choose the points on the interval, are there optimal ones to choose?" The answer is a resounding *yes*. The general form of Gauss integration is

$$\int_{-1}^{1} f(x)\,dx = \sum_{i=1}^{n} w_i f(x_i)$$ (4.19)

where both w_i and x_i are chosen so that for a given n, the rule is exact for polynomials up to and including degree $2n - 1$. The w_i are called the *weights*. Note the fixed range of integration from -1 to 1. Thus, if the range is a to b, a change of variables is required. That is, if the *original variable* is z and the range is $[a, b]$, then set

$$x = \frac{2z - a - b}{b - a}$$ (4.20)

The Gauss formula for $n = 2$ from Equation 4.19 becomes

$$\int_{-1}^{1} f(x)\,dx = w_1 f(x_1) + w_2 f(x_2)$$

Make this exact for polynomials of degree 0, 1, 2, and 3 ($2n - 1$) as follows:

$$\int_{-1}^{1} 1\,dx = 2 = w_1 + w_2$$

$$\int_{-1}^{1} x\,dx = 0 = w_1 x_1 + w_2 x_2$$

$$\int_{-1}^{1} x^2\,dx = \frac{2}{3} = w_1 x_1^2 + w_2 x_2^2$$

$$\int_{-1}^{1} x^3\,dx = 0 = w_1 x_1^3 + w_2 x_2^3$$

This gives four equations and four unknowns, which can be solved to give

$$x_1 = {-1}\big/{\sqrt{3}} \quad x_2 = {1}\big/{\sqrt{3}} \quad w_1 = 1 \quad w_2 = 1$$

Thus,

$$\int_{-1}^{1} f(x)\,dx = f\left(\frac{-1}{\sqrt{3}}\right) + f\left(\frac{1}{\sqrt{3}}\right)$$

TABLE 4.3

Gauss Points for $n = 2, 3, ..., 6$

	Gauss Points and Weights	
n	$\pm x$	w_i
2	0.57735	1
3	0	0.88889
	0.7746	0.55556
4	0.33998	0.65215
	0.86114	0.34785
5	0	0.56889
	0.53847	0.47863
	0.90618	0.23693
6	0.23862	0.46791
	0.66121	0.36076
	0.93247	0.17132

Error analysis is not so simple. The above formula is *exact* if $f(x)$ is a cubic polynomial (or a simpler one). A rule of thumb is that the order of accuracy of Gauss integration is twice that of equally spaced methods using the same number of data points.

Gauss quadrature formulas for higher n can be derived in a similar manner, but only the final results are shown here. Table 4.3 shows Gauss points for values of n up to 6.

Example 4.3: Using Gauss Integration

Consider the same problem as in Example 4.2 using four-point Gauss integration. First, perform a change of variables as indicated in Equation 4.20. Let

$$y = \frac{2x - 2}{2} = x - 1 \text{ or } x = y + 1$$

The restated problem is then

$$I = \int_{-1}^{1} \pi \left(1 + \left(\frac{y+1}{2} \right)^2 \right)^2 dy$$

Note that x has been replaced with $y + 1$, dx with dy, and the limits of integration changed accordingly. The four-point Gauss calculations are shown in the following Excel spreadsheet:

Four Point Gauss Quadrature			
y	$f(y)$	Weight	$f(y)*wt$
−0.86114	3.17196	0.34785	1.10338
−0.33998	3.86313	0.65215	2.51932
0.33998	6.59507	0.65215	4.30094
0.86114	10.93838	0.34785	3.80497
		Integral =	**11.72861**

This example illustrates the power of Gauss integration. A four-point formula gives essentially the exact result. [This is due to the fact that $f(x)$ is a quartic function, and the four-point formula is exact for polynomials up to degree 7.]

In summary, the trapezoidal rule is easy to implement and usually accurate enough. Furthermore, it can be used with nonequally spaced data (such as real experimental data). When there is a limit to the number of "sample" points, but they can be placed at will, then Gauss integration is often the best choice. For example, suppose a restriction is that only four samples can be obtained on a process during a test whose duration is 1 h. When during the hour should the samples be collected? The answer to this is left as an exercise.

4.3　CURVE FITTING FOR INTEGRATION

Another approach to numerical integration is to "fit" the data to a particular function form and then do the integration analytically. Sophisticated curve fitting methods are covered in Chapter 7. For now, Excel's graphing capability allows the fitting of simple functions to data. Once the fitting function has been determined, it can be integrated analytically. This is demonstrated in Example 4.4 (along with the trapezoidal rule).

Example 4.4: Integrating Fermentation Data

The data shown in the following spreadsheet represent the rate of evolution of CO_2 and the take-up rate of O_2 during a fermentation reaction. It is important in the study of fermentation processes to obtain the net amount of these gases used and released. This is accomplished by *integrating* the rates over time. The fourth and fifth columns in the spreadsheet show the results of integrating the data using the trapezoidal rule, and the results are

Total CO_2 evolution = 168.3450 g
Total O_2 evolution = 145.5200 g

In this case, h is fixed by the available experimental data.

Time(h)	CO₂ Rate (g/h)	O₂ Rate (g/h)	Trap CO₂ Evolution	Trap O₂ Evolution	CO₂ Evol Curve Fit	O₂ Evol Curve Fit
140	15.72	15.49	0.00	0.00	0.00	0.00
141	15.53	16.16	15.63	15.83	15.66	16.11
142	15.19	15.35	30.99	31.58	31.58	31.71
143	16.56	15.13	46.86	46.82	47.78	46.90
144	16.21	14.20	63.25	61.49	64.26	61.74
145	17.39	14.23	80.05	75.70	81.04	76.32
146	17.36	14.29	97.42	89.96	98.15	90.70
147	17.42	12.74	114.81	103.48	115.59	104.97
148	17.60	14.74	132.32	117.22	133.39	119.21
149	17.75	13.68	150.00	131.43	151.56	133.49
150	18.95	14.51	168.35	145.52	170.12	147.88

A graph of the experimental data is shown in Figure 4.3 along with a curve fit quadratic polynomial equation. These equations were determined by right clicking on one of the experimental points and choosing Add Trendline, which then displays the window shown in Figure 4.4. The Polynomial button was selected along with the order 2. This produced the following curve fit equations:

$$CO_2 = 0.0082t^2 - 2.0567t + 142.76$$

$$O_2 = 0.0385t^2 - 11.335t + 851.48$$

Upon integrating these equations from $t = 140$ to $t = 150$, there results (see columns 6 and 7 of the spreadsheet)

Total CO_2 evolution = 170.12 g
Total O_2 evolution = 147.8 g

These results compare favorably with those obtained from the trapezoidal rule.

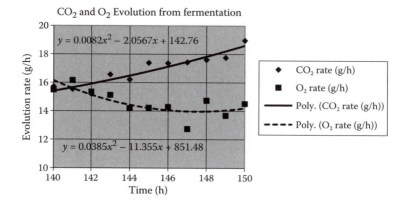

FIGURE 4.3   Graph with data points and trendline with equations displayed.

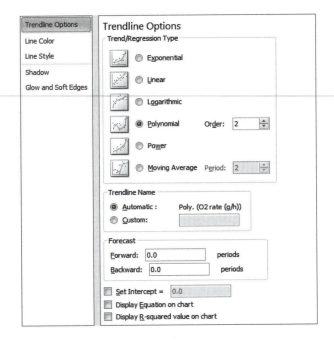

FIGURE 4.4 Trendline options window.

EXERCISES

Exercise 4.1: The heat capacity at constant pressure is defined as

$$C_p = \left(\frac{\partial H}{\partial T} \right)_p$$

where C_p is the heat capacity at constant pressure, H is the molar enthalpy, and T is temperature. The following table shows heat capacity versus temperature data for carbon dioxide (http://webbook.nist.gov/chemistry/fluid/).

Temperature (°C)	Enthalpy (kJ/mol)
100	25.186
150	27.254
200	29.408
250	31.640
300	33.942
350	36.307
400	38.732
450	41.209
500	43.736
550	46.307

600	48.919
650	51.568
700	54.252
750	58.966
800	59.710

a. Use finite differences to compute the heat capacity of carbon dioxide at each of the given temperatures. Be sure to use the "end-point" formulas for the first and last entries.

b. Graph the enthalpy data and curve-fit it with a suitable polynomial. Then, calculate the heat capacity at each temperature using analytical differentiation. Compare these results with those of part a.

c. Compare the results for C_p with those from the *nist* database (these values appear in the data table of Exercise 4.2). Be sure to use consistent units for comparison.

Exercise 4.2: The enthalpy required to heat n moles of a gas from T_1 to T_2 can be found by integrating the heat capacity at constant pressure over the temperature range. The following table lists heat capacity data for CO_2 (also from the *nist* database):

Temperature (°C)	Cp (J/mol*K)
100	40.461
150	42.256
200	43.881
250	45.355
300	46.695
350	47.917
400	49.034
450	50.055
500	50.989
550	51.843
600	52.624
650	53.339
700	53.994
750	54.593
800	55.144

a. Calculate the enthalpy of one mole of CO_2 over the temperature range given in the table using the trapezoidal rule with the following formula (n is the number of moles):

$$\Delta H = n \int_{T_1}^{T_2} C_p(T)\,dT$$

b. Curve fit the heat capacity data using an appropriate polynomial. Then find ΔH (800) (with a reference temperature of 100°C) by integrating the resulting function analytically. What is the percentage error between the two methods?

c. Compare your results for ΔH with those from the *nist* database (these values appear in the data table of Exercise 4.1). Be sure to use consistent units and reference temperature when making comparisons.

Exercise 4.3: Evaluate the following integral using three- and four-point Gauss integrations:

$$erf(p) = \frac{2}{\sqrt{\pi}} \int_0^p e^{-x^2}\,dx$$

To get a value for p, generate a random integer between 0 and 300 using the RANDBETWEEN function. Then let p = the random integer divided by 100 (this gives a floating point number between 0 and 3). Be sure to make a "copy" of p because the random number generator will keep changing it every time a mouse or keyboard command is given.

This is the familiar "error" function (Gaussian normal probability distribution), and the exact value when $p = 1$ is 0.84270073517 (this can be verified by using the Excel function ERF(p)). Remember to change variables so that the interval of integration is from –1 to 1.

What is the percent error produced by the Gauss method for the value of p generated?

Exercise 4.4: Evaluate the following integral using
a. The trapezoidal rule. Experiment with the Δx increment to produce good results.
b. Gauss quadrature. Use 2-, 3-, and 4-point formulas and compare results.

$$I = \int_0^1 \sqrt{x}\,dx$$

Exercise 4.5: For the function $x^2 \cos x$; $0 \le x \le 1$,
a. Calculate the numerical derivative of this function at each point using Δx values of 0.1 and 0.01. Be sure to use end-point formulas at $x = 0$ and $x = 1$. Compare the numerical derivatives at each point with the exact values.

b.   Find the integral of this function using the trapezoidal rule over the same range. Calculate the % error of the integral at $x = 1$ for both Δx values.

Exercise 4.6: For the function $f(x) = \sqrt{x}\sin^2(x)$, do the following operations:

1.   Make a table of the function between $a = 0.4$ and $b = 1.6$ using a Δx of 0.1.
2.   Generate a column of random numbers between 0 and 1 using the Excel function RAND(). Make a *copy* of the random numbers in another column (i.e., Paste by Value). This is so the random numbers do not keep changing.
3.   In the next column, generate numbers according to the formula (*Rand* $- 0.5)/10$. In other words, subtract 0.5 from the column of random numbers and divide the result by 10. This constitutes a column of "noise" to be added to the original function values.
4.   Generate a column of numbers by adding to $f(x)$ from part 1 the noise values of part 3.
5.   Compute the numerical first derivative, $f'(x)$, from the original function values (no noise added).
6.   Compute the numerical first derivative, $f'(x)$, from the noisy function values (after noise has been added).
7.   From a graph of noisy $f(x)$ values versus x, add a trendline and have the equation displayed on the graph.
8.   Differentiate the trendline equation *analytically* and evaluate it for each value of x.
9.   Produce a graph of the three columns of derivative calculations (steps 6, 7, and 8) versus x.
10. Comment on the agreement (or disagreement) between the three derivative estimates.

Exercise 4.7: Shown below are compressibility data for nitrogen:

| | Compressibility Factor, z | | |
Pressure (atm)	0°C	25°C	50°C
0	1.000	1.000	1.000
10	0.996	0.998	1.000
50	0.985	0.996	1.004
100	0.984	1.004	1.018
200	1.036	1.057	1.072
300	1.134	1.146	1.154
400	1.256	1.254	1.253
600	1.524	1.495	1.471
800	1.798	1.723	1.697

a. The pure component fugacity of a substance can be computed from

$$\ln\left(\frac{f}{P}\right) = \int_0^P \frac{z-1}{P}\,dP$$

where z is the compressibility factor $\left(P\hat{V}/RT\right)$, P is the pressure, R is the gas constant, T is the absolute temperature, and $\hat{V}$ is the specific volume.

Using the data in the table, compute the pure component fugacity of nitrogen at 0°C, 25°C, and 50°C. Since the data are not equally spaced, use of the trapezoidal rule is suggested to perform the numerical integration.

b. The pure component enthalpy relative to zero enthalpy at 0 atm and 25°C is given by

$$h(P, 25°C) = \int_0^P -\frac{RT^2}{P}\left(\frac{\partial z}{\partial T}\right)_P\,dP$$

The derivative of z with respect to T at constant P is required within the integrand. Since data are available at 0°C, 25°C, and 50°C, this derivative can be estimated using a centered difference approximation at each pressure. Compute h from this formula for nitrogen at all pressures given. Note that at zero pressure, the integrand is indeterminate (0/0). Using L'Hopital's rule, the value of the integrand is given by

$$-RT^2\left(\frac{\partial^2 z}{\partial T \partial P}\right)_{P=0}$$

The derivative in this expression can be evaluated from the data by first evaluating $\left(\frac{\partial z}{\partial T}\right)_P$ at $P = 0$, 200, and 400 atm and then using the "left end-point" second-order correct finite difference formula for the first derivative (to get the derivative with respect to P). The resulting value of the integrand is –1.296 cal/gmol-atm (this should be verified).

Exercise 4.8: A process engineer is performing tests on a unit that has been giving problems. A crucial measurement is the *average* concentration of a particular component in the feed stream to the unit. The analytical method available for determining the concentration of this key component is very expensive and time consuming, such that the budget allows only four samples to be drawn for the purpose of determining the average concentration. Further, the experiment will take place over a 2-h period. Suggest *when* the samples should be taken and how the best possible average concentration can be determined.

Hint: One way to compute an average of a function, $f(t)$, is to integrate over the time interval and divide by the interval width. That is,

$$Average(f(t)) = \frac{1}{t_{max}} \int_0^{t_{max}} f(t)\,dt$$

5 Ordinary Differential Equations (Initial Value Problems)

5.1 INTRODUCTION

Many differential equations defy analytical solution. Still others are such that analytical solutions are onerous. In these cases, a numerical solution is usually the best (and sometimes the only) option. In this chapter, several methods are presented for solving single or multiple ordinary differential equation(s) numerically. Here, only initial value problems (IVPs) are considered, where all necessary information is given at the origin of the independent variable (usually time or distance). The coverage of methods is not complete. Only the more popular methods used in engineering problem solving are considered. These include the Euler, backward Euler, trapezoidal, and Runge–Kutta (RK) methods.

5.1.1 General Statement of the Problem

The IVP involving first-order ordinary differential equation(s) can be written as follows:

$$\frac{dy}{dt} = f(y,t), \quad y(0) = y_0 \tag{5.1}$$

where $f(y, t)$ is an n-vector function of the n-vector y. t is the independent variable, and y_0 is an n-vector of the initial condition(s). When $n = 1$, there is a single ordinary differential equation (ODE).

5.2 EULER-TYPE METHODS

5.2.1 Euler's Method for Single ODE

The simplest numerical method for solving one ODE is called the Euler method and is based on approximating the derivative with a forward difference approximation to the derivative as follows:

$$\frac{y_{n+1} - y_n}{h} \cong f(y_n, t_n) \tag{5.2}$$

99

where

$$y_{n+1} = y(t_n + h)$$
$$y_n = y(t_n) \tag{5.3}$$
$$h = \Delta t$$

The Euler approximation can be written in terms of a "recurrence" relation as follows:

$$y_{n+1} = y_n + hf(y_n, t_n) \tag{5.4}$$

Starting with the initial condition, this recurrence formula can be used to "step forward in time" as follows:

$$y_1 = y_0 + hf(y_0, 0)$$
$$y_2 = y_1 + hf(y_1, h)$$
$$\vdots \tag{5.5}$$
$$y_n = y_{n-1} + hf(y_{n-1}, (n-1)h)$$

This kind of sequential calculation is called an *explicit* method.

Example 5.1: Euler Method for an ODE-IVP

$$\frac{dy}{dt} = -20y + 7\exp(-0.5t), \quad y(0) = 5 \tag{5.6}$$

The analytical solution is obtained easily using an integrating factor to give

$$y = 5e^{-20t} + (7/19.5)(e^{-0.5t} - e^{-20t}) \tag{5.7}$$

Using the Euler method with $h = 0.01$, the first few steps of the calculations are as follows:

$$y_1 = 5 + 0.01(-20 * 5 + 7 * \exp(0)) = 4.07000$$
$$y_2 = 4.07 + 0.01(-20 * 4.07 + 7 * \exp(-0.005)) = 3.32565$$

Figure 5.1 summarizes the calculations for t up to 0.1 using $h = 0.01$. The values from the analytical solution are also shown for comparison.

t	y	y exact
0.00	5.00000	5.00000
0.01	4.07000	4.15693
0.02	3.32565	3.46638
0.03	2.72982	2.90068
0.04	2.25282	2.43721
0.05	1.87087	2.05745
0.06	1.56497	1.74622
0.07	1.31990	1.49109
0.08	1.12352	1.28191
0.09	0.96607	1.11033
0.10	0.83977	0.96956

FIGURE 5.1   Euler method results for $h = 0.01$.

From the table, it can be seen that the disagreement between the numerical solution and analytical solution grows with t at a significant rate. This suggests that a smaller value for h is required. The percentage error for $h = 0.01$, 0.001, and 0.0001, respectively, is shown in the graph in Figure 5.2.

It can be observed that the error decreases approximately one order of magnitude for every order of magnitude decrease in h. This demonstrates "empirically" that this method is of $\mathcal{O}(h)$. One example is not the proof of this error behavior for Euler's method, but it is persuasive, and in practice, this is borne out to be the case.

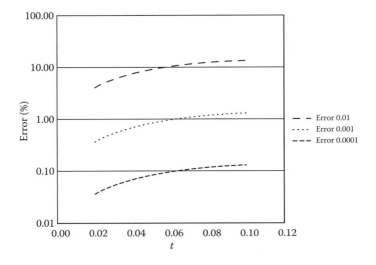

FIGURE 5.2   Error using the Euler method when changing h.

5.2.2 STABILITY OF NUMERICAL SOLUTIONS OF ODEs

It can be instructive to study the simple ODE-IVP

$$\frac{dy}{dt} = -ay; \quad y(0) = 1 \quad a > 0 \tag{5.8}$$

whose exact solution is $y(t) = \exp(-at)$. $1/a$ is called the system *time constant*. Applying the Euler method, there results

$$y_{n+1} = (1 - ah)y_n \tag{5.9}$$

Since the exact solution decreases exponentially with t, it is clear that the approximate solution should also decrease continuously. Therefore, it is necessary that

$$(1 - ah) < 1 \ or \ 0 < ah < 1 \qquad \text{(note: } ah \text{ cannot be negative)}$$

Further, if $1 < ah < 2$, then the solution *alternates sign* at each step. And, if $ah > 2$, the *magnitude* of the solution increases at each step and it *oscillates*—this is called *instability*.

This is a new kind of error. Truncation error exists when h is too large, round-off error occurs when h is too small, and the numerical solution exhibits instability depending on the product of the system time constant and h.

Although this analysis has been performed only for the simplest of ODEs, in practice, this same behavior is often observed. Obviously, some care must be given to the selection of h to avoid unacceptable errors of any kind.

5.2.3 EULER BACKWARD METHOD

Consider a backward finite difference approximation for the derivative

$$\frac{y_{n+1} - y_n}{h} \cong f(y_{n+1}, t_{n+1}) \tag{5.10}$$

This equation is "implicit" in y_{n+1} since it appears on both sides of the equation. It can be shown that this method is also $\mathcal{O}(h)$.

Applying this approximation to the exponential test problem given by Equation 5.8, the following recurrence results:

$$y_{n+1} = \frac{1}{(1 + ah)} y_n \tag{5.11}$$

This solution decreases monotonically with t for *any* positive value of h. While it is said to be *unconditionally stable*, it still suffers significant truncation error. Also, if $f(y, t)$ is not linear, then a nonlinear equation must be solved at each time step.

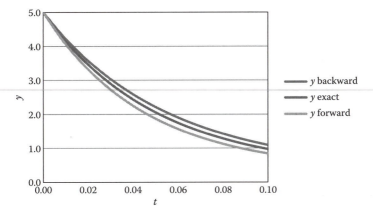

FIGURE 5.3   Comparison of Euler with backward Euler method.

Example 5.2: Backward Euler Method

The following steps show the application of the backward Euler method to the ODE-IVP of Equation 5.6:

$$\frac{y_{n+1} - y_n}{h} = -20y_{n+1} + 7\exp(-0.5t_{n+1})$$

$$(1 + 20h)y_{n+1} = y_n + 7h\exp(-0.5t_{n+1}) \tag{5.12}$$

$$y_{n+1} = \frac{y_n + 7h\exp(-0.5t_{n+1})}{(1 + 20h)}$$

Shown in Figure 5.3 is a comparison of the Euler and backward Euler methods for this problem. Only results for $h = 0.01$ are shown. For this example (and it often happens that) one of these methods errs on one side of the solution and the other on the other side.

Here, the Euler method "undershoots" the exact solution, while the backward Euler method "overshoots" it. This suggests that an average of the two methods might give better results, and this is indeed that case (see Section 5.2.4).

5.2.4   TRAPEZOIDAL METHOD (MODIFIED EULER)

This method applies a centered difference approximation for the derivative and an average value for f on the right-hand side. The finite difference analog is "centered about the ½ point." That is, the differential equation is discretized at $t_{n+1/2}$, and $f(y_{n+1/2}, t_{n+1/2})$ is approximated by the average of the end-point values. Therefore, this method is an average of the forward and backward Euler methods, and the discrete approximation appears in the following equation:

$$\frac{y_{n+1} - y_n}{h} = \frac{1}{2}[f(y_{n+1}, t_{n+1}) + f(y_n, t_n)] \tag{5.13}$$

This equation is implicit in y_{n+1}. If f is nonlinear, then a nonlinear equation must be solved at each time step. It can be shown that this method is $\mathcal{O}(h^2)$.

Looking again at the exponential test problem given by Equation 5.8 and applying the trapezoidal method, the following recurrence results:

$$y_{n+1} = \frac{(1-ah/2)}{(1+ah/2)} y_n \qquad (5.14)$$

This equation is stable for $0 < ah/2 < 1$ or $0 < ah < 2$.

Example 5.3: Trapezoidal Method

Applying the trapezoidal method to the ODE-IVP of Equation 5.6 leads to the following:

$$\frac{y_{n+1} - y_n}{h} = \frac{1}{2}[-20y_{n+1} + 7\exp(-0.5t_{n+1}) - 20y_n + 7\exp(-0.5t_n)]$$

$$(1+10h)y_{n+1} = y_n + \frac{h}{2}[7\exp(-0.5t_{n+1}) + 7\exp(-0.5t_n) - 20y_n] \qquad (5.15)$$

$$y_{n+1} = \frac{y_n + \frac{h}{2}[7\exp(-0.5t_{n+1}) + 7\exp(-0.5t_n) - 20y_n]}{(1+10h)}$$

Figure 5.4 shows results for $h = 0.01$ and $h = 0.001$. It can be observed that the error is reduced by two orders of magnitude when h is decreased by only

$y(0) =$	5			Percent	
t	$y(h = 0.01)$	$y(h = 0.001)$	y exact	error(0.01)	error(0.001)
0	5	5	5	0	0
0.01	4.15439	4.15691	4.15693	0.061286	0.000609
0.02	3.46220	3.46633	3.46638	0.120305	0.001197
0.03	2.89556	2.90063	2.90068	0.176501	0.001756
0.04	2.43163	2.43716	2.43721	0.229239	0.002282
0.05	2.05173	2.05739	2.05745	0.277817	0.002766
0.06	1.74060	1.74616	1.74622	0.321490	0.003202
0.07	1.48573	1.49104	1.49109	0.359504	0.003582
0.08	1.27689	1.28186	1.28191	0.391146	0.003899
0.09	1.10572	1.11029	1.11033	0.415804	0.004146
0.1	0.96536	0.96952	0.96956	0.433033	0.004319

FIGURE 5.4   Trapezoidal method results.

one order of magnitude. This empirical observation is evidence that the method is $\mathcal{O}(h^2)$.

5.2.5   ACCURACY OF EULER-TYPE METHODS

It has been stated that the Euler and backward Euler methods are $\mathcal{O}(h)$ and that the trapezoidal method is $\mathcal{O}(h^2)$. These results can be shown more convincingly for the exponential test problem given by Equation 5.8, the exact solution of which is the negative exponential. This can be written in the following incremental form:

$$y(t) = \exp(-at)$$

$$y_{n+1} = y_n \exp(-ah) = \left(1 - ah + \frac{1}{2}(ah^2) - \frac{1}{6}(ah)^3 + \cdots\right)y_n \qquad (5.16)$$

The original Euler method has the recurrence (see Equation 5.9)

$$y_{n+1} = (1 - ah)y_n$$

which represents the first two terms in the series solution. The error term, therefore, is proportional to h^2, but this error is made at *each step*. The global error at the end of many steps is proportional to h itself, so the entire process is $\mathcal{O}(h)$.

For the backward Euler, the recurrence is

$$y_{n+1} = \frac{1}{(1+ah)}\, y_n = (1 - (ah) + (ah)^2 - (ah)^3 + \cdots)y_n \qquad (5.17)$$

This equation, when compared to the exact series expansion, is accurate to the first two terms, and the term involving h^2 is the magnitude of the error. So, once again, each step is $\mathcal{O}(h^2)$ while the global error is $\mathcal{O}(h)$.

For the trapezoidal method,

$$y_{n+1} = \frac{(1 - ah/2)}{(1 + ah/2)}\, y_n = \left(1 - (ah) + \frac{1}{2}(ah)^2 - \frac{1}{4}(ah)^3 + \cdots\right)y_n \qquad (5.18)$$

This equation is in agreement with the first three terms of the exact expansion, so the step error is $\mathcal{O}(h^3)$, while the global error is $\mathcal{O}(h^2)$.

5.3   RK METHODS

These are among the most popular methods for solving ODE-IVPs. They are *explicit* and are based on the idea of using intermediate points in each major time step (note: even though the independent variable often is time, it can be distance, volume, or some other quantity). Recall that any explicit method can encounter stability issues if the time step is not carefully chosen.

Consider the general ODE-IVP

$$\frac{dy}{dt} = f(y,t), \quad y(0) = y_0 \tag{5.19}$$

To calculate y_{n+1} at $t_{n+1} = t_n + h$ with a known value of y_n, Equation 5.19 can be integrated over the interval $[t_n, t_{n+1}]$ as

$$y_{n+1} = y_n + \int_{t_n}^{t_{n+1}} f(y,t)\, dt \tag{5.20}$$

RK methods are derived by applying a numerical integration method to the integral on the right-hand side.

5.3.1   SECOND-ORDER RK METHOD

It is straightforward to derive the recurrence relations for the second-order RK method. Suppose that the trapezoidal rule is used to evaluate the integral on the right-hand side in Equation 5.20:

$$\int_{t_n}^{t_{n+1}} f(y,t)\, dt = \frac{1}{2} h[f(y_n,t_n) + f(y_{n+1},t_{n+1})] \tag{5.21}$$

Since y_{n+1} is not known, consider approximating it by $f(\bar{y}_{n+1}, t_{n+1})$ where $\bar{y}_{n+1}$ is a "first estimate" for y_{n+1} calculated by the forward Euler method:

$$\bar{y}_{n+1} = y_n + hf(y_n,t_n)$$

$$y_{n+1} = y_n + \frac{1}{2} h[f(y_n,t_n) + f(\bar{y}_{n+1},t_{n+1})] \tag{5.22}$$

A standard computational notation is as follows:

$$k_1 = hf(y_n,t_n)$$

$$k_2 = hf(y_n + k_1, t_{n+1})$$

$$y_{n+1} = y_n + \frac{1}{2}[k_1 + k_2] \tag{5.23}$$

Without proof, this method is $\mathcal{O}(h^2)$.

5.3.2 FOURTH-ORDER RK METHOD

The fourth-order RK method can be derived in a manner similar to that for the second-order method. The basic idea, once again, is to subdivide the interval h and to use successive approximations to y_{n+1}. The final y_{n+1} is a weighted average of the individual approximations. The resulting equations are

$$k_1 = hf(y_n, t_n)$$

$$k_2 = hf\left(y_n + \frac{k_1}{2}, t_n + \frac{h}{2}\right)$$

$$k_3 = hf\left(y_n + \frac{k_2}{2}, t_n + \frac{h}{2}\right) \tag{5.24}$$

$$k_4 = hf(y_n + k_3, t_n + h)$$

$$y_{n+1} = y_n + \frac{1}{6}[k_1 + 2k_2 + 2k_3 + k_4]$$

This method can be shown to be $\mathcal{O}(h^4)$ and is perhaps the most popular method for solving ODE-IVPs numerically.

Example 5.4: Second- and Fourth-Order RK Methods

Suppose chemical A is in solution in a perfectly stirred tank and its concentration is $C_A^0 (\text{g/L})$. The constant volumetric flow into and out of the tank is F (L/min) and the tank volume is V(L). A mass balance on component A leads to

$$V\frac{dC_A}{dt} = -FC_A \tag{5.25}$$

The analytical solution is

$$C_A = C_A^0 e^{-tF/V} \tag{5.26}$$

Calculations for the second-order RK method are shown in Figure 5.5. Also shown are the analytical (exact) solution and that produced by the Euler method using the same $h = 0.1$ min.

It is obvious that the second-order RK method produces significantly better results than those of the Euler method. Figure 5.5 also illustrates that it is straightforward to implement this method in Excel®. Similar calculations using the fourth-order RK method are shown in Figure 5.6.

From Figure 5.6, it can be seen that the fourth-order RK method produces essentially exact results for this problem. As previously stated, the overall error associated with the fourth-order RK method is $\mathcal{O}(h^4)$. The method is easy to implement in Excel since it is an *explicit* method.

dt =	0.1	Tau =	0.5		
Time	RK2	k_1	k_2	Exact	Euler
0	1.0000	−0.2000	−0.1600	1.0000	1.0000
0.1	0.8200	−0.1640	−0.1476	0.8187	0.8000
0.2	0.6642	−0.1328	−0.1196	0.6703	0.6400
0.3	0.5380	−0.1076	−0.0968	0.5488	0.5120
0.4	0.4358	−0.0872	−0.0784	0.4493	0.4096
0.5	0.3530	−0.0706	−0.0635	0.3679	0.3277
0.6	0.2859	−0.0572	−0.0515	0.3012	0.2621
0.7	0.2316	−0.0463	−0.0417	0.2466	0.2097
0.8	0.1876	−0.0375	−0.0338	0.2019	0.1678
0.9	0.1519	−0.0304	−0.0274	0.1653	0.1342
1	0.1231	−0.0246	−0.0222	0.1353	0.1074

FIGURE 5.5   Second-order RK results for mixing tank.

dt =	0.1	Tau =	0.5				
Time	RK4	k_1	k_2	k_3	k_4	Exact	Euler
0	1.0000	−0.2000	−0.1800	−0.1820	−0.1636	1.0000	1.0000
0.1	0.8187	−0.1637	−0.1474	−0.1490	−0.1339	0.8187	0.8000
0.2	0.6703	−0.1341	−0.1207	−0.1220	−0.1097	0.6703	0.6400
0.3	0.5488	−0.1098	−0.0988	−0.0999	−0.0898	0.5488	0.5120
0.4	0.4493	−0.0899	−0.0809	−0.0818	−0.0735	0.4493	0.4096
0.5	0.3679	−0.0736	−0.0662	−0.0670	−0.0602	0.3679	0.3277
0.6	0.3012	−0.0602	−0.0542	−0.0548	−0.0493	0.3012	0.2621
0.7	0.2466	−0.0493	−0.0444	−0.0449	−0.0403	0.2466	0.2097
0.8	0.2019	−0.0404	−0.0363	−0.0367	−0.0330	0.2019	0.1678
0.9	0.1653	−0.0331	−0.0298	−0.0301	−0.0270	0.1653	0.1342
1	0.1353	−0.0271	−0.0244	−0.0246	−0.0221	0.1353	0.1074

FIGURE 5.6   Fourth-order RK results for mixing tank.

5.4   STIFF ODEs

Stiffness often refers to a system with a very short time constant. For the system model

$$y' = -ay + s(t); \; y(0) = y_0 \tag{5.27}$$

the time constant is $1/|a|$; if a is large, the time constant is small. Note that the units of a are time^{-1}.

The solution to this equation can be written as

$$y(t) = y_0 e^{-at} + e^{-at} \int_0^t s(x) e^{ax} \, dx \qquad (5.28)$$

Even if a is very large, if $s(t)$ is a slowly varying function, the system responds slowly. To follow this slow response numerically is problematic since the stability of the method depends only on a and has nothing to do with $s(t)$. A very small time step might be required, even though the overall system response is a slowly varying function. This is one example of a phenomenon called *stiffness*.

Another example of stiffness occurs with a system (two or more) of ODEs each with greatly different time constants. For example, consider the system

$$y' = -y + z + 3$$
$$z' = -10^7 z + y \qquad (5.29)$$

The very short time constant in the second equation requires a very small h to follow both y and z in time. Problems of instability can easily arise. To solve stiff problems, the most typical approach is to use an *implicit* method, which is known to exhibit excellent stability properties (recall the backward Euler method). Special software packages are available for solving stiff systems. Fortunately, many of the straightforward chemical engineering problems encountered in practice do not yield stiff systems, but when difficulties arise, stiffness might well be the culprit.

5.5   SOLVING SYSTEMS OF ODE-IVPs

The following is a system of ODE-IVPs:

$$\frac{dy_1}{dt} = f_1(y_1, y_2, \cdots, y_n, t); \qquad y_1(0) = y_{10}$$

$$\frac{dy_2}{dt} = f_2(y_1, y_2, \cdots, y_n, t); \qquad y_2(0) = y_{20} \qquad (5.30)$$

$$\vdots$$

$$\frac{dy_n}{dt} = f_n(y_1, y_2, \cdots, y_n, t); \qquad y_n(0) = y_{n0}$$

The only difference between solving a single ODE-IVP and solving a system of them is that all variables and functions become vectors. This is illustrated in Example 5.5.

Example 5.5: Solving a System of ODE-IVPs

Suppose the following chemical reactions take place in a continuous stirred tank reactor (CSTR):

$$A \underset{k_2}{\overset{k_1}{\Longleftrightarrow}} B \underset{k_4}{\overset{k_3}{\Longleftrightarrow}} C \qquad (5.31)$$

where the rate constants are as follows:

$$k_1 = 1 \text{ min}^{-1}, \ k_2 = 0 \text{ min}^{-1}, \ k_3 = 2 \text{ min}^{-1}, \ k_4 = 3 \text{ min}^{-1}$$

The initial charge to the reactor is all A, so the initial conditions are (in mol/L)

$$C_{A_0} = 1 \quad C_{B_0} = 0 \quad C_{C_0} = 0$$

An unsteady-state mass balance on each component leads to the following set of ODEs:

$$\frac{dC_A}{dt} = -k_1 C_A + k_2 C_B$$

$$\frac{dC_B}{dt} = k_1 C_A - k_2 C_B - k_3 C_B + k_4 C_C \qquad (5.32)$$

$$\frac{dC_C}{dt} = k_3 C_B - k_4 C_C$$

The following spreadsheet displays a solution to this system using the Euler method for time only up to 0.13 min to conserve space:

$k_1 =$	1	$k_2 =$	0
$k_3 =$	2	$k_4 =$	3
$h =$	0.01		
Time	CA	CB	CC
0.00	1.0000	0.0000	0.0000
0.01	0.9900	0.0100	0.0000
0.02	0.9801	0.0197	0.0002
0.03	0.9703	0.0291	0.0006
0.04	0.9606	0.0382	0.0012
0.05	0.9510	0.0471	0.0019
0.06	0.9415	0.0556	0.0028
0.07	0.9321	0.0639	0.0038
0.08	0.9227	0.0720	0.0050
0.09	0.9135	0.0798	0.0063
0.10	0.9044	0.0873	0.0077
0.11	0.8953	0.0946	0.0092
0.12	0.8864	0.1017	0.0108
0.13	0.8775	0.1085	0.0125

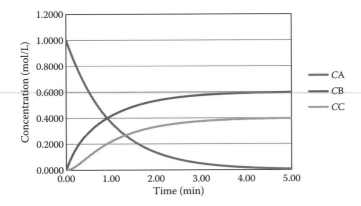

FIGURE 5.7 Euler solution for three simultaneous ODE-IVPs.

Shown in Figure 5.7 is the solution for time up to 5 min at which time a steady state has been essentially reached. It can be observed that component A decreases monotonically while B and C increase.

5.6 HIGHER-ORDER ODEs

Consider a general second-order ODE-IVP of the form

$$\frac{d^2 y}{dt^2} = f(y, y', t) \quad y(0) = y_0 \quad y'(0) = y_0' \tag{5.33}$$

Since all of the methods discussed apply only to first-order equations, one way to solve those of the form of Equation 5.33 is to convert them into two simultaneous first-order equations. This is easily accomplished by defining a *new variable z* as follows:

$$\frac{dy}{dt} = z; \quad y(0) = y_0 \tag{5.34}$$

Equation 5.34 is a first-order ODE-IVP involving both y and z. If this equation is differentiated with respect to t and substituted into Equation 5.33, there results

$$\frac{dz}{dt} = f(y, z, t); \quad z(0) = y'(0) \tag{5.35}$$

Equations 5.34 and 5.35 constitute a system of *coupled* ODE-IVPs. This system can be solved numerically by any of the methods previously discussed.

Example 5.6: Second-Order System Response

A linear second-order dynamic system is defined by the ODE-IVP

$$\tau^2 \frac{d^2y}{d\theta^2} + 2\tau\zeta \frac{dy}{d\theta} + y = f(\theta)$$

(5.36)

where
   τ = the system *time constant*
   ζ = the damping factor
   $f(\theta)$ = the system *input* or *forcing* function

and the initial value of y and y' are given.
   By defining a dimensionless time as $t = \theta/\tau$ and assuming a constant input of 1 (a unit step function occurring at time zero), the model equation becomes

$$\frac{d^2y}{dt^2} + 2\zeta \frac{dy}{dt} + y = 1$$

(5.37)

The only parameter of this model is the damping factor. The damping factor determines if the system is overdamped ($\zeta > 1$), underdamped or oscillatory ($\zeta < 1$), or critically damped ($\zeta = 1$).
   In order to solve this model equation numerically with, for example, the Euler method, it must be converted into two first-order equations as follows:
   Define a new variable z such that

$$\frac{dy}{dt} = z; \, y(0) = y_0$$

(5.38)

Differentiating this expression with respect to t and substituting into the original model equation, there results

$$\frac{dz}{dt} = 1 - 2\zeta z - y; \, z(0) = y'(0)$$

(5.39)

Equations 5.38 and 5.39 are two first-order ODE-IVPs that can be solved by any of the methods previously described. Calculations using the Euler method with $h = 0.1$ and $\zeta = 0.5$ are shown in the following spreadsheet for t up to 2. The formulas for the right-hand side of the equations for y and z are shown in bold type.

	A	B	C	D	E	F
1	DT =	0.1	Zeta =	0.5		
2	Time	y	z			
3	0	0	0	=B3+$B$1*C3		
4	0.1	0	0.1	=C3+$B$1*(1-2*$D$1*C3-B3)		
5	0.2	0.01	0.19			
6	0.3	0.029	0.27			
7	0.4	0.056	0.3401			
8	0.5	0.09001	0.40049			
9	0.6	0.130059	0.45144			
10	0.7	0.175203	0.49329			
11	0.8	0.224532	0.526441			
12	0.9	0.277176	0.551344			
13	1	0.33231	0.568492			
14	1.1	0.38916	0.578411			
15	1.2	0.447001	0.581654			
16	1.3	0.505166	0.578789			
17	1.4	0.563045	0.570393			
18	1.5	0.620084	0.557049			
19	1.6	0.675789	0.539336			
20	1.7	0.729723	0.517824			
21	1.8	0.781505	0.493069			
22	1.9	0.830812	0.465611			
23	2	0.877373	0.435969			

A graph of the system output appears in Figure 5.8. Note that the system response is oscillatory and *overshoots* before settling to the final steady-state value of 1.

Example 5.7: A VBA Program for the Euler Method

In certain circumstances, it can be more efficient to write a VBA program to solve ODE-IVPs numerically. Such cases include times when it is convenient to simply alter a subroutine subprogram that defines the right-hand side function and also when there is a large system of ODE-IVPs involved. This example shows the development of a VBA program to implement Euler's method for any number of simultaneous ODEs. First, a program Specification is given that defines the Excel interface to the program followed by an Algorithm Design, Coding in VBA, and Testing using a problem previously solved using only Excel.

SPECIFICATION

The following "mock-up" of an Excel spreadsheet for solving any number of ODE-IVPs using the Euler method represents a program specification:

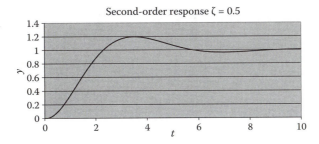

FIGURE 5.8 Response of a second-order system to a step response.

	A	B	C	D	E	F
1	2	N	0.01	h		100	TimeMax
2	Time	y1	y2	y3	...	yN
3	0	Initial Conditions Go Here				
4	0.01	y11	y21	y31	...	yN1
5	0.02	y12	y22	y32	...	yN2
6	⋮	⋮	⋮	⋮	⋮	⋮
7	TimeMax	y1N	y2N	y3N	...	yNN
8						
9	Cell A1 contains the number of simultaneous ODE IVPs to
10	solve. Cell C1 holds the time step and cell E1 the maximum
11	time required. The second row contains appropriate labels
12	(generic ones are shown here). Cells labeled y11, y21, etc.
13	contain values produced by the program. The program
14	"reads" N, h and MaxTime as well as the initial conditions
15	(cells B3 ...). The program computes, one row at a time, the
16	values of y and outputs them to the appropriate row of the
17	spreadsheet.

ALGORITHM DESIGN

The structure chart of Figure 5.9 shows an algorithm design for the main program and for a subroutine to calculate the right-hand side functions. A dictionary of variables is also given.

VBA CODE

A listing of the VBA code for this program is shown below:

```
Option Base 1
Option Explicit
Sub EulerMain()
'Main routine for solving a set of ODE's using the Euler Method.
    Dim y() As Double                   'size N
    Dim f() As Double                   'size N
    Dim h As Double                     'fixed time step
    Dim T1 As Double                    'intermediate time
    Dim TimeMax As Double               'maximum integration time
    Dim Time As Double                  'time counter
    Dim ActRow As Integer               'row counter
    Dim i As Integer                    'a counter
    Dim N As Integer                    'number of state variables
    ' N = number of state variables and equations
    ' DeltaT = Step size in independent variable
    Worksheets(ActiveSheet.Name).Activate

    N = ActiveSheet.Cells(1, 1).Value           'The number of ODEs
    ReDim y(1 To N) As Double                    'size N
    ReDim f(1 To N) As Double                   'size N
    h = ActiveSheet.Cells(1, 3).Value           'get DeltaT from spreadsheet
    TimeMax = ActiveSheet.Cells(1, 5).Value     'get TimMax from spreadsheet
    Time = ActiveSheet.Cells(3, 1).Value        'get initial value of time
    For i = 1 To N
       y(i) = ActiveSheet.Cells(3, 1 + i).Value    'get initial y values
    Next i
    ActRow = 4                                  'calculated output starts here
    While (Time <= TimeMax)
            'Here are the Euler calculations for one time step
        Call FCalc(Time, y(), N, f())
        For i = 1 To N
            y(i) = y(i) + h * f(i)
        Next i
        Time = Time + h         'Display the results on the spreadsheet
        ActiveSheet.Cells(ActRow, 1) = Time     'put time on spreadsheet
        For i = 1 To N
            ActiveSheet.Cells(ActRow, i + 1).Value = y(i) 'output
        Next i
        ActRow = ActRow + 1
    Wend                                        'repeat for all times
    End Sub
```

Name	Usage
N	The number of simultaneous ODEs to solve
h	The time step (or step for any independent variable)
TimeMax	The maximum time over which the solution is sought
y	N-length vector of dependent variables
ActRow	Pointer to current row in spreadsheet
Time	Current value of time
f	N-length vector of right side functions
Func	Subroutine to calculate f given y and N
Zeta	Damping factor for second order system

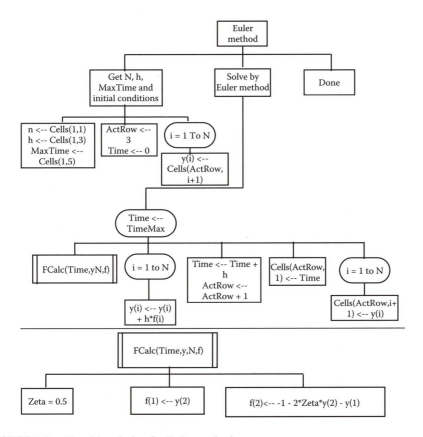

FIGURE 5.9 Algorithm design for Euler method.

```
Sub FCalc(Time, y, N, f)

'This is the "right hand side" function
'for a second order system step response

Dim Zeta As Double
Zeta = 0.5

f(1) = y(2)
f(2) = 1 - 2 * Zeta * y(2) - y(1)

End Sub
```

PROGRAM EXECUTION

The following spreadsheet shows program execution up to Time = 1. These results are identical to those of Example 5.6.

2	= N	0.1	= h	10	= TimeMax
T	y	z			
0	0.0000	0.0000			
0.1	0.0000	0.1000			
0.2	0.0100	0.1900			
0.3	0.0290	0.2700			
0.4	0.0560	0.3401			
0.5	0.0900	0.4005			
0.6	0.1301	0.4514			
0.7	0.1752	0.4933			
0.8	0.2245	0.5264			
0.9	0.2772	0.5513			
1	0.3323	0.5685			

EXERCISES

Exercise 5.1: Dynamic flow in a tank can be modeled by making a mass balance on the fluid in the tank (Figure 5.10). The nature of the resulting ODE-IVP depends on the model used for the outlet valve. If a linear valve is

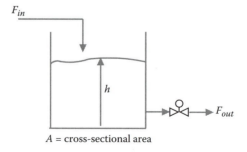

A = cross-sectional area

FIGURE 5.10 Dynamic flow in a tank schematic.

assumed, then the model ODE is linear. A more accurate valve description makes the outlet flow a nonlinear function of the height. Both of these cases are considered in that which follows.

Mass balance:

$$d(\text{mass in tank at any time})/dt = \text{Input} - \text{Output} + \text{Generation}$$

Assume a pure (single) component of constant density fluid:

$$\frac{d\rho V}{dt} = F_{in} - F_{out} \tag{5.40}$$

where
  ρ = density
  V = volume
  F = mass flow rate

In terms of cross-sectional area and height,

$$\frac{d\rho Ah}{dt} = F_{in} - F_{out} \tag{5.41}$$

Since density and area are constant,

$$\frac{dh}{dt} = \frac{F_{in} - F_{out}}{\rho A} \tag{5.42}$$

F_{out} can be
  a. Directly proportional to h: $F_{out} = C_v h$ in which case the model equation is

$$\frac{dh}{dt} = \frac{F_{in} - C_v h}{\rho A} \tag{5.43}$$

  b. A function of h, such as $F_{out} = C_v h^{1/2}$ and the model becomes

$$\frac{dh}{dt} = \frac{F_{in} - C_v \sqrt{h}}{\rho A} \tag{5.44}$$

Additional pertinent data are as follows:
  • Tank diameter = 6 ft
  • Density of liquid = 62.5 lb/ft^3
  • Valve constant C_v = 250 lb/h-ft for the linear case

- Valve constant $C_v = 500$ lb/h-ft$^{0.5}$ for the nonlinear case
- Initial height = 4 ft
- $F_{in} = 1000$ lb/h (initial steady-state inlet flow)
- t = time (h)

F_{in} undergoes a change at time 0+ to 1400 lb/h until time 10 h, after which it returns to 1000 lb/h.

a. Solve the linear ODE (F_{out} proportional to h) using Euler's method. Start with a Δt of 0.5 and again with $\Delta t = 0.05$ to determine an acceptable Δt. Use an IF function to generate the proper values for F_{in}.

b. Repeat part a for the nonlinear case (F_{out} proportional to $h^{1/2}$).

Exercise 5.2: Solve the ODEs of Exercise 5.1 using the second-order RK method. Find acceptable values of h when using the second-order RK method.

Exercise 5.3: Solve the ODEs of Exercise 5.1 using the Euler VBA program as given in Example 5.7.

Exercise 5.4: Suppose the following chemical reactions take place in a continuous stirred tank reactor (CSTR):

$$A \overset{k_1}{\underset{k_2}{\Longleftrightarrow}} B$$

where the rate constants are as follows:

$$k_1 = 1 \text{ min}^{-1}, k_2 = 0.5 \text{ min}^{-1}$$

The charge to the reactor is all component A, so initially, the concentrations within the reactor are

$$C_{A_0} = 1 \quad C_{B_0} = 0 \text{ (gmol/L)}$$

An unsteady-state mass balance on each component leads to the following set of ODEs:

$$\frac{dC_A}{dt} = -k_1 C_A + k_2 C_B$$

$$\frac{dC_B}{dt} = k_1 C_A - k_2 C_B$$

(5.45)

Solve this ODE-IVP system using the Euler method in Excel with a time step of 0.1 min and repeat with a time step of 0.01 min. If a significant difference results (compare values at a specific time to see if they are different), repeat with a time step of 0.001, etc. until a satisfactory time step is found. Integrate out to time = 2 min. Plot the solutions in each case and compare them. Report the percent difference in the solutions at time = 1 min.

Perform experiments by changing the rate constants in one of the solutions to see how Excel immediately recalculates the solution and updates the graph.

Exercise 5.5: Solve the ODEs of Exercise 5.4 using the second-order RK method. Compare results for $\Delta t = 0.1$ min and for $\Delta t = 0.01$ min.

Exercise 5.6: Solve the ODEs of Exercise 5.4 using the Euler VBA program as given in Example 5.7. Use time steps of 0.1, 0.01, and 0.001 min, respectively, and deduce which is the most appropriate for an acceptable solution.

Exercise 5.7: Solve the ODEs of Exercise 5.7 using the Euler VBA program as given in Example 5.7. Use time steps of 0.1, 0.01, and 0.001 min and compare the results to find an acceptable solution.

Exercise 5.8: Implement the second-order RK method for *any number* of simultaneous ODE-IVPs in VBA. Use Example 5.7 as a model for designing, coding, and testing the program. Solve the problem of Exercise 5.7 to test the program.

Exercise 5.9: Penicillin Fermentation

A model for a batch reactor in which penicillin is produced by fermentation has been derived as follows (Constantantinides et al. 1970) for cell production and penicillin synthesis, respectively:

$$\frac{dy_1}{dt} = b_1 y_1 - \frac{b_1}{b_2} y_1^2 \qquad y_1(0) = 0.03$$

$$\frac{dy_2}{dt} = b_3 y_1 \qquad y_2(0) = 0.0$$

(5.46)

where
  y_1 = dimensionless concentration of cell mass
  y_2 = dimensionless concentration of penicillin
  t = dimensionless time, $0 \le t \le 1$

Experiments have determined that

$$b_1 = 13.1$$

$$b_2 = 0.94$$

$$b_3 = 1.71$$

a. Solve this ODE-IVP using the Euler method and Excel (not VBA).
b. Solve this ODE-IVP using the second-order RK method and Excel (not VBA).
c. Solve this ODE-IVP using the Euler VBA program as given in Example 5.7.
d. Solve this ODE-IVP using the second-order RK VBA program from Exercise 5.8.

In all cases, experiment with the time step to assure an accurate solution. Also, graph y_1 and y_2 versus time with appropriate annotations.

Exercise 5.10: Bioreaction Kinetics
The "Monod" model for bioreaction kinetics can be expressed as

$$\frac{ds}{dt} = -\frac{ksx}{k_s + s} \qquad s(0) = s_o, x(0) = x_o \qquad (5.47)$$
$$\frac{dx}{dt} = y\frac{ksx}{k_s + s} - bx$$

where
s = Growth limiting substrate concentration (ML^{-3})
x = Biomass concentration (ML^{-3})
k = Maximum specific uptake rate of the substrate (T^{-1}) = 5
k_s = Half saturation constant for growth (ML^{-3}) = 20
y = Yield coefficient (MM^{-1}) = 0.05
b = Decay coefficient (T^{-1}) = 0.01

Initial conditions are s_o = 1000 and x_o = 100.
a. Solve this ODE-IVP using the Euler method and Excel (not VBA).
b. Solve this ODE-IVP using the second-order RK method and Excel (not VBA).
c. Solve this ODE-IVP using the Euler VBA program as given in Example 5.7.
d. Solve this ODE-IVP using the second-order RK VBA program from Exercise 5.8.

An appropriate maximum time over which to integrate these equations must be determined by experimentation. Also, experiment with the time step to guarantee good results. Graph s and x versus time with appropriate annotations (do this with data for only an appropriate time step).

Exercise 5.11: Implement the fourth-order RK method for *any number* of simultaneous ODE-IVPs in VBA.

The only thing a user must change to use the program is the Subroutine to calculate the "right-hand side" functions for the ODEs. The user must also specify the required input data. A *suggested* user interface associated with the Excel spreadsheet that interacts with the VBA program is as shown in Example 5.7.

In what follows, some hints are given to assist in preparing the VBA program.

Recall the algorithm for using the fourth-order RK method with only one ODE:

$$k_1 = hf(y_n, t_n)$$

$$k_2 = hf\left(y_n + \frac{k_1}{2}, t_n + \frac{h}{2}\right)$$

$$k_3 = hf\left(y_n + \frac{k_2}{2}, t_n + \frac{h}{2}\right) \tag{5.48}$$

$$k_4 = hf(y_n + k_3, t_n + h)$$

$$y_{n+1} = y_n + \frac{1}{6}[k_1 + 2k_2 + 2k_3 + k_4]$$

To generalize this to N simultaneous ODEs, make each k, y, and f an array with subscripts from 1 to N. Here are suggested ReDim statements:

```
ReDim k1(N) as double
ReDim k2(N) as double
ReDim k3(N) as double
ReDim k4(N) as double
ReDim y(N) as double
ReDim f(N) as double
ReDim z(N) as double
```

Use a subroutine (such as Sub FCalc of Example 5.7) that has as input the current time, the y array, and the number of equations (N), and it returns N right-hand-side functions (f). This subroutine must be called *four times per time step*. Note that y, k_1, k_2, k_3, and k_4 are N-length vectors.

1. Once at the base point $f(y_n, t_n)$

2. A first time at the half-way point $f\left(y_n + \frac{k_1}{2}, t_n + \frac{h}{2}\right)$

3. A second time at the half-way point $f\left(y_n + \frac{k_2}{2}, t_n + \frac{h}{2}\right)$

4. And finally at the end point $f(y_n + k_3, t_n + h)$

It is suggested to define another array (call it z) and use this for the first argument in the subroutine calls at the half-way and end points. Also, define a second time variable (call it Ttemp). For example, for the first call at the half-way point, use the sequence

```
z(i) = y(i) + k₁(i)/2          i = 1, 2,..., N
Ttemp = Time + h/2
Call FCalc(Ttemp, z(), N, f())
k₂(i) = h*f(i)                 i = 1, 2,..., N
```

It is strongly urged to sketch out the logic of the code and to "run through" this logic carefully before actually starting to write the code.

Finally, test the resulting VBA program by solving the ODE-IVP of Exercise 5.10. Experiment with the time step to be sure that an acceptable solution is being generated. Beware of the results for a time step of 0.1.

6 Ordinary Differential Equations (Boundary Value Problems)

6.1 INTRODUCTION

So-called boundary value problems (BVPs) occur most often when the system model is a second-order ODE and the known information is available for two different values of the independent variable. For example, consider the following ODE problem:

$$\frac{d^2y}{dx^2} = f(x, y) \quad y(x_1) = y_1 \quad y(x_2) = y_2 \tag{6.1}$$

This problem differs from ODE-IVP since two initial conditions are not given. The two values of the independent variable where information is available are usually at a physical boundary of the system, and the problem is referred to as a *boundary value problem*. Here is a more specific example:

Example 6.1: Heat Conduction in a Rod

A copper rod of length 1 m is placed between two tanks, one containing boiling water and the other containing ice. The rod is exposed to the air. The mathematical model for this system can be expressed as follows:

$$\frac{d^2T}{dx^2} = \frac{4h}{Dk}(T - T_a) \tag{6.2}$$

where
  h = Heat transfer coefficient between rod and air = 50 W/(m² K)
  D = Diameter of the rod = 4 cm = 0.04 m
  k = Thermal conductivity of the rod = 390 W/(m K)
  T_a = Air temperature = 25°C

The *boundary* conditions are

$$T(0) = 100°C$$
$$T(1) = 0°C$$

The methods covered in Chapter 5 cannot be used directly for this problem. The standard procedure would be to convert Equation 6.2 into two first-order ODEs, which would require two initial conditions. In the present example, $0 \leq x \leq 1$, and only one condition is available at $x = 0$. Therefore, some strategy must be used so that available methods can be applied.

6.2 SHOOTING METHOD

The *shooting method*, as the name implies, makes a guess at a second initial condition and then applies one of the numerical methods covered in Chapter 5 to "shoot" at the far boundary. Based on the error in the result from the known second boundary condition, the guessed initial condition is adjusted until the condition is satisfied. This idea is now applied to the problem of Example 6.2.

Example 6.2: Shooting Method for Heat Conduction in a Rod

First, the second-order ODE is transformed into two first-order ODEs. Define $F = \dfrac{dT}{dx}$.

Then, the converted problem becomes

$$\frac{dF}{dx} = \frac{4h}{Dk}(T - T_a)$$
$$\frac{dT}{dx} = F$$
$$T(0) = 100 \tag{6.3}$$
$$F(0) = ?$$
$$T(1) = 0$$

Since $F(0)$ is not known, a guess is made and the problem is solved as an IVP. The value of T at $x = 1$ is then checked; if it is too low, $F(0)$ is increased (aimed too low); if too high, $F(0)$ is decreased (aimed too high).

At the Excel® level, the Goal Seek tool can be used to converge the second boundary condition. This process is illustrated in the following spreadsheet where the initial guess was $F(0) = -100$ (the derivative must be negative if the temperature is to decrease).

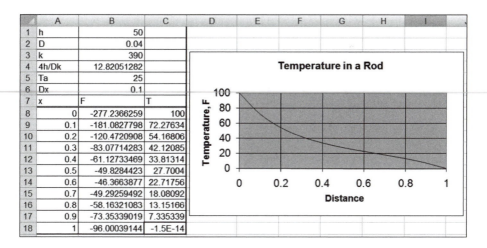

	A	B	C	D	E	F	G	H	I
1	h	50							
2	D	0.04							
3	k	390							
4	4h/Dk	12.82051282							
5	Ta	25							
6	Dx	0.1							
7	x	F	T						
8	0	-277.2366259	100						
9	0.1	-181.0827798	72.27634						
10	0.2	-120.4720908	54.16806						
11	0.3	-83.07714283	42.12085						
12	0.4	-61.12733469	33.81314						
13	0.5	-49.8284423	27.7004						
14	0.6	-46.3663877	22.71756						
15	0.7	-49.29259492	18.08092						
16	0.8	-58.16321083	13.15166						
17	0.9	-73.35339019	7.335339						
18	1	-96.00039144	-1.5E-14						

The lower right-hand cell contains the value of $T(1)$, which should be zero. It was driven to (nearly) zero using Goal Seek by varying $F(0)$. The initial "guess" was −100 for $F(0)$ and Goal Seek found the value −277.237.

6.3 SPLIT BVPs USING FINITE DIFFERENCES

Another approach to solving BVPs is to discretize the ODEs using finite differences. The boundary conditions are applied directly, and the resulting set of equations are solved simultaneously. This approach is well suited to *linear* problems (ones that lead to a set of linear algebraic equations). When the problem leads to nonlinear equations, a tool (such as Solver) must be used. There are also other methods that have been developed for nonlinear problems, but these are beyond the scope of the present discussion (an excellent discussion is given in Riggs 1994). Note that the shooting method can be used without undue difficulty on nonlinear problems.

Example 6.3: Finite Difference Solution for Heat Conduction in a Rod

Consider the same problem as in Example 6.2, which used the shooting method. Dividing the line from 0 to 1 into equal increments of Δx and writing the finite difference form of the equation using a central difference approximation, at the ith node, there results

$$\frac{T_{i-1} - 2T_i + T_{i+1}}{\Delta x^2} = \frac{4h}{Dk}(T_i - T_a) \tag{6.4}$$

with the conditions

$$T_0 = 100$$
$$T_{n+1} = 0$$

There are n interior points. The relationship between n and Δx is

$$\Delta x = \frac{1}{n+1} \tag{6.5}$$

Writing the equation at $i = 1$ gives (let $4h/Dk = a$, and recall that $T_0 = 100$)

$$
\begin{aligned}
\frac{T_0 - 2T_1 + T_2}{\Delta x^2} &= a(T_1 - T_a) \\
-2T_1 + T_2 &= a(T_1 - T_a)\Delta x^2 - 100 \\
-(2 + a\Delta x^2)T_1 + T_2 &= -a\Delta x^2 T_a - 100 \\
(2 + a\Delta x^2)T_1 - T_2 &= a\Delta x^2 T_a + 100
\end{aligned} \tag{6.6}
$$

Let

$$
\begin{aligned}
b &= (2 + a\Delta x^2) \\
c &= a\Delta x^2 T_a
\end{aligned} \tag{6.7}
$$

Then the equation for $i = 1$ becomes

$$bT_1 - T_2 = c + 100 \tag{6.8}$$

Writing the equation for $i = 2$ gives

$$
\begin{aligned}
\frac{T_1 - 2T_2 + T_3}{\Delta x^2} &= a(T_2 - T_a) \\
T_1 - 2T_2 + T_3 &= a(T_2 - T_a)\Delta x^2 \\
T_1 - (2 + a\Delta x^2)T_2 + T_3 &= -a\Delta x^2 T_a \\
-T_1 + (2 + a\Delta x^2)T_2 - T_3 &= a\Delta x^2 T_a \\
-T_1 + bT_2 - T_3 &= c
\end{aligned} \tag{6.9}
$$

Each succeeding equation is like this one, until $i = n$, in which case (recall that $T_{n+1} = 0$) the following results:

$$
\begin{aligned}
\frac{T_{n-1} - 2T_n + T_{n+1}}{\Delta x^2} &= a(T_n - T_a) \\
T_{n-1} - 2T_n + 0 &= a(T_n - T_a)\Delta x^2 \\
T_{n-1} - (2 + a\Delta x^2)T_n &= -a\Delta x^2 T_a \\
-T_{n-1} + (2 + a\Delta x^2)T_n &= a\Delta x^2 T_a \\
-T_{n-1} + bT_n &= c
\end{aligned} \tag{6.10}
$$

Writing all of the equations simultaneously gives a system of *linear* algebraic equations, which can be solved in Excel using the SYSLIN function (or any other method). Note that the matrix of the linear system for problems like this is tridiagonal (has nonzero elements on the diagonal and just above and below the diagonal). These can be solved by SYSLIN, but it is much more efficient to use special numerical methods that take advantage of the "sparseness" of the matrix. The function SYSLIN3 does exactly this and was used to produce the results shown in the following spreadsheet:

h	50.000										
D	0.040										
k	390.000										
4h/Dk	12.821										
Ta	25.000										
n	9.000	Dx =	0.100								
a =	12.821										
b	2.128										
c	3.205										

A =

									c	T
2.128	-1.000								103.205	76.966
-1.000	2.128	-1.000							3.205	60.594
	-1.000	2.128	-1.000						3.205	48.785
		-1.000	2.128	-1.000					3.205	40.025
			-1.000	2.128	-1.000				3.205	33.192
				-1.000	2.128	-1.000			3.205	27.409
					-1.000	2.128	-1.000		3.205	21.935
						-1.000	2.128	-1.000	3.205	16.068
							-1.000	2.128	3.205	9.056

x	T
0.000	100.000
0.100	76.966
0.200	60.594
0.300	48.785
0.400	40.025
0.500	33.192
0.600	27.409
0.700	21.935
0.800	16.068
0.900	9.056
1.000	0.000

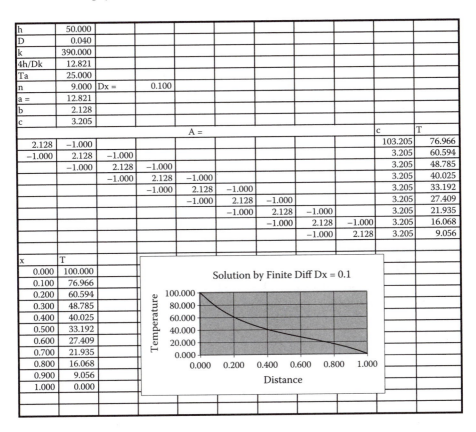

Solution by Finite Diff Dx = 0.1

6.4 MORE COMPLEX BOUNDARY CONDITIONS WITH ODE-BVPs

Consider again the problem of heat transfer in a rod, where the ODE is

$$\frac{d^2T}{dx^2} = \frac{4h}{Dk}(T - T_a) \qquad (6.11)$$

However, now assume that the left boundary is still exposed to boiling water, but that the right side is insulated (no heat flow). From Fourier's law of heat conduction,

$$q = -hA \frac{dT}{dx} \tag{6.12}$$

If $q = 0$, this implies that $\frac{dT}{dx} = 0$. So, now the boundary conditions for the problem are

$$T(0) = 100$$
$$\frac{dT(1)}{dx} = 0 \tag{6.13}$$

The shooting method can be used to solve this problem with no difficulty by transforming the second-order ODE into two first-order ones with one of the variables being the temperature gradient. That is, define F such that

$$\frac{dT}{dx} = F$$
$$\frac{dF}{dt} = \frac{4h}{Dk}(T - T_a) \tag{6.14}$$

In the shooting method, the procedure is to assume a value for $F(0)$ and "shoot" for a value $F(1) = 0$.

The finite difference method can also be applied. The only thing that changes is the "last" equation (Equation 6.10). Recall the finite difference equation when $i = n$:

$$\frac{T_{n-1} - 2T_n + T_{n+1}}{\Delta x^2} = \alpha(T_n - T_a) \tag{6.15}$$

where

$$a = \frac{4h}{Dk}$$

The right-hand boundary condition can be written using the right-hand second-order correct finite difference formula (the third equation in Table 4.2). This is the second-order correct backward difference formula applied at the right boundary:

$$\frac{3T_{n+1} - 4T_n + T_{n-1}}{2\Delta x} = 0$$
$$T_{n+1} = \frac{4T_n - T_{n-1}}{3} = \frac{4}{3}T_n - \frac{1}{3}T_{n-1} \tag{6.16}$$

Substituting for T_{n+1} results in

$$\frac{T_{n-1} - 2T_n + T_{n+1}}{\Delta x^2} = a(T_n - T_a)$$

$$T_{n-1} - 2T_n + \frac{4}{3}T_n - \frac{1}{3}T_{n-1} = a\Delta x^2(T_n - T_a) \tag{6.17}$$

$$\frac{2}{3}T_{n-1} - \left(\frac{2}{3} + a\Delta x^2\right)T_n = -a\Delta x^2 T_a$$

So, the last of the n equations is *special* and accounts for the insulated boundary condition.

Another more complex boundary condition involves what is often called (incorrectly) the *radiation* boundary condition. For example, at the end of the rod, there might be heat transfer to the surroundings as follows:

$$-kA\frac{dT(1)}{dx} = hA[T(1) - T_a] \text{ or}$$

$$\frac{dT(1)}{dx} = -\frac{h}{k}[T(1) - T_a] \tag{6.18}$$

As with the simpler insulated boundary condition, this more complex one presents no particular difficulty when solving using the shooting method together with Goal Seek. When applying finite differences, the equation representing the boundary ($i = n$) must be altered.

EXERCISES

Exercise 6.1: Consider the problem of heat transfer in a fin with variable thermal conductivity. For a rectangular fin of thickness $2B$ and length L, it can be shown (with suitable assumptions) that the governing ODE is

$$\frac{d}{dx}\left(k\frac{dT}{dx}\right) = \frac{h}{B}(T - T_a) \tag{6.19}$$

where

$T = T(x)$ is the temperature in the fin (°F)

$h =$ Heat transfer coefficient between fin and air [Btu/(h ft^2 °F)]

$B =$ The half thickness of the fin, 0.02 in. (note: inches)

$L =$ Length of the fin, 1.5 in. (note: inches)
$k =$ Thermal conductivity of the fin [Btu/(h ft °F)]
$T_a =$ Air temperature $= 90°F$
$T_w =$ Temperature of the wall to which the fin is affixed $= 450°F$
$h = 40$ Btu/(h ft² °F)

The thermal conductivity of the fin varies with distance as follows:

$$k = k_0 (1 + x) \tag{6.20}$$

$$k_0 = 60 \text{ Btu/(h ft °F)}$$

The boundary conditions are

$$T(0) = T_w$$
$$\frac{dT(L)}{dy} = 0 \text{ (insulated)}$$

Defining a dimensionless distance, $y = x/L$, and differentiating the first term in the ODE (using the equation for k as a function of x), the result is (should be verified)

$$k \frac{d^2T}{dy^2} + k_0 L \frac{dT}{dy} = \frac{hL^2}{B}(T - T_a) \tag{6.21}$$

with boundary conditions

$$T(0) = T_w$$
$$\frac{dT(1)}{dy} = 0 \text{ (insulated)}$$

a. Convert the ODE to two first-order ODEs and solve this problem using the Euler method together with the shooting method.
b. Change the second boundary condition to $T(L) = T_a$ (90°F) and repeat the solution.
c. Change the right-hand boundary condition (B.C.) to the "radiation," which can be expressed as

$$\frac{dT(1)}{dy} = -\frac{hL}{k}[T(1) - T_a]$$

Exercise 6.2: Consider the problem of diffusion and reaction in a cylindrical pore (e.g., in a solid catalyst) where component A reacts at the walls of the cylinder according to

$$A \xrightarrow{r} B \qquad\qquad (6.22)$$

where

$$r = kC_A^2 \text{ (second-order reaction)} \qquad\qquad (6.23)$$

In this system, component A diffuses into the pore due to lower concentration of A inside the pore than at the pore mouth. Since B is produced by the reaction, the concentration of B inside the pore is larger than at the inlet, causing diffusion of B out of the pore. At the inlet of the pore ($x = 0$), the concentration is C_{A0}. The end of the pore ($x = L$) is assumed to be sealed, so there is no flux of A at $x = L$. The mathematical model for this system can be expressed as follows:

$$D_A \frac{d^2 C_A}{dx^2} = kC_A^2$$
$$C_A(0) = C_{A0} \qquad\qquad (6.24)$$
$$\frac{dC_A(L)}{dx} = 0$$

The second boundary condition is the "no flux at $x = L$" condition. This is a split BVP that can be solved using the shooting method. Here are some data for the problem:

k = 0.01 L/(gmol s) (rate constant)
C_{A0} = 1.0 gmol/L (inlet concentration)
D_A = 1 × 10⁻³ cm²/s (diffusivity)
L = 1 cm (length of pore)

Prepare an Excel spreadsheet to solve this problem using the Euler method. Use Goal Seek to implement the shooting method. Be sure to get a reasonable value of the unknown boundary condition (by experimentation) before invoking Goal Seek.

Exercise 6.3: Consider the same problem as in Exercise 6.2, but with a first-order reaction:

$$A \xrightarrow{r} B \qquad\qquad (6.25)$$

where

$$r = kC_A \text{ (first-order reaction)} \tag{6.26}$$

The model equations then become

$$D_A \frac{d^2 C_A}{dx^2} = kC_A$$

$$C_A(0) = C_{A0} \tag{6.27}$$

$$\frac{dC_A(L)}{dx} = 0$$

Apply the finite difference method to solve this problem (first introduce a dimensionless distance $y = x/L$). Use the same data as for the second-order reaction case. For the right-hand boundary condition, do not forget to use a second-order correct finite difference form.

Exercise 6.4: Consider heat transfer in a counter-current heat exchanger as depicted in Figure 6.1.

When the flow direction of the two streams is opposite, counter-current flow exists and the system equations take the form

$$\frac{dT'}{dx} = -\frac{U_i \pi D_i}{m' C_p'} (T' - T)$$

$$\quad \text{with } T'(0) \text{ and } T(L) \text{ known} \tag{6.28}$$

$$\frac{dT}{dx} = \frac{U_i \pi D_i}{m C_p} (T' - T)$$

where

T = Temperature on the shell (outer) side
T' = Temperature on the tube (inner) side
U_i = Overall heat transfer coefficient based on D_i
D_i = Inside diameter of the inner pipe (tube side)
m = Mass flow rate on the shell side
m' = Mass flow rate on the tube side

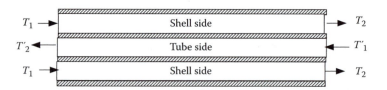

FIGURE 6.1   Double pipe heat exchanger schematic.

C_p = Fluid heat capacity on the shell side
C_p' = Fluid heat capacity on the tube side
L = Length of the heat exchanger

Pertinent data are as follows:

$$\frac{U_i \pi D_i}{m' C_p'} = 1.2$$

(6.29)

$$\frac{U_i \pi D_i}{m C_p} = 0.5$$

Assume that the length of the exchanger = 1 and then consider the boundary conditions

$$T'(0) = 180;\ T(1) = 70$$

In order to integrate the two ODEs from $x = 0$ to $x = 1$, $T(0)$ must be guessed and the integration performed to see if the target of 70 is hit. In effect, the single nonlinear equation $T(1) - 70 = 0$ must be solved.

Use the Euler method along with Goal Seek to solve this problem (at the Excel level). Begin using $\Delta x = 0.1$; find $T(0)$ that satisfies the right-hand boundary condition. Then reduce Δx until comparable temperature profiles result. Finally, plot the temperature profiles with appropriate annotations on the graph. Assume that temperatures are in degrees Celsius.

Exercise 6.5: The transient BVP for a rectangular fin can be stated as follows:

$$\frac{\partial^2 T}{\partial x^2} - \beta^2 (T - T_a) = \frac{1}{\alpha} \frac{\partial T}{\partial t}$$

$$0 \le x \le 1$$

$$T(0,t) = T_1$$

$$T(1,t) = T_2$$

$$T(x,0) = 0$$

α and β are constants.
Define a dimensionless temperature and time as follows:

$$\theta = \frac{T - T_a}{T_1 - T_a} \qquad \tau = \alpha t$$

The BVP then becomes

$$\frac{\partial^2 \theta}{\partial x^2} - \beta^2 \theta = \frac{\partial \theta}{\partial \tau}$$

The new boundary conditions are

$$\theta(0,t) = 1$$
$$\theta(1,t) = \frac{T_2 - T_a}{T_1 - T_a}$$
$$\theta(x,0) = \frac{-T_a}{T_1 - T_a}$$

The steady-state model results when the time derivative is zero:

$$\frac{\partial^2 \theta}{\partial x^2} - \beta^2 \theta = 0 \quad \theta(0) = 1 \quad \theta(1) = \frac{T_2 - T_a}{T_1 - T_a}$$

The steady-state solution can be shown to be

$$\theta_{ss} = \frac{\sinh[\beta(1-x)]}{\sinh \beta} + \left(\frac{T_2 - T_a}{T_1 - T_a} \right) \frac{\sinh \beta x}{\sinh \beta}$$

For simplicity, take $T_a = 0$, $T_1 = 1$, $T_2 = 0$, and $\beta = 4$.
a.   Solve the steady-state dimensionless problem using the shooting method and the Euler method. Check the results with the analytical solution.
b.   Solve the steady-state dimensionless problem using finite differences. Again, compare the results with the analytical solution.

Exercise 6.6: (*Note: This problem involves significant VBA programming and complex logic.*) Solve the ODE-BVP of Equation 6.11 with boundary conditions given by Equation 6.13 using the VBA program of Example 5.7 (Euler method). The requisite ODE and boundary conditions are repeated below:

$$\frac{d^2 T}{dx^2} = \alpha(T - T_a) \qquad (6.30)$$

$$T(0) = 100$$

$$\frac{dT(1)}{dx} = 0 \qquad (6.31)$$

where
$$\alpha = 12.8$$
$$T_a = 25$$

While Goal Seek can be invoked from VBA, it is not possible to use it in the context of the Euler program. Therefore, it is necessary to include VBA code to implement one of the methods described in Chapter 1 to solve a single nonlinear equation. In particular, use the *secant method* to do this. The logic changes required in the Example 5.4 VBA program must be carefully thought out, but in general, the steps required are as follows:
a. Execute the current VBA program logic for one initial guess for $T'(0)$.
b. Run the current VBA program logic for a second initial guess for $T'(0)$.
c. Use the results of steps a and b to produce a new guess for $T'(0)$ and use the logic of the secant method to proceed (replace one of the initial guesses and repeat this step until convergence is achieved). The process is converged when $T'(1)$ is close to zero.

Exercise 6.7: Solve the problem of Exercise 6.2 using the instructions given in Exercise 6.6. That is, use the Euler VBA program of Example 5.4 together with the secant method. As with Exercise 6.6, this problem requires significant logic planning and VBA programming.

REFERENCE

Riggs, J.B., *An Introduction to Numerical Methods for Chemical Engineers*, 2nd ed., Texas Tech University Press, Lubbock, TX. pp. 241–282 (1994).

7 Regression Analysis and Parameter Estimation

7.1 INTRODUCTION AND THE GENERAL METHOD OF LEAST SQUARES

The problem of regression analysis is to find coefficients (parameters) of a function that are believed to properly represent a set of experimental data *and* to perform statistical analyses to confirm (or deny) that the function gives a good fit to the data. Without the additional statistical analysis, just finding the parameters of some candidate function is called *curve fitting*. Curve fitting, while useful in certain circumstances, is not as powerful as regression analysis. Consider data as represented in Figure 7.1.

It is assumed that x, the independent variable, is error free (this might be time or temperature, for example), and y, the dependent variable, contains experimental error. In chemical and biomolecular engineering applications, theoretical knowledge often exists of the function that should "fit" the data. If the deviations (errors) between the data and the fitting function are statistically distributed with a normal distribution with zero mean and *constant variance*, then it can be shown that a proper way to find the unknown coefficients of the function is to minimize the sum of squares of the errors. It is common nomenclature to call the errors "residuals," which are defined as follows:

$$r_i = y_i^{calc} - y_i^{data}; i = 1, 2, \ldots, n_d \tag{7.1}$$

where n_d is the number of data points, y_i^{calc} is the value of y calculated from the fitting function, and y_i^{data} is the associated data value. The function to be minimized is then

$$q = \frac{1}{2} \sum_{i=1}^{n_d} r_i^2 \tag{7.2}$$

This is, therefore, called the *method of least squares*. If the fitting function can be represented as

$$y_i^{calc} = f_i(c_1, c_2, \ldots, c_n; x_1, x_2, \ldots, x_m) \, i = 1, 2, \ldots, n_d \tag{7.3}$$

where the x_i are the "independent variables" (e.g., time, temperature, distance, etc. in which there are no significant errors) and y is the "dependent variable" that contains

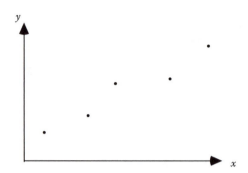

FIGURE 7.1 Data with experimental error in the dependent variable.

random errors (perhaps due to measurement error or some other error source). The c_i are unknown model parameters to be determined. Note that the unknowns to be determined are not represented by x. It takes some time to change the usual mind-set that x is the unknown.

To minimize Equation 7.2, differentiate with respect to the unknown parameters and set the result to zero (seeking a stationary point):

$$\frac{\partial q}{\partial c_k} = \sum_{i=1}^{n_d} r_i \frac{\partial r_i}{\partial c_j} = 0 \quad j = 1, 2, \ldots, n \tag{7.4}$$

If r_i is expanded into a Taylor's series about a point c^0 (this is an "initial guess" of the c's), then

$$r_i = r_i(c^0) + \sum_{k=1}^{n} \frac{\partial r_i(c^0)}{\partial c_k} \cdot \Delta c_k \tag{7.5}$$

where

$$\Delta c_k = c_k - c_k^0 \tag{7.6}$$

Substituting Equation 7.5 into Equation 7.4 gives

$$\sum_{i=1}^{n_d} \left(r_i + \sum_{k=1}^{n} \frac{\partial r_i}{\partial c_k} \right) \frac{\partial r_i}{\partial c_j} \cdot \Delta c_k = 0 \quad j = 1, 2, \ldots, n \tag{7.7}$$

In Equation 7.7, it is understood that the residuals and their derivatives are evaluated at c^0 (the initial guess).

By expanding Equation 7.7 and defining

$$Z_k = \frac{\partial f(c^0)}{\partial c_k} \tag{7.8}$$

the following set of equations in matrix–vector form result where the *summations are over all data points*:

$$\begin{bmatrix} \sum Z_1 Z_1 & \sum Z_1 Z_2 & \cdots & \sum Z_1 Z_n \\ \sum Z_2 Z_1 & \sum Z_2 Z_2 & \cdots & \sum Z_2 Z_n \\ \vdots & \vdots & \ddots & \vdots \\ \sum Z_n Z_1 & \sum Z_n Z_2 & \cdots & \sum Z_n Z_n \end{bmatrix} \begin{bmatrix} \Delta c_1 \\ \Delta c_2 \\ \vdots \\ \Delta c_n \end{bmatrix} = - \begin{bmatrix} \sum r_i Z_1 \\ \sum r_i Z_2 \\ \vdots \\ \sum r_i Z_n \end{bmatrix} \tag{7.9}$$

A shorthand representation of Equation 7.9 is

$$G\Delta c = -b \tag{7.10}$$

where G is the left-hand side matrix, Δc is the vector of unknowns, and b is the right-hand side vector. Since Equations 7.9 and 7.10 are applicable to any fitting function, these are often called the normal equations of the general method of least squares. The symmetric matrix G is referred to here as the *Gauss–Newton matrix*, but it goes by several other names in the literature.

7.2 LINEAR REGRESSION ANALYSIS

If the Zs are constant (not dependent on the cs), Equation 7.10 is a linear system and the unknown coefficients can be easily determined. Note that a "linear regression" does not require the fitting function to be linear (a straight line); the only requirement is that the Zs are constants (the function is linear in its parameters).

Another way of getting the coefficients for *a limited number of fitting functions* is to use X–Y scatter graphs in Excel® and add a trendline.

7.2.1 STRAIGHT LINE REGRESSION

Consider "straight line" regression where theory dictates that the data should lie on a straight line (or in the absence of a theoretical justification, a plot of the data suggests that a straight line should suffice). The fitting function is then

$$y_i^{calc} = c_1 + c_2 x_i \tag{7.11}$$

To calculate the regression coefficients using Equation 7.9, proceed as follows:

$$y_i^{calc} = c_1 + c_2 x_i$$

$$Z_1 = \frac{\partial y^{calc}}{\partial c_1} = 1 \quad Z_2 = \frac{\partial y^{calc}}{\partial c_2} = x_i \tag{7.12}$$

$$G = \begin{bmatrix} \sum_{i=1}^{10}(1)(1) & \sum_{i=1}^{10}(1)(x_i) \\ \sum_{i=1}^{10}(x_i)(1) & \sum_{i=1}^{10}(x_i)(x_i) \end{bmatrix} = \begin{bmatrix} 10 & \sum_{i=1}^{10} x_i \\ \sum_{i=1}^{10} x_i & \sum_{i=1}^{10}(x_i)^2 \end{bmatrix}$$

If the initial guess is taken to be zero (this is always suggested for linear problems), then

$$\Delta c_k = c_k \tag{7.13}$$

and

$$r_i = -y_i^{data} \text{ (since } y_i^{calc} = 0 \text{ when } c_1 = c_2 = 0) \tag{7.14}$$

The equations to solve for the unknown coefficients become

$$\begin{bmatrix} 10 & \sum_{i=1}^{10} x_i \\ \sum_{i=1}^{10} x_i & \sum_{i=1}^{10}(x_i)^2 \end{bmatrix} \begin{bmatrix} c_1 \\ c_2 \end{bmatrix} = + \begin{bmatrix} \sum_{i=1}^{10} y_i^{data} \\ \sum_{i=1}^{10} y_i^{data} x_i \end{bmatrix} \tag{7.15}$$

The following Excel spreadsheet shows the calculation of all quantities required in these equations:

Example 7.1: Evaporation Coefficient Correlation

Figure 7.2 displays data for an evaporation coefficient for different air velocities over a pool of liquid.

Shown in Figure 7.3 is a plot of the original data produced with Excel. Note the regression equation that is displayed on the graph, which was calculated by Excel by adding a "trendline." To add the trendline, after having plotted the data as an X–Y scatter plot, right click the mouse on any data point. This brings up a menu of choices, one of which is Add trendline. A number of choices appear for the fitting function. After selecting the fitting function, click on the Options tab; some check boxes appear at the bottom. One of these says "Display Equation on Chart," and the other says "Display R-squared Value on Chart." Check both of these boxes, and the result should look like that shown in Figure 7.2. The quantity R^2 is covered in more detail later.

x Air velocity cm/sec	y Evap. coeff mm^2/sec
20	0.18
60	0.37
100	0.35
140	0.78
180	0.56
220	0.75
260	1.18
300	1.36
340	1.17
380	1.65

FIGURE 7.2 Evaporation coefficient data.

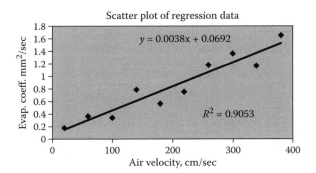

FIGURE 7.3 Scatter plot of evaporation coefficient data.

	x Air velocity cm/sec	y Evap. coeff mm^2/sec	x^2	$x\,y$	y^2
	20	0.18	400	3.6	0.0324
	60	0.37	3600	22.2	0.1369
	100	0.35	10000	35	0.1225
	140	0.78	19600	109.2	0.6084
	180	0.56	32400	100.8	0.3136
	220	0.75	48400	165	0.5625
	260	1.18	67600	306.8	1.3924
	300	1.36	90000	408	1.8496
	340	1.17	115600	397.8	1.3689
	380	1.65	144400	627	2.7225
sums	2000	8.35	532000	2175.4	9.1097
Equations:	10	2000	8.35		
	2000	532000	2175.4		
G Inv.	0.40303	−0.001515	c1=	0.069242	
	−0.001515	7.576E-06	c2=	0.003829	

Note the following in the spreadsheet:

$$
G = \begin{bmatrix} 10 & 2000 \\ 2000 & 532000 \end{bmatrix} \quad -b = \begin{bmatrix} 8.35 \\ 21.75 \end{bmatrix} \quad c = \begin{bmatrix} 0.069242 \\ 0.003829 \end{bmatrix} \quad (7.16)
$$

The cs are identical to those found by Excel when a trendline was added to the plot of the data. The cs were determined using the inverse of the matrix G. The G matrix and its inverse have other powerful uses as well, as will be covered in later discussions.

One might (justifiably) ask why learn all of these details when Excel produced the desired results so easily. The reasons are twofold. First, it is always useful to know the details behind any automatic computations. Second, the computational details are required when the problem is nonlinear (they are not done automatically by Excel).

7.2.2 CURVILINEAR REGRESSION

The next example considers data that obviously cannot be represented by a straight line. The fitting function is often chosen as a polynomial or other function that can "bend" to better represent the data. It is important that although the fitting function is nonlinear in the dependent variable (x), it remains linear in the unknown coefficients (c).

Example 7.2: Curvilinear Regression

Consider the data given in Figure 7.4. A scatter plot of the data is shown in Figure 7.5. From the plot, it does not appear that a straight line can represent the data adequately. In the absence of a physical system model (conservation of mass, energy, or momentum principles), the usual procedure is to simply seek a function that will represent the data in an appropriate manner (this can be highly subjective—even very approximate models are better than a pure guess about the function).

The function next up the ladder of complexity from a straight line is a quadratic, which can be written as follows:

$$y^{calc} = c_1 + c_2 x + c_3 x^2 \tag{7.17}$$

Varnish additive, g	Drying time, hours
0	12.0
1	10.5
2	10.0
3	8.0
4	7.0
5	8.0
6	7.5
7	8.5
8	9.0

FIGURE 7.4   Drying time data.

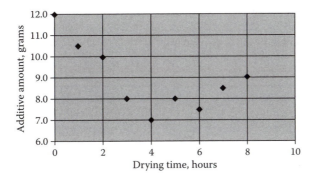

FIGURE 7.5   Effect of additive on drying time.

To set up the least-squares (normal) equations, proceed by finding the Zs (the derivatives of y^{calc} with respect to the unknown coefficients):

$$Z_1 = 1$$
$$Z_2 = x$$
$$Z_3 = x^2$$

(7.18)

Since the Zs are not dependent on the cs, this is still a linear regression problem. Taking $c^0 = 0$ (i.e., all initial guesses of the cs $= 0$), the normal equations result from Equation 7.9 as follows:

$$r_i = -y_i^{data} \text{ (since } y^{calc}(x^0) = 0)$$

(7.19)

$$
\begin{bmatrix}
\sum 1 & \sum x & \sum x^2 \\
\sum x & \sum x^2 & \sum x^3 \\
\sum x^2 & \sum x^3 & \sum x^4
\end{bmatrix}
\begin{bmatrix}
c_1 \\
c_2 \\
c_3
\end{bmatrix}
=
\begin{bmatrix}
\sum y^{data} \\
\sum x \cdot y^{data} \\
\sum x^2 \cdot y^{data}
\end{bmatrix}
$$

(7.20)

Shown below is an Excel spreadsheet that performs the indicated calculations:

x	y	x^2	x^3	x^4	yx	yx^2
0	12.0	0	0	0	0	0
1	10.5	1	1	1	10.5	10.5
2	10.0	4	8	16	20	40
3	8.0	9	27	81	24	72
4	7.0	16	64	256	28	112
5	8.0	25	125	625	40	200
6	7.5	36	216	1296	45	270
7	8.5	49	343	2401	59.5	416.5
8	9.0	64	512	4096	72	576
36	80.5	204	1296	8772	299	1697

So, the normal equations become

$$
\begin{bmatrix}
9 & 36 & 204 \\
36 & 204 & 1296 \\
204 & 1296 & 8772
\end{bmatrix}
\begin{bmatrix}
c_1 \\
c_2 \\
c_3
\end{bmatrix}
=
\begin{bmatrix}
80.5 \\
299 \\
1697
\end{bmatrix}
$$

(7.21)

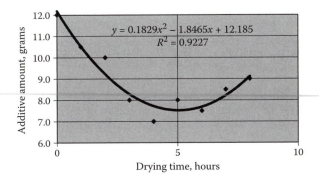

FIGURE 7.6 Effect of additive on drying time with trendline.

These linear algebraic equations can be solved using any of the methods previously discussed. The solution is shown below in terms of the approximating function:

$$y = 12.2 - 1.85x + 0.183x^2 \qquad (7.22)$$

This equation together with the correlation coefficient can be found using Excel's graphing capabilities. Simply plot the data (points only) and add a trendline (Figure 7.6). Use the appropriate Option to show the equation and R^2. The coefficients for the quadratic trendline are identical to those calculated previously.

This example will be amplified upon discussion after having introduced the "statistical part" of regression analysis.

7.3 HOW GOOD IS THE FIT FROM A STATISTICAL PERSPECTIVE?

There are several tools to help visualize if the chosen function is a good one. Five of these will be covered:

- Residual plots
- Correlation coefficient
- Parameter standard deviations
- Parameter confidence intervals
- Parameter t-ratios

7.3.1 RESIDUAL PLOTS

When the residuals ($y^{calc} - y^{data}$) versus x are plotted, these values should distribute themselves somewhat evenly about zero, and their magnitude should be approximately constant (these were the basic assumptions for the least-squares method).

7.3.2 CORRELATION COEFFICIENT

A quantitative measure of the "goodness of fit" is provided by the correlation coefficient:

$$R^2 = 1 - \frac{\sum\limits_{i=1}^{n_d} r_i^2}{\sum\limits_{i=1}^{n_d} (y_i - \bar{y})^2} \tag{7.23}$$

where

$$\bar{y} = \frac{1}{n_d} \sum\limits_{i=1}^{n_d} y_i$$

An R^2 value near 1 indicates that all residuals are small and that the fit is "good." An elaborate treatment regarding R^2 and the analysis of variance for regression analysis is beyond the scope of the present discussion.

7.3.3 PARAMETER STANDARD DEVIATIONS

The parameter (regression coefficient) variances (and thus standard deviations) can be estimated from the matrix G of Equation 7.10. The following is stated without proof:

$$\text{var}(c) = S = s_e^2 G^{-1} \tag{7.24}$$

where

$$s_e^2 = \frac{\sum\limits_{i=1}^{n_d} r_i^2}{v} \tag{7.25}$$

and v is the number of degrees of freedom (see Equation 7.26).

Variance [var(c)] represent the parameter variances. S is sometimes called the variance–covariance matrix. Variance is a measure of the spread of expected values of random variables belonging to a specific probability distribution. As has been mentioned previously, the validity of the least-squares method for determining regression parameters is based on errors in the data having a normal (Gaussian) distribution (the familiar bell-shaped curve) with zero mean and constant variance. The values of the parameters determined from data with such normal errors are, in a

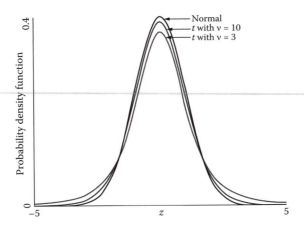

FIGURE 7.7 Plot of the *t*-distribution.

sense, average values—they too are subject to errors. If the variance of the errors in the data were known, then the parameters themselves would be normally distributed. However, it is rarely the case that this variance is known with certainty. Therefore, the variance must be estimated. The matrix S of Equation 7.24 provides an estimate of the variances of the individual parameters (the diagonal elements of S) as well as the off-diagonal covariances between pairs of parameters (no further treatment of the covariances is given here).

When the variances must be estimated, the probability distribution of the parameters is the *t*-distribution. The *t*-distribution is similar to the normal distribution, but it has an additional argument called the degrees of freedom, which is the difference between the number of data points and the number of parameters:

$$\text{Degrees of freedom} = \nu = n_d - n \tag{7.26}$$

The *t*-distribution probability density function looks very much like that of the normal distribution, but it is "shorter" and has a longer tail. Figure 7.7 displays the *t*-distribution with 3 and 10 degrees of freedom as well as the normal distribution. The *t*-distribution with an infinite number of degrees of freedom is coincident with the normal distribution.

7.3.4 PARAMETER CONFIDENCE INTERVALS

As previously stated, the regression parameters have the *t*-distribution. Again, without proof, the following inequality can be written:

$$c_i - t_{1-\alpha/2} s_e \sqrt{S_{ii}} < \bar{c}_i < c_i + t_{1-\alpha/2} s_e \sqrt{S_{ii}} \tag{7.27}$$

where $\bar{c}_i$ is the true value of c_i.

7.3.5 USING T-RATIOS (T-STATISTICS) FOR INDIVIDUAL PARAMETER SIGNIFICANCE

When choosing an arbitrary function to fit a set of data (such as the quadratic in Example 7.2), it might be asked if all three terms in the equation are needed or if it can be simplified by dropping perhaps the linear term (the one involving $c_2 x$). To answer this question, statistics gives the answer. It can be shown that the ratio of the optimal parameter values divided by their standard deviations (as determined by Equation 7.7) has a t-distribution with $n_d - n - 1$ degrees of freedom. That is,

$$t_i = \frac{c_i}{s_i} \tag{7.28}$$

Looking at a table of the t-distribution (Google it), it will be noticed that for $\alpha = 0.025$ (two-sided confidence of 95%), for more than 3 degrees of freedom, the tabulated values are all near 2. So, for convenience, it can be stated as an informal test of the null hypothesis that a parameter's contribution is insignificant (i.e., its value is zero) if the calculated value $|t_i| < 2$ and rejected if the calculated value $|t_i| \geq 2$. This can be done more formally by using the precise value of the t-distribution with the proper number of degrees of freedom. In Excel, this can be found using the function TINV(α, v). Note that α and not $\alpha/2$ is used with the TINV function. This is because the function returns what is called the two-tailed t-value; α is split between the negative tail and the positive tail of the t-distribution. Only a much more detailed study of the statistics associated with these arguments would explain these concepts fully, but this is beyond the present discussion. Suffice it to say that if a parameter t-ratio (Equation 7.28) is less than about 2, it can be concluded that the parameter is no different than zero and can be removed from the correlating equation.

Example 7.3: Applying Statistics to the Problem of Example 7.2

Consider again the quadratic function fit to the drying time data of Example 7.2. In what follows, each of the statistical tools for determining the quality of the "fit" is demonstrated.

RESIDUAL PLOT

The residual plot for Example 7.2 (quadratic curve fit) appears in Figure 7.8.

With only 9 data points, the data distribute nicely about zero, and the magnitude does not (visually) appear to be a function of x. It can be concluded that the fit is adequate. Note that a single point with a very large residual is a candidate "outlier" and might be omitted if this can be justified (poor experimental procedure, other extenuating circumstances, etc.). However, the arbitrary exclusion of outliers must be avoided.

CORRELATION COEFFICIENT

From Equation 7.23, the R^2 value for Example 7.2 can be calculated as 0.9227. This value is identical to that shown in the trendline graph in Figure 7.4.

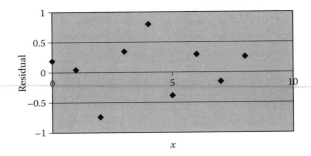

FIGURE 7.8 Residual plot for Example 7.2.

PARAMETER STANDARD DEVIATIONS

The G matrix for Example 7.2 was

9	36	204
36	204	1296
204	1296	8772

The associated inverse matrix is

0.660606	−0.30909	0.030303
−0.30909	0.224459	−0.02597
0.030303	−0.02597	0.003247

x	residual	resid^2
0	0.185	0.034225
1	0.0389	0.001513
2	−0.7414	0.549674
3	0.3441	0.118405
4	0.7954	0.632661
5	−0.3875	0.150156
6	0.2954	0.087261
7	−0.1559	0.024305
8	0.2586	0.066874
	sum	1.665074
	se^2	0.277512

0.183326	−0.085777	0.008409
−0.085777	0.06229	−0.007208
0.008409	−0.007208	0.000901

Parameter	Variance	Standard deviation
c_1	0.1833	0.4282
c_2	0.0623	0.2496
c_3	0.0009	0.0300

FIGURE 7.9 Parameter standard deviations.

The final results are shown in Figure 7.9.

To obtain the parameter variances, the diagonal elements must be multiplied by s_e^2 (see Equation 7.25). s_e^2 is the sum of squares of residuals divided by the degrees of freedom (number of data points – number of parameters; 6 in the present case). Shown below is the calculation of s_e^2 followed by the matrix whose diagonal elements are the parameter variances:

PARAMETER CONFIDENCE INTERVALS (95%)

From Equation 7.27, the 95% parameter confidence intervals are shown in the following inequalities:

$$c_1: 11.633 < 12.185 < 12.737$$

$$c_2: -2.168 < -1.847 < -1.525 \tag{7.29}$$

$$c_3: 0.1442 < 0.1829 < 0.2216$$

An examination of these confidence intervals reveals that the uncertainty in all three parameters is not unreasonably large.

PARAMETER t-RATIOS

From Equation 7.28, the t-ratios for the three parameters are shown in Figure 7.10.
The t-ratios for c_1, c_2, and c_3 have absolute values considerably greater than 2. Therefore, it can be concluded that all parameters are significant and must be retained in the correlating equation.

Parameter	t-ratio
c_1	28.4582
c_2	-7.3986
c_3	6.0932

FIGURE 7.10 Parameter t-ratios.

Example 7.4: Another Curvilinear Regression Problem

Consider fitting a polynomial to the data in Figure 7.11.

A typical first step when choosing a fitting function (in the absence of one based on theory) is to prepare a graph of the data. When using Excel, a variety of trendline types can also be helpful in arriving at a suitable functional form. Shown in Figure 7.12 is a plot of the data together with both quadratic and cubic polynomial fitting functions.

The graph in Figure 7.12 suggests clearly that a cubic fitting function is superior to a quadratic one. A cubic fitting function is as shown by the following equation:

$$y_{calc} = c_1 + c_2 v + c_3 v^2 + c_4 v^3 \qquad (7.30)$$

The derivative of Equation 7.29 with respect to each of the four parameters gives

$$Z_1 = 1, \; Z_2 = v, \; Z_3 = v^2, \; Z_4 = v^3 \qquad (7.31)$$

v	y Data
0.00	1.766
0.25	2.478
0.50	3.690
0.75	6.397
1.00	6.649
1.25	10.045
1.50	12.924
1.75	15.957
2.00	17.008
2.25	21.196
2.50	24.113
2.75	25.570
3.00	28.258
3.25	32.129
3.50	32.494
3.75	34.031
4.00	34.088
4.25	32.974
4.50	31.815
4.75	30.647
5.00	26.050
5.25	23.453
5.50	17.694
5.75	9.444
6.00	1.734

FIGURE 7.11 Data for curvilinear regression.

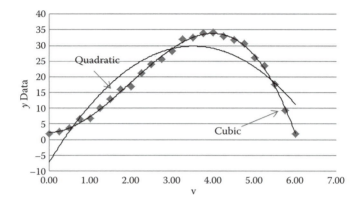

FIGURE 7.12 Curvilinear data showing quadratic and cubic fitting functions.

The normal equations [see Equation 7.10] are as shown in Equation 7.32:

$$
\begin{bmatrix}
25.00 & 75.00 & 306.25 & 1406.25 \\
75.00 & 306.25 & 1406.25 & 6886.80 \\
306.21 & 406.25 & 6886.80 & 35126.95 \\
1406.25 & 6886.80 & 35126.95 & 184262.87
\end{bmatrix}
\begin{bmatrix}
c_1 \\
c_2 \\
c_3 \\
c_4
\end{bmatrix}
=
\begin{bmatrix}
482.60 \\
1697.26 \\
6779.00 \\
29158.68
\end{bmatrix}
\qquad (7.32)
$$

Solving these equations gives

c_1	2.2241
c_2	0.1829
c_3	5.8748
c_4	−0.9855

The gross sum of squares of residuals = 0.5191.

The inverse of G is as follows:

0.4810	−0.5869	0.1915	−0.0182
−0.5869	1.0442	−0.3946	0.0407
0.1915	−0.3946	0.1604	−0.0173
−0.0182	0.0407	−0.0173	0.0019

The parameter standard deviations and t-ratios are shown in Figure 7.13.

	Base value	Std. dev	t-ratios
c_1	2.2241	0.4997	4.451
c_2	0.1829	0.7363	0.248
c_3	5.8747	0.2886	20.357
c_4	−0.9855	0.0316	−31.2

FIGURE 7.13 Parameter values, standard deviations, and t-ratios.

	Base value	Std. dev.	t-ratios
c_1	2.327	0.274	8.489
c_2	5.944	0.075	79.209
c_3	−0.993	0.013	−76.610

FIGURE 7.14 Revised parameter values, standard deviations, and t-ratios.

From Figure 7.13, it can be deduced that c_2 is poorly determined. Upon removing the term involving c_2, the new model equation is given by

$$y = c_1 + c_2 x^2 + c_3 x^3 \tag{7.33}$$

When the computations are repeated using the revised fitting function, the optimal coefficients, standard deviations, and t-ratios are as shown in Figure 7.14.

Now, all t-ratios are well above 2, and Equation 7.33 gives an adequate representation of the data.

As a note of warning, when turned out that two of the original parameters had t-ratios less than 2, it was appropriate to remove only the one with the smallest (in absolute value) t-ratio. Once the *worst actor* removed, the revised t-ratio of the *other offender* was greater than 2.

7.4 REGRESSION USING EXCEL'S® REGRESSION ADD-IN

Now that all of the computations for regression analysis the "hard way" (e.g., do it yourself) have been covered, a *little secret* can be revealed: Excel will do almost everything that has been discussed as long as the problem is one of linear regression. To invoke this package, go to Data/Data Analysis/Regression. The following spreadsheet displays the original data and columns for other terms to be included in the original fitting function (Equation 7.30). Also shown is the window that is presented by the Regression Add-In:

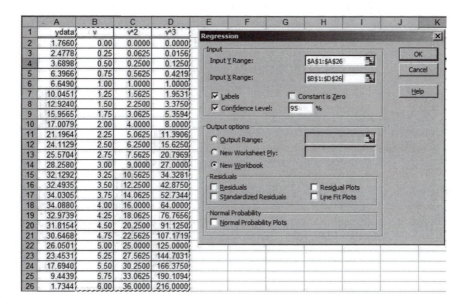

For the Input Y Range, the first column was selected, and for the Input X Range, the next three columns were identified. Since the first row contains labels, the "Labels" box was checked. Also checked was the Confidence Level box and 95% for the confidence level. For the Output options, the New Workbook radio button was checked, and the following is the output from the Regression Add-In:

	A	B	C	D	E	F	G
1	SUMMARY OUTPUT						
2							
3	Regression Statistics						
4	Multiple R	0.998277703					
5	R Square	0.996558373					
6	Adjusted R Square	0.996066712					
7	Standard Error	0.72049303					
8	Observations	25					
9							
10	ANOVA						
11		df	SS	MS	F	Significance F	
12	Regression	3	3156.587363	1052.196	2026.922	5.16918E-26	
13	Residual	21	10.90131434	0.51911			
14	Total	24	3167.488678				
15							
16		Coefficients	Standard Error	t Stat	P-value	Lower 95%	Upper 95%
17	Intercept	2.224111248	0.499705233	4.450846	0.000221	1.184917331	3.2633052
18	v	0.182815879	0.736261921	0.248303	0.806312	-1.348324599	1.7139564
19	v^2	5.874718908	0.28858153	20.35722	2.63E-15	5.274580765	6.4748571
20	v^3	-0.985488667	0.031585686	-31.2005	4.45E-19	-1.051174697	-0.919803

The analysis of variance (ANOVA) portion is not covered here. The other results are identical to those that were performed "by hand." Note that the t Stat (same as t-ratio) for v is less than 2, so the associated term (involving v) was deleted and the analysis was repeated. To do this, the second data column was not included in the X-Range in the Regression Add-In window. Results when doing that are as follows:

	A	B	C	D	E	F	G
1	SUMMARY OUTPUT						
2							
3	Regression Statistics						
4	Multiple R	0.998272642					
5	R Square	0.996548269					
6	Adjusted R Square	0.996234475					
7	Standard Error	0.704960337					
8	Observations	25					
9							
10	ANOVA						
11		df	SS	MS	F	gnificance F	
12	Regression	2	3156.555358	1578.278	3175.807	8.29E-28	
13	Residual	22	10.9333197	0.496969			
14	Total	24	3167.488678				
15							
16		Coefficients	Standard Error	t Stat	P-value	Lower 95%	Upper 95%
17	Intercept	2.32685819	0.274098785	8.489123	2.17E-08	1.758412	2.895304
18	v^2	5.943797784	0.07503901	79.20944	1.6E-28	5.788176	6.099419
19	v^3	-0.992608943	0.012956692	-76.6098	3.32E-28	-1.01948	-0.96574

Once again, the results are identical with the "hand calculations." One might ask why the hand calculations were explained when Excel Add-In would perform all of the requisite computations? The reasons are twofold: (1) so the power and results from the Regression Add-In can be fully appreciated, and (2) the Regression Add-In works only for *linear* regression, where the Zs in Equation 7.9 do not involve the unknown parameters. For nonlinear regression, the calculations must be done "by hand." Nonlinear regression is covered in Chapter 9.

7.5 NUMERICAL DIFFERENTIATION AND INTEGRATION REVISITED

When data to be differentiated or integrated contain random errors (usually experimental errors), it can be advantageous to first fit the data to an appropriate function using regression techniques. Then, the fitted function can be differentiated or integrated analytically. Example 7.5 illustrates this approach.

Example 7.5: Mean Heat Capacity of Gaseous Propane

The mean heat capacity of a gas is determined by the relationship

$$\bar{C}_p = \frac{\int_{T_{ref}}^{T} C_p \, dT}{T - T_{ref}} \tag{7.34}$$

The data of Figure 7.15 are for gaseous propane.

Point number	Temperature (K)	Heat capacity (kJ/kg-mol-K)
1	50	34.16
2	100	41.3
3	150	48.79
4	200	56.07
5	273.15	68.74
6	300	73.93
7	400	94.01
8	500	112.59
9	600	128.7
10	700	142.67
11	800	154.77
12	900	163.35
13	1000	174.6
14	1100	182.67
15	1200	189.74
16	1300	195.85
17	1400	201.21
18	1500	205.89

FIGURE 7.15 Heat capacity data for propane gas (reference temperature = 50 K).

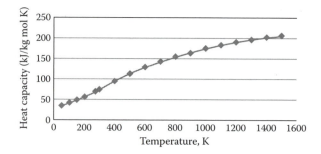

FIGURE 7.16 Plot of heat capacity of propane (reference temperature = 50 K).

A plot of the data is shown in Figure 7.16.

Because of the curvature, initially a cubic polynomial fit to the data was tried. Two parameters had *t*-ratios less than 2, and the coefficient of the cubic term was the *worst actor.* This led to a quadratic fit with the following result:

$$C_p = -6.17 \times 10^{-5}T^2 + 0.2175T + 18.1968 \qquad (7.35)$$

Integrating this function between limits of 50 and 1500 K and dividing by the temperature range gives a mean heat capacity of 138.79 kJ/(kg mol K). Performing the integration by the trapezoidal rule on the same data gives a result of 139.70. Since the data are quite smooth (small experimental error), it is to be expected that these results will be similar.

EXERCISES

Exercise 7.1: For the data shown below:

v	y
0	1.766
0.25	2.478
0.5	3.690
0.75	6.397
1	6.649

a. Fit the data to a quadratic polynomial. As a first step, plot the data and add a quadratic polynomial trendline—this will indicate what the coefficients will be when they are calculated. Do these calculations *by hand*; determine c_1, c_2, c_3, the parameter standard deviations, the *t*-ratios, and the R^2 value (which can also be checked from the trendline).

b. Repeat the calculations of part a using the Excel Data Analysis/ Regression Add-In.

c. Based on the results of parts a and b, if any of the *t*-ratios suggest an alteration in the fitting function, change the function and recalculate all necessary quantities using the Excel Data Analysis/Regression Add-In.

d. Fit the data to a fourth-degree polynomial using an Excel trendline *only*. Comment on the results (Is it good, and if so why? Is it bad, if so why?).

Exercise 7.2: Consider the following data:

x	y
0	0.998
0.1	1.061
0.2	1.050
0.3	1.111
0.4	1.298
0.5	1.482
0.6	1.751
0.7	2.211
0.8	2.658
0.9	3.262
1	3.965

a. Plot the data in the usual way.
b. Plot the data again, but with no connecting line.
c. Add a trendline to the data; display the equation and the R^2 value.
d. Use the Excel Regression Add-In to fit the data to a cubic polynomial: $y = c_1 + c_2x + c_3x^2 + c_4x^3$. Based on the calculated t-ratios (called t-stat in the results table), should all four constants be retained? If not, use the Excel Regression Add-In to redo the regression with the most significant terms in the fitting polynomial. Continue this until all t-ratios are satisfactory.
e. (Optional) Now that the answers are known, redo step d *by hand* using the appropriate equations and formulas. Formulate the linear equations to be solved, and calculate the value of R^2, the parameter standard deviations (called the standard error by the Excel Add-In), and the t-ratios. Again, repeat this step until all t-ratios are satisfactory.

Exercise 7.3: The Clausius–Clapyron equation relates the latent heat of vaporization to temperature and vapor pressure according to the following equation:

$$\ln(p) = -\frac{\Delta H_v}{RT} + k \tag{7.36}$$

Data for water are shown below (units are K for T and mmHg for p). Use linear regression to find the heat of vaporization of water; the published value is –540 cal/g. The heat of vaporization (/R) is the slope of the line described by Equation 7.36.

1/T	ln(P)
0.002755	6.2649
0.002740	6.3404
0.002725	6.4149
0.002710	6.4886
0.002695	6.5615
0.002681	6.6333
0.002667	6.7043
0.002653	6.7743
0.002639	6.8436
0.002625	6.9121
0.002611	6.9797

Exercise 7.4: The data shown below are to be fitted to the function

$$y^{calc} = c_1 + c_2 \frac{\ln x}{x^2} + c_3 e^{-x} \tag{7.37}$$

Consider the following data (these are "artificial" and were generated to suit this problem):

x	y Data
1	–0.3680
1.1	–0.2540
1.2	–0.1740
1.3	–0.1170
1.4	–0.0749
1.5	–0.0429
1.6	–0.0183
1.7	0.0009
1.8	0.0161
1.9	0.0282
2	0.0380
2.1	0.0458
2.2	0.0521
2.3	0.0572
2.4	0.0612
2.5	0.0645
2.6	0.0671
2.7	0.0690
2.8	0.0705
2.9	0.0716
3	0.0723

a. Find the three parameters c_1, c_2, and c_3 using the Excel Data Analysis Regression Add-In. If any of the parameters are insignificant based on the t-ratios, repeat the calculations using the reduced model.

b. Do the calculations for the three parameter models *by hand*. Your results should be the same as those from part a.

Exercise 7.5: The following are vapor–liquid equilibrium data for the binary system SO_2–water at 20°C.

x (mole fr. SO$_2$ in liquid)	y (mole fr. SO$_2$ in vapor)
0	0
5.62E–05	0.000685
0.00014	0.00158
0.00028	0.00421
0.000422	0.00763
0.000564	0.0112
0.000842	0.01855
0.001403	0.0342
0.001965	0.0513
0.00279	0.0775
0.0042	0.121
0.00698	0.212
0.01385	0.443
0.0206	0.682
0.0273	0.917

Fit these data to an appropriate polynomial form. First, plot the data and successively fit a linear, quadratic, cubic, and quartic *trendline*, noting the R^2 value each time; when the R^2 no longer improves, use the Excel Regression Add-On to find the parameters and statistical indicators. Verify that the polynomial chosen does indeed satisfy the usual statistical criteria. In all cases, the constant term must be zero since the (0, 0) data point is without error.

8 Partial Differential Equations

8.1 INTRODUCTION

The most frequently occurring partial differential equations (PDEs) involving two independent variables are as follows:

Parabolic:

$$\frac{\partial u}{\partial t} = \frac{\partial^2 u}{\partial x^2} \quad \text{one-dimensional unsteady state} \tag{8.1}$$

Elliptic:

$$\frac{\partial^2 u}{\partial y^2} + \frac{\partial^2 u}{\partial x^2} = 0 \quad \text{two-dimensional steady state} \tag{8.2}$$

Hyperbolic:

$$\frac{\partial^2 u}{\partial t^2} + \frac{\partial^2 u}{\partial x^2} = 0 \quad \text{vibration of a string} \tag{8.3}$$

In this chapter, only parabolic and elliptic PDEs are considered.

8.2 PARABOLIC PDEs

The equation

$$\frac{\partial u}{\partial t} = \alpha \frac{\partial^2 u}{\partial x^2} \tag{8.4}$$

describes many phenomena in chemical and biomolecular engineering. Examples are one-dimensional, unsteady-state heat conduction (where α is the thermal diffusivity), fluid flow (where α is kinematic viscosity), and molecular diffusion (where

α is the molecular diffusivity). The heat conduction examples are most widely used since they are easy to visualize.

Recall Example 6.1 involving heat conduction in a thin rod of length 1 (when the perimeter is insulated). For transient operation (unsteady state), this system can be described by

$$\frac{\partial u}{\partial \theta} = \frac{\partial^2 u}{\partial x^2} \qquad (8.5)$$

where α is the thermal diffusivity of the rod material and θ is a dimensionless time defined as

$$\theta = \alpha t \qquad (8.6)$$

This kind of problem is a PDE-IVP. It is a boundary value problem in x and an initial value problem in θ. It remains to specify initial and boundary conditions. For simplicity, consider the following:

$$u(0,\theta) = 0; \; u(1,\theta) = 1; \; u(x,0) = 1 \quad \text{(1 everywhere except } x = 0) \qquad (8.7)$$

Physically, this represents heat transfer in a one-dimensional object (a rod or slab) with unit initial temperature. At time 0, the temperature at the left boundary undergoes a step change to zero while the right boundary is maintained at a temperature of 1. This problem can be solved analytically to give

$$u = x + \frac{2}{\pi} \sum_{n=1}^{\infty} \frac{1}{n} \exp(-n^2 \pi^2 \theta) \sin(n\pi x) \qquad (8.8)$$

The following spreadsheet shows the calculation of *one* temperature at the center-line ($x = 0.5$) and for $\theta = 0.1$:

n	n^2 Pi^2 Theta	n Pi x	exp(−B)	sin C	Term	Sum
1	0.9869605	1.57080	0.372708	1	0.23727	0.23727
2	3.9478419	3.14159	0.019296	−5E–08	0.00000	0.23727
3	8.8826442	4.71239	0.000139	−1	−0.00003	0.23724
4	15.7913675	6.28319	1.39E–07	9E–08	0.00000	0.23724
5	24.6740117	7.85398	1.92E–11	1	0.00000	0.23724

Since $x = 0.5$, the solution at this point is $u = 0.73724$. This exercise gives a *point* that can be compared to that produced by any numerical solution.

To solve such problems numerically, a popular approach is to substitute finite difference analogs for the derivatives and solve the resulting algebraic equations. There are many ways in which this can be done, a few of which are summarized below:

1. Centered difference in x, forward difference in θ
2. Centered difference in x, backward difference in θ
3. Centered difference in x, centered difference in θ

8.2.1 Explicit Method (Centered Difference in x, Forward Difference in θ)

Solution by centered difference in x, forward difference in θ, is now illustrated. This approach suffers the same shortcomings as the Euler method for ODEs (the truncation error is $\mathcal{O}(\Delta x)$). By replacing the derivatives with the appropriate finite difference analogs (see Table 4.2) as follows:

$$\frac{u_{i,j+1} - u_{i,j}}{\Delta\theta} = \frac{u_{i-1,j} - 2u_{i,j} + u_{i+1,j}}{\Delta x^2}; \; j = 0,1,\ldots,M; \; i = 0,1,2,\ldots,N+1 \qquad (8.9)$$

M should be large enough to reach the steady-state solution.

This corresponds to a two-dimensional "grid" with θ on the vertical axis and x on the horizontal axis. Subscript j is an index for the θ axis while i is an index for the x-axis. When $i = 0$, the left boundary condition applies; when $i = N + 1$, the right boundary condition applies. The process begins with u *known* when $j = 0$ ($\theta = 0$ or time = zero) for all i; these are the initial conditions (zero at all points—Equation 8.7). The only *unknown* in this equation is $u_{i,j+1}$. Solving for this unknown gives

$$u_{i,j+1} = u_{i,j} + \frac{\Delta\theta}{\Delta x^2}(u_{i-1,j} - 2u_{i,j} + u_{i+1,j}) \qquad (8.10)$$

$\Delta\theta$ and Δx must be determined so that reasonably accurate results occur (if possible). A typical value for N is 19, but a larger value might be required. This corresponds to Δx of 0.05.

This method is the simplest approach to solving the problem numerically. It can be implemented directly in Excel®, or a VBA Macro can be written.

Example 8.1: Transient Heat Conduction in a Rod

Shown below are the first few rows of an Excel program to solve Equation 8.10 together with the initial and boundary conditions given in Equation 8.7:

	A	B	C	D	E	F	G	H	I	J	K	L	M	N	O	P	Q	R	S	T	U	V
1	Dtheta	0.001	Dx	0.05	N	20	Dtheta/Dx^2		0.4													
2	x																					
3	Theta	0	0.05	0.1	0.15	0.2	0.25	0.3	0.35	0.4	0.45	0.5	0.55	0.6	0.65	0.7	0.75	0.8	0.85	0.9	0.95	1
4	0	0	1	1	1	1	1	1	1	1	1	1	1	1	1	1	1	1	1	1	1	1
5	0.001	0	0.6000	1	1	1	1	1	1	1	1	1	1	1	1	1	1	1	1	1	1	1
6	0.002	0	0.5200	0.8400	1	1	1	1	1	1	1	1	1	1	1	1	1	1	1	1	1	1
7	0.003	0	0.4400	0.7760	0.9360	1	1	1	1	1	1	1	1	1	1	1	1	1	1	1	1	1
8	0.004	0	0.3984	0.7056	0.8976	0.9744	1	1	1	1	1	1	1	1	1	1	1	1	1	1	1	1
9	0.005	0	0.3619	0.6595	0.8515	0.9539	0.9898	1	1	1	1	1	1	1	1	1	1	1	1	1	1	1

Values of $\Delta\theta$ of 0.01, 0.001, and 0.0001 were tried. $\Delta\theta = 0.001$ gave nearly the same results as $\Delta\theta = 0.0001$. The value of u corresponding to $\theta = 0.1$ and $x = 0.5$ was 0.7361. This compares favorably with the analytically computed value of 0. 0.7372. When $\Delta\theta = 0.001$, the numerically computed value was 0.7371. Given the simplicity of this method, these results are remarkably good. It must be pointed out, however, that when $\Delta\theta = 0.01$, the solution was unstable (wildly fluctuating temperatures were the result).

8.2.2 CENTERED DIFFERENCE IN x, BACKWARD DIFFERENCE IN θ

In the interest of keeping this discussion brief, this method will not be implemented. The required algorithms are similar to what is discussed next. A potential significant advantage of this approach is that it has excellent stability attributes. For more information, Google it.

8.2.3 CRANK–NICHOLSON METHOD (CENTERED DIFFERENCE IN x, CENTERED DIFFERENCE IN θ)

The third method (called the Crank–Nicholson method) applies some innovation in that the finite difference analog is "centered about a fictitious half-way point" as shown in Figure 8.1.

Here are the finite difference analogs for the terms in the PDE; the unknowns are at level $j + 1$ for any i:

$$\frac{\partial u}{\partial t} = \frac{u_{i,j+1} - u_{i,j}}{\Delta\theta} + \mathcal{O}(\Delta\theta^2) \quad \text{(centered about } j + 1/2) \tag{8.11}$$

$$\frac{\partial^2 u}{\partial x^2} = \frac{u_{i-1,j+1/2} - 2u_{i,j+1/2} + u_{i+1,j+1/2}}{\Delta x^2} + \mathcal{O}(\Delta x^2) \quad \text{(centered about } i) \tag{8.12}$$

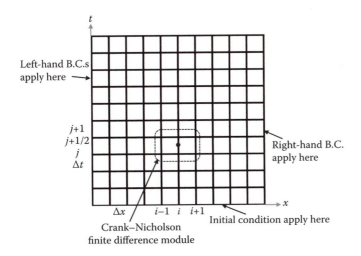

FIGURE 8.1 Crank–Nicholson finite difference nomenclature.

The $j + 1/2$ terms are replaced with its *average at the j and j + 1 points*:

$$u_{i,j+1/2} = \frac{1}{2}(u_{i,j} + u_{i,j+1}) \tag{8.13}$$

With these substitutions, Equation 8.5 is replaced by the following finite difference equations:

$$-u_{i-1,j+1} + 2\left(1 + \frac{\Delta x^2}{\Delta\theta}\right)u_{i,j+1} - u_{i+1,j+1} = u_{i-1,j} - 2\left(1 - \frac{\Delta x^2}{\Delta\theta}\right)u_{i,j} + u_{i+1,j} \tag{8.14}$$

Note that the unknowns are at level $j + 1$, so all terms on the right-hand side are known. When the finite difference equation is applied for all i, a tridiagonal system of equations results. Such systems can be solved very efficiently (as compared to full-matrix linear systems) using a method called the Thomas algorithm (see below). The Matrix.xla function SYSLINT can also be used.

Example 8.2: Crank–Nicholson Method

The algorithm for solving the example problem can be summarized as follows:

1. Start with all us at the initial conditions.
2. Set up the tridiagonal set of equations for time $\Delta\theta$. Apply the boundary conditions for $i = 0$ and $i = N + 1$.
3. Use the Thomas algorithm or SYSLINT to find all us at $\Delta\theta$.
4. Repeat the process for 2 $\Delta\theta$, 3 $\Delta\theta$, and so forth.

The initial and boundary conditions of Equation 8.7 must be introduced. Let the vector v represent the present (known or level j) temperatures and u be the unknown (level $j + 1$) temperatures. The finite difference equations are as follows (note that the initial values for v are all 1). Recall that the left boundary value is 0 while the right boundary value is 1.

$$-0 + 2\left(1 + \frac{\Delta x^2}{\Delta\theta}\right)u_1 - u_2 = v_0 - 2\left(1 - \frac{\Delta x^2}{\Delta\theta}\right)v_1 + v_2 \quad \text{(left boundary)}$$

$$-u_1 + 2\left(1 + \frac{\Delta x^2}{\Delta\theta}\right)u_2 - u_3 = v_1 - 2\left(1 - \frac{\Delta x^2}{\Delta\theta}\right)v_2 + v_3$$

$$-u_2 + 2\left(1 + \frac{\Delta x^2}{\Delta\theta}\right)u_3 - u_4 = v_2 - 2\left(1 - \frac{\Delta x^2}{\Delta\theta}\right)v_3 + v_4 \tag{8.15}$$

$$\vdots$$

$$-u_{N-1} + 2\left(1 + \frac{\Delta x^2}{\Delta\theta}\right)u_N = v_{N-1} - 2\left(1 - \frac{\Delta x^2}{\Delta\theta}\right)v_N + 1 \quad \text{(right boundary)}$$

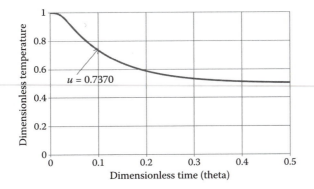

FIGURE 8.2   Centerline temperature ($\Delta\theta = 0.1$, $\Delta x = 0.05$).

This is a tridiagonal system. Once solved for all us, replace v with u and repeat for any desired number of time steps.

Figure 8.2 is a graph of the temperature when $x = 0.5$ (the centerline). The numerical solution at the centerline (when $\Delta\theta = 0.1$ and $\Delta x = 0.05$) for $\theta = 0.1$ is 0.7370, which is the same as the infinite series solution to three significant figures. Accuracy to additional significant figures is easily obtained by reducing $\Delta\theta$. It is highly significant that a value of $\Delta\theta = 0.01$ is one or two orders of magnitude larger than was required with the Euler in time method.

8.3 THOMAS ALGORITHM FOR TRIDIAGONAL SYSTEMS

The Thomas algorithm is well known in numerical mathematics as an implementation of Gaussian elimination applied specifically to tridiagonal systems. If each equation is of the form

$$c_i x_{i-1} + d_i x_i + e_i x_{i+1} = b_i \tag{8.16}$$

applying Gaussian elimination to the system results in the following algorithm:

1. $\beta_1 = d_1$

2. $\gamma_1 = \dfrac{b_1}{\beta_1}$

3. $\beta_i = d_i - \dfrac{c_i \cdot e_{i-1}}{\beta_{i-1}} \quad i = 2,3,\cdots,n$

4. $\gamma_i = \dfrac{b_i - c_i \cdot \gamma_{i-1}}{\beta_i} \quad i = 2,3,\cdots,n$ (8.17)

5. $x_n = \gamma_n$

6. $x_i = \gamma_i - \dfrac{e_i \cdot x_{i+1}}{\beta_i} \quad i = n-1, n-2, \cdots, 1$

Example 8.3: Illustration of the Thomas Algorithm

A three-equation, three-unknown example is as follows:

$$\begin{bmatrix} 2 & 1 & 0 \\ 1 & 2 & 1 \\ 0 & 1 & 2 \end{bmatrix} \begin{bmatrix} x_1 \\ x_2 \\ x_3 \end{bmatrix} = \begin{bmatrix} 3 \\ 4 \\ 3 \end{bmatrix} \quad \text{(note that the matrix is tridiagonal)}$$

The steps of the Thomas algorithm are shown below:

$$\beta_1 = 2 \quad \gamma_1 = 3/2$$

$$i = 2: \quad \beta_2 = 2 - \frac{1 \cdot 1}{2} = 3/2$$

$$\gamma_2 = \frac{4 - 1(3/2)}{3/2} = \frac{8/2 - 3/2}{3/2} = \frac{5/2}{3/2} = 5/3$$

$$i = 3: \quad \beta_3 = 2 - \frac{1 \cdot 1}{3/2} = 6/3 - 2/3 = 4/3$$

$$\gamma_3 = \frac{3 - 1(5/3)}{4/3} = \frac{9/3 - 5/3}{4/3} = \frac{4/3}{4/3} = 1$$

$$x_3 = 1$$

$$i = 2: \quad x_2 = \frac{5}{3} - \frac{1 \cdot 1}{3/2} = \frac{5}{3} - \frac{2}{3} = 1$$

$$i = 1: \quad x_1 = \frac{3}{2} - \frac{1 \cdot 1}{2} = \frac{3}{2} - \frac{1}{2} = 1$$

Example 8.4: VBA Program for the Crank–Nicholson Method

A VBA program that implements the Crank–Nicholson method for the problem of Equation 8.5 with the initial and boundary conditions of Equation 8.7 is shown below. Equation 8.15 forms the basis for the tridiagonal system to solve at each time step.

```
Sub CrankNicholsonZLOR()
Dim DeltaX As Double
Dim DeltaT As Double
Dim Ratio As Double
Dim NumXSteps As Long
Dim MaxTime As Double
Dim Time As Double
Dim TOld() As Double
Dim TNew() As Double

Dim C() As Double
Dim D() As Double
Dim E() As Double
Dim B() As Double
Dim X() As Double
Dim I, J As Long

DeltaT = ActiveSheet.Cells(2, 3).Value  'get time step
DeltaX = ActiveSheet.Cells(3, 3).Value  'get distance step
Ratio = DeltaX ^ 2 / DeltaT

MaxTime = ActiveSheet.Cells(2, 5).Value 'get maximum time
NumXSteps = Round(1 / DeltaX) - 1
Time = 0                                'initialize time
ReDim TOld(1 To NumXSteps + 2) As Double
ReDim TNew(1 To NumXSteps + 2) As Double
ReDim C(1 To NumXSteps) As Double        'lower diagonal coefficients
ReDim D(1 To NumXSteps) As Double        'diagonal coefficients
ReDim E(1 To NumXSteps) As Double        'upper diagonal coefficients
ReDim B(1 To NumXSteps) As Double        'right hand side vector
ReDim X(1 To NumXSteps) As Double        'unknown temperatures at new time step
          'Get the initial temperatures from the spreadsheet
I = 6
For J = 1 To NumXSteps + 2
  TOld(J) = ActiveSheet.Cells(I, J + 1)
Next J

Time = 0
While Time <= MaxTime
 'Set up the tridiagonal system coefficients
  For J = 2 To NumXSteps + 1
    C(J - 1) = -1
    D(J - 1) = 2 * (1 + Ratio)
    E(J - 1) = -1
    B(J - 1) = TOld(J - 1) - 2 * (1 - Ratio) * TOld(J) + TOld(J + 1)
  Next J
  B(NumXSteps) = B(NumXSteps) + 1
  Call Thomas(NumXSteps, C, D, E, B, X)

  TNew(1) = TOld(1)
  For J = 1 To NumXSteps
    TNew(J + 1) = X(J)
  Next J
  TNew(NumXSteps + 2) = TOld(NumXSteps + 2)

  Time = Time + DeltaT   'this is the "new" time
 'display the temperatures at the new time
  I = I + 1                        'move down one row
  ActiveSheet.Cells(I, 1) = Time
  For J = 1 To NumXSteps + 2
    ActiveSheet.Cells(I, J + 1) = TNew(J)
    TOld(J) = TNew(J)
  Next J
Wend
```

```
Sub Thomas (N, C, D, E, B, X)
Dim Beta () As Double
Dim Gamma () As Double
ReDim Beta (1 To N) As Double
ReDim Gamma (1 To N) As Double
Dim I As Long
Dim J As Long

Beta (1)  = D(1)
Gamma (1)  = B(1) / Beta(1)
For I = 2 To N
   Beta (I)  = D(I)  -  C(I) * E(I - 1) / Beta (I - 1)
    Gamma (I)  =  (B(I)  -  C(I) * Gamma (I - 1)) / Beta (I)
Next I
X (N)  = Gamma (N)
For I = 2 To N
   J = N - I + 1
   X (J)  = Gamma (J)  -  E(J) * X(J + 1) / Beta (J)
Next I

End Sub
```

The spreadsheet shown on the next page illustrates the program user interface and output. Note, particularly, the value for $\theta = 0.1$ and $x = 0.5$. To three significant figures, this value is the same as that provided by the analytical solution.

To review, the Crank–Nicholson method involves writing second-order correct finite difference approximations by using a fictitious "halfway point" in time. To resolve the unknowns at the halfway point, the average between the old time points and the new time points is used. This leads to a tridiagonal system of linear equations to be solved at each time step. The resulting algorithm is second-order correct in both distance and time. Using the same example, the time step required by the Crank–Nicholson method was two orders of magnitude greater than when the Euler in time method was used.

Heat transfer in a slab

Dt = 0.01 Dx = 0.05 Tmax = 1

Crank–Nicholson

Left boundary zero Right boundary at one IC = 1

Distance, dimensionless

Time	0	0.05	0.1	0.15	0.2	0.25	0.3	0.35	0.4	0.45	0.5	0.55	0.6	0.65	0.7	0.75	0.8	0.85	0.9	0.95	1
0	0	1	1	1	1	1	1	1	1	1	1	1	1	1	1	1	1	1	1	1	1
0.01	0	3E-12	0.5	0.75	0.875	0.938	0.969	0.984	0.992	0.996	0.998	0.999	1	1	1	1	1	1	1	1	1
0.02	0	0.333	0.333	0.5	0.667	0.792	0.875	0.927	0.958	0.977	0.987	0.993	0.996	0.998	0.999	0.999	1	1	1	1	1
0.03	0	0.074	0.352	0.472	0.579	0.683	0.774	0.847	0.9	0.936	0.96	0.976	0.986	0.992	0.995	0.997	0.998	0.999	1	1	1
0.04	0	0.193	0.243	0.395	0.523	0.624	0.71	0.782	0.842	0.888	0.923	0.949	0.967	0.979	0.987	0.992	0.995	0.997	0.998	0.999	1
0.05	0	0.09	0.271	0.365	0.467	0.568	0.656	0.731	0.794	0.845	0.886	0.918	0.942	0.96	0.973	0.982	0.989	0.993	0.996	0.998	1
0.06	0	0.138	0.208	0.335	0.438	0.529	0.612	0.686	0.751	0.806	0.851	0.888	0.917	0.94	0.957	0.97	0.98	0.987	0.992	0.996	1
0.07	0	0.09	0.224	0.309	0.404	0.495	0.577	0.65	0.714	0.77	0.818	0.858	0.891	0.918	0.939	0.956	0.969	0.979	0.987	0.994	1
0.08	0	0.11	0.187	0.294	0.384	0.467	0.546	0.618	0.682	0.739	0.789	0.831	0.867	0.897	0.921	0.941	0.957	0.971	0.982	0.991	1
0.09	0	0.086	0.194	0.274	0.362	0.444	0.52	0.59	0.654	0.711	0.762	0.806	0.844	0.876	0.903	0.926	0.945	0.962	0.976	0.988	1
0.1	0	0.094	0.171	0.264	0.345	0.423	0.497	0.566	0.629	0.686	**0.737**	0.782	0.822	0.856	0.886	0.912	0.934	0.953	0.97	0.985	1
0.11	0	0.081	0.173	0.249	0.33	0.406	0.478	0.545	0.607	0.663	0.715	0.761	0.802	0.838	0.87	0.898	0.922	0.944	0.964	0.982	1

Additional header annotations appearing in the table: under "Left boundary zero" → 0.25; under "Right boundary at one" → 0.125, 0.25.

8.4   METHOD OF LINES

When using finite differences to solve the unsteady heat conduction problem, another approach involves writing finite difference equations at each grid point (node) only for the spatial variables while leaving the time derivative intact. This leads, generally, to a large number of simultaneous ODEs, which can be solved by, for example, a Runge–Kutta method. However, one must be careful since this set of ODEs can be *stiff.* Consider the same one-dimensional, unsteady state heat conduction problem as solved in Examples 8.1 and 8.2. This problem is solved by the method of lines in the next example.

Example 8.5: Method of Lines

$$\frac{\partial u}{\partial \theta} = \frac{\partial^2 u}{\partial x^2}$$

$$u(0,t) = 0; \quad u(1,t) = 1 \quad u(x,0) = 1$$

(8.18)

Applying a second-order correct finite difference analog for the spatial derivative term at "line" i gives

$$\frac{du_i}{dt} = \frac{u_{i-1} - 2u_i + u_{i+1}}{\Delta x^2}; \quad i = 1,2,\cdots n$$

(8.19)

At the left boundary, $u_{i-1} = 0$ because of the left boundary condition. So, the ODE when $i = 1$ becomes

$$\frac{du_1}{dt} = \frac{-2u_1 + u_2}{\Delta x^2}$$

(8.20)

For $2 \le i \le n - 1$, the ODEs are

$$\frac{du_i}{dt} = \frac{u_{i-1} - 2u_i + u_{i+1}}{\Delta x^2}; \quad i = 2,3,\cdots n-1$$

(8.21)

And the last equation ($i = n - 1$) is as follows because the right-hand boundary condition is 1:

$$\frac{du_n}{dt} = \frac{u_{n-1} - 2u_n + 1}{\Delta x^2}$$

(8.22)

Shown below is a VBA subprogram `FCalc` that would be called by an ODE solver such as a Runge–Kutta method. The subprogram implements the right-hand-side functions of Equations 8.20 through 8.22.

```
Sub FCalc(Time, y, N, RHS)

Dim i As Long

'This is the "right hand side" function for the heat conduction problem
'Left boundary insultate, right boundary at zero, IC = 1

RHS(1) = (-2 * y(1) + y(2)) / 0.05 ^ 2
For i = 2 To N - 1
  RHS(i) = (y(i - 1) - 2 * y(i) + y(i + 1)) / 0.05 ^ 2
Next i
RHS(N) = (y(N - 1) - 2 * y(N) + 1) / 0.05 ^ 2

End Sub
```

The results for this implementation are essentially identical to the results shown in Example 8.2.

8.5　SUCCESSIVE OVERRELAXATION FOR ELLIPTIC PDEs

Consider Laplace's equation in two dimensions (this is an elliptic equation):

$$\frac{\partial^2 \phi}{\partial x^2} + \frac{\partial^2 \phi}{\partial y^2} = 0 \tag{8.23}$$

This equation represents many steady-state phenomena in two dimensions, such as temperature distributions, laminar flow distributions, and voltage distributions.

Example 8.6: Relaxation Method for an Elliptic Equation

Consider a rectangular (thin) flat plate as shown in Figure 8.3 with the given conditions along each edge.

Substituting second-order correct finite difference analogs into Equation 8.23, there results

$$\frac{\phi_{i-1,j} - 2\phi_{i,j} + \phi_{i+1,j}}{\Delta x^2} + \frac{\phi_{i,j-1} - 2\phi_{i,j} + \phi_{i,j+1}}{\Delta y^2} = 0 \tag{8.24}$$

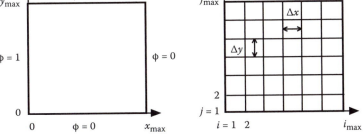

FIGURE 8.3　Nomenclature: finite difference representation of heat transfer in a flat plate.

Solving for $\phi_{i,j}$ gives

$$\phi_{i,j} = \frac{\Delta x^2(\phi_{i,j-1} + \phi_{i,j+1}) + \Delta y^2(\phi_{i-1,j} + \phi_{i+1,j})}{2\Delta x^2 + 2\Delta y^2} \qquad (8.25)$$

If $\Delta x = \Delta y$, this can be written as the average of the four surrounding temperatures as in Equation 8.26.

$$\phi_{i,j} = \frac{\phi_{i,j-1} + \phi_{i,j+1} + \phi_{i-1,j} + \phi_{i+1,j}}{4} \qquad (8.26)$$

To apply this *recurrence* relation, start by guessing the φ values at all nodes and then sweep through all *i* and *j* updating with the above "average of 4" formula. This process is repeated until there is only a small change from one sweep (iteration) to the next or for a fixed number of iterations (easier). The process is called *relaxation*. The relaxation method is a procedure for solving simultaneous equations by guessing a solution and then reducing the errors that result by *successive approximations* until all the errors are less than some specified amount. The VBA program shown below implements the relaxation method of Equation 8.26 and for the boundary conditions shown in Figure 8.3. The program is set to 20 iterations; this is much easier testing to see if the temperatures cease to change. Results are shown in Figure 8.4.

1	0	0	0	0
1	0.4654	0.2322	0.0997	0
1	0.6308	0.3649	0.1673	0
1	0.6949	0.4309	0.2053	0
1	0.7201	0.4605	0.2235	0
1	0.7270	0.4691	0.2290	0
1	0.7207	0.4612	0.2240	0
1	0.6960	0.4322	0.2060	0
1	0.6320	0.3663	0.1681	0
1	0.4663	0.2332	0.1003	0
1	0	0	0	0

FIGURE 8.4 Relaxation results after 20 iterations.

```
Sub SOR()
'This program applies the method of Successive Over-Relaxation
'    to the solution of an elliptic PDE.
Dim n As Integer 'number of x-grid intervals;
                 'unknowns are for i = 2, 3, ..., n-1.
                 'Boundary values are when i = 1 or n
Dim m As Integer 'number of y-grid intervals;
                 'unknowns are for j = 2, 3, ..., m-1.
                 'Boundary values are when i = 1 or m
Dim Phi() As Double   'dimensionless temperture within the grid
'First, read the BCs and initial guesses from the spreadsheet
Dim Dx, Dy As Double    'increments in x- and y-directions
Dim i, j, k As Integer
Dim SRF, PhiOld As Double

n = Cells(1, 2)
m = Cells(1, 4)
Dx = 1 / (n + 1)
Dy = 1 / (m + 1)
SRF = Cells(1, 6)
ReDim Phi(n, m)
For i = 1 To n
  For j = 1 To m
    Phi(i, j) = Cells(i + 2, j + 1)
  Next j
Next i
'Apply SOR
For k = 1 To 20
For i = 2 To n - 1
  For j = 2 To m - 1
    PhiOld = Phi(i, j)
    Phi(i, j) = (Dx ^ 2 * (Phi(i, j - 1) + Phi(i, j + 1)) + _
                Dy ^ 2 * (Phi(i - 1, j) + Phi(i + 1, j))) / _
                (2 * Dx ^ 2 + 2 * Dy ^ 2)
    Phi(i, j) = PhiOld + SRF * (Phi(i, j) - PhiOld)
  Next j
Next i
Next k
'Output the results
For i = 2 To n - 1
  For j = 2 To m - 1
    Cells(i + 2, j + 1) = Phi(i, j)
  Next j
Next i
End Sub
```

Because of symmetry, it is known that the values 0.2883 and 0.2924 should be the same, but even after 15 iterations, they are relatively far apart. To improve this, there is the method of successive overrelaxation or the SOR method. The idea here is that on each sweep, the newly calculated value is not used directly; instead an interpolation/extrapolation formula as shown by the following equation is used:

$$\phi_{New} = \phi_{Old} + SRF(\phi_{Calc} - \phi_{Old}) \tag{8.27}$$

where *SRF* is the *successive relaxation factor*. If *SRF* = 1, the new value is equal to the calculated one. If *SRF* < 1, the formula interpolates, and if *SRF* > 1, it extrapolates. Typically "good" values for *SRF* are between 1.2 and 1.5.

Example 8.7: Successive Overrelaxation

When Equation 8.27 is used with an SRF of 1.5 and 20 iterations, the results are as shown in Figure 8.5. Note that these data are symmetric to the precision shown.

Much more time could be spent on PDEs of various kinds and with a variety of boundary conditions. Time does not allow this, but with the background gained so far, the reader should be able to comprehend more advanced numerical analysis textbooks and research papers to learn about solving other kinds of problems. Want to know more? Google it!

1	0	0	0	0
1	0.4666	0.2335	0.1005	0
1	0.6327	0.3671	0.1686	0
1	0.6971	0.4335	0.2068	0
1	0.7223	0.4631	0.2251	0
1	0.7290	0.4714	0.2304	0
1	0.7223	0.4631	0.2251	0
1	0.6971	0.4335	0.2068	0
1	0.6327	0.3671	0.1686	0
1	0.4666	0.2335	0.1005	0
1	0	0	0	0

FIGURE 8.5   Successive overrelaxation results for 20 iterations.

EXERCISES

Exercise 8.1: Rework the problem of Example 8.1 using the simple explicit method and the following initial and boundary conditions:

a. $u(x, 0) = 1$; $u(0, t) = u(1, t) = 0$
b. $u(x, 0) = 1$; $u(0, t) = 0$; $du(1, t)/dx = 0$ (right side insulated)
c. $u(x, 0) = x$ (linear initial temperature profile); $u(0, t) = 1$; $u(1, t) = 0$
d. $u(x, 0) = \sin(\pi x)$; $u(0, t) = u(1, t) = 0$
e. $u(x, 0) = x$; $du(0, t)/dx = du(1, t)/dx = 0$

Exercise 8.2: Rework Exercise 8.1 using the Crank–Nicholson method.

Exercise 8.3: Rework Exercise 8.1 using the method of lines.

Exercise 8.4: The boundary value problem for a rectangular fin is as follows:

$$\frac{\partial^2 T}{\partial x^2} - \beta^2 (T - T_a) = \frac{1}{\alpha}\frac{\partial T}{\partial t} \tag{8.28}$$

$$0 \le x \le 1$$

$$T(0,t) = T_1$$

$$T(1,t) = T_2 \tag{8.29}$$

$$T(x,0) = 0$$

Define a dimensionless temperature and time as follows:

$$u = \frac{T - T_a}{T_1 - T_a} \quad \tau = \alpha t \tag{8.30}$$

The boundary value problem then becomes

$$\frac{\partial^2 u}{\partial x^2} - \beta^2 u = \frac{\partial u}{\partial \tau} \tag{8.31}$$

The BCs become

$$u(0,t) = 1$$

$$u(1,t) = \frac{T_2 - T_a}{T_1 - T_a} \tag{8.32}$$

$$u(x,0) = \frac{-T_a}{T_1 - T_a}$$

The steady-state solution can be shown to be

$$u_{ss} = \frac{\sinh[\beta(1-x)]}{\sinh\beta} + \left(\frac{T_2 - T_a}{T_1 - T_a}\right)\frac{\sinh\beta x}{\sinh\beta} \tag{8.33}$$

For simplicity, take $T_a = 0$, $T_1 = 1$, $T_2 = 0$, and $\beta = 4$.

The steady-state problem was described in Exercise 6.5. The present exercise involves solving the transient problem by the methods presented in this chapter. Be sure that in each case, the eventual profile corresponds to the steady-state solution.

a. Solve the transient (PDE) dimensionless problem using the explicit method.

b. Solve the transient (PDE) dimensionless problem using the Crank–Nicholson method.

c. Solve the transient (PDE) dimensionless problem using the method of lines.

Exercise 8.5: Resolve the example elliptic PDE of Equation 8.23 using the following boundary conditions and with an SOR factor of 1.4. Use a grid such that $\Delta x = \Delta y$.

$d\varphi(0, y)/dy = 0$	Insulated left boundary
$\varphi(1, y) = 1$	Right side at 1
$d\varphi(x, 0)/dx = 0$	Insulated bottom boundary
$\varphi(x, 1) = 0$	Top side at 0

Experiment with different SOR factors and compare the results.

9 Linear Programming, Nonlinear Programming, Nonlinear Equations, and Nonlinear Regression Using Solver

9.1 INTRODUCTION

Solver is a powerful tool that is a standard Excel® Add-In. It can produce solutions to many different kinds of problems, among which are the following:

- Linear programming
- Nonlinear programming
- Nonlinear regression
- Nonlinear sets of equations

Solver can also be used for simpler problems such as solving one nonlinear equation (such as those discussed in Chapter 1). It is a much more powerful tool than Goal Seek and might be used in instances where Goal Seek fails. For most problem types (except most notably linear programming), there is no guarantee that Solver will find a solution. An initial guess must be provided, and then Solver attempts to find better and better estimates for the unknowns until it finds conditions that indicate that a solution has been found, or it gives an error message stating failure.

9.2 LINEAR PROGRAMMING

Linear programming (LP) involves problems with an objective function (often profit or loss) and constraints (equality and inequality) that typically specify the availability (or lack thereof) of resources. The problem is to maximize (or minimize) the objective function while satisfying all of the constraints. As the name implies, the objective function and constraints are *linear* functions of the unknown variables. In this context, the word *programming* does not refer to a computer program but to the action or process of scheduling something such as assigning people to jobs or how to allocate resources. Many important engineering (and other) problems can be formulated as linear programs.

Example 9.1: A Simple LP Problem

Consider the following LP in two variables:

$$\text{Maximize } y = 2x_1 + 4x_2$$

subject to

$$x_2 - x_1 \leq 4$$
$$x_1 + x_2 \leq 8$$

(9.1)

It is typical to also impose so-called nonnegativity constraints. These are not essential but "traditional." This can always be made the case by change of variables, but some lower limit is most often imposed. The nonnegativity constraints are expressed as

$$x_1 \geq 0$$
$$x_2 \geq 0$$

(9.2)

A graphical representation of this LP problem is shown in Figure 9.1. The solid lines represent the two constraints, and the arrows point into the feasible region (where both constraints are satisfied). Lines representing the objective function (y) are dashed. It is obvious from the figure that the maximum value of y on (or within) the feasible region is 28. This same result can be found by simply solving the two constraints as equalities.

In the late 1940s, George Dantzig perfected the so-called *simplex algorithm* for solving LP problems. This effort was initiated during World War II but was kept secret until 1946. The simplex method starts with an initial guess of the origin (called the

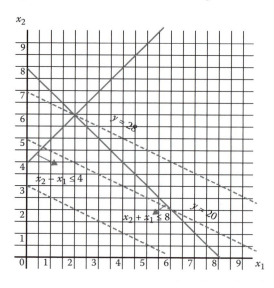

FIGURE 9.1 Graphical interpretation of a linear programming problem.

basic solution since it always satisfies all of the constraints). It then *visits* vertices based on those that give the most improvement in the objective function. It is actually a bit more complex than that, but this is the general idea. For a long time, researchers were surprised by the repeated successes of the simplex algorithm. It was finally proven that any LP problem can be solved in what is called *polynomial time* (Google it).

Solver can handle LPs very robustly. Before invoking Solver, a spreadsheet must be set up to represent the objective function and constraints. A typical spreadsheet setup for the example problem appears below:

	A	B	C
1	x1	x2	
2	0	0	
3	y =	0	=2*A2+4*B2
4	Constraint 1	0	=B2-A2
5	Constraint 2	0	=A2+B2

The actual formulas for the cells in column B are shown in bold in column C. The initial values for x_1 and x_2 have been set to zero; other values could be used, but selecting the origin as initial values is the usual procedure.

To start Solver, use the menu Data/Solver. The following window shows the Solver Parameters filled in for the example problem. Note that cell B3 contains the value of the objective function, and the Max button is selected. The By Changing Variable Cells has the range indicator for x_1 and x_2. The constraints were included by using the Add button and entering the appropriate values. Note that the box next to Make Unconstrained Variables Non-Negative is checked.

Importantly, the Select a Solving Method shows that the Simplex LP method has been chosen.

When the Solve button is selected, the spreadsheet changes to the following:

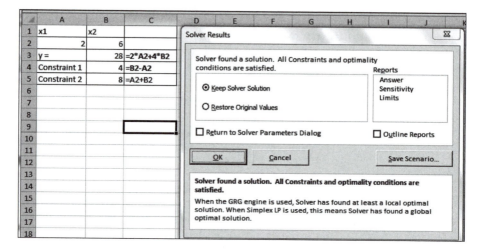

The Solver Results window indicates if a solution has been found that satisfies all constraints and optimality conditions. By selecting OK, the Solver Results window disappears, and the optimal values of x_1 and x_2 remain displayed on the spreadsheet (hitting cancel causes all spreadsheet values to revert to the initial ones).

The next example is a more realistic one of interest to chemical engineers. It is a very simplified version of what might be used by oil refinery management in order to optimize plant operation.

Example 9.2: Refinery Linear Program

Suppose that a refinery has four different types of crude oil (in a tank farm) available. In order to optimize plant operation (maximize profit), the plant management can process different amounts of the four crudes. Perhaps the hardest part of any LP problem is collecting the data and constructing the objective function and constraints. Figure 9.2 shows the product fractions that result from refining each

Crude number		1	2	3	4
Fractional product	Gasoline	0.15	0.55	0.3	0.25
	Diesel	0.25	0.2	0.3	0.35
	Aviation fuel	0.25	0.15	0.25	0.15
	Lube oil	0.25	0.05	0	0.1
	Losses	0.01	0	0.05	0.15
Availability 1000 Bbl/day		20	25	35	30

FIGURE 9.2 Crude oil product fractions and availability.

crude. Also shown is the availability of each crude oil. Figure 9.3 contains data regarding processing costs, sales price, and market demand.

Let x_1, x_2, x_3, and x_4 represent the amount of each crude to be processed (1000 Bbl/day). The objective function is the profit generated by processing the four crude oils in the optimal proportions. Profit for each crude is calculated as the amount of the crude (in barrels/day) times the profit margin (selling price − processing cost) times the fraction yield for each product. For example, for crude no. 1, the net profit is calculated as

$$x_1[(56 - 52)(0.15) + (49 - 45)(0.25) + (62 - 55)(0.25)$$
$$+ (70 - 60)(0.25)] * 1000 = 5850x_1 \tag{9.3}$$

Using similar calculations for the other crudes, the objective function becomes

$$y = 5850x_1 + 5050x_2 + 5150x_3 + 4450x_4 \tag{9.4}$$

There are three kinds of constraints: crude availability, market demands, and nonnegativity.

Crude availability constraints:

$$x_1 \le 20 \; x_2 \le 25 \; x_3 \le 35 \; x_4 \le 30 \tag{9.5}$$

Demand constraints:

$$0.15x_1 + 0.55x_2 + 0.30x_3 + 0.25x_4 \le 38$$
$$0.25x_1 + 0.20x_2 + 0.30x_3 + 0.35x_4 \le 18$$
$$0.25x_1 + 0.15x_2 + 0.25x_3 + 0.15x_4 \le 30 \tag{9.6}$$
$$0.25x_1 + 0.05x_2 + 0.00x_3 + 0.10x_4 \le 5$$

Nonnegativity constraints:

$$x_1 \ge 0 \; x_2 \ge 0 \; x_3 \ge 0 \; x_4 \ge 0 \tag{9.7}$$

Product	Processing cost $/barrel	Selling price $/barrel	Minimum daily demand 1000 Bbl/day
Gasoline	52	56	38
Diesel	45	49	18
Aviation fuel	55	62	30
Lube oil	60	70	5

FIGURE 9.3   Financial and daily demand data for products.

The following spreadsheet shows the setup for this LP problem. Note that the "initial guess" for all variables is zero.

	A	B	C	D	E	F
1	x1	x2	x3	x4		
2	0	0	0	0		
3						
4	Constraints					
5	LHS	RHS	comment			
6	0	38	LE	gasoline		
7	0	18	LE	diesel		
8	0	30	LE	aviation fuel		
9	0	5	LE	lube oil		
10	0	20	LE	x1 upper		
11	0	25	LE	x2 upper		
12	0	35	LE	x3 upper		
13	0	30	LE	x4 upper		
14	Obj. func.	0	=5850*A2+5050*B2+5150*C2+4450*D2			

When Solver is invoked, the spreadsheet changes to the following:

	A	B	C	D	E	F
1	x1	x2	x3	x4		
2	15	25	30.83333	0		
3						
4	Constraints					
5	LHS	RHS	comment			
6	25.25	38	.LE.	gasoline		
7	18	18	.LE.	diesel		
8	15.20833	30	.LE.	aviation fuel		
9	5	5	.LE.	lube oil		
10	15	20	.LE.	x1 upper		
11	25	25	.LE.	x2 upper		
12	30.83333	35	.LE.	x3 upper		
13	0	30	.LE.	x4 upper		
14	Obj. Func.	372791.7	=5850*A2+5050*B2+5150*C2+4450*D2			

Therefore, it is optimal to process 15,000 Bbl/day of crude 1, 25,000 Bbl/day of crude 2, 30,833 Bbl/day of crude 3, and *none* of crude 4. In typical refinery operations, the constraining data change often, so when the LP is run on another day, the optimal values are apt to change.

9.3 NONLINEAR PROGRAMMING

Nonlinear programming (NLP), as the name implies, is similar to LP, but the objective function or constraints can be nonlinear functions. There are no algorithms (like the simplex method) that guarantee a solution for NLP problems. Many methods have been developed, and Solver has one of these built in (called Generalized Reduced Gradient). The subject of NLP is quite complex and far beyond what can be covered here. NLP is introduced by way of a simple example. Even the simplest of chemical and biomolecular engineering NLP problems can be too complex to warrant coverage here.

Example 9.3: An NLP Problem

Consider the following NLP:

$$\text{Minimize } y = 2x_1^2 + 2x_2^2 + x_3^2 - 2x_1x_2 - 4x_1 - 6x_2 \qquad (9.8)$$

$$\text{subject to}$$

$$x_1 + x_2 + x_3 = 2$$

$$x_1^2 + 5x_2 = 5 \tag{9.9}$$

$$x_i \geq 0, \quad i = 1,2,3$$

Clearly, the objective function and the second (equality) constraint are nonlinear since they involve quadratic terms. For such problems, a good initial guess is often necessary if Solver is to find a solution. It is sometimes necessary to try several initial guesses before a proper solution can be found. In this case, the initial guess of [0 1 1] was tried. The following spreadsheet shows a setup for this problem:

	A	B	C	D
1	x1	x2	x3	
2	0	1	1	
3	y =	-3		
4	Constraint 1	2 =		2
5	Constraint 2	5 =		5

The associated Solver window is as shown below–note that the Select a Solving Method displays GRG Nonlinear:

After selecting Solve, the spreadsheet then appears as follows:

	A	B	C	D
1	x1	x2	x3	
2	0.991125466	0.804	0.205	
3	y =	-7.08		
4	Constraint 1	2	=	2
5	Constraint 2	5	=	5

The solution shown may or may not be the global optimum. Also, there could be more than one solution (even a simple quadratic equation usually has two solutions). The only way to discover this is to try several other initial guesses. This is left as an additional exercise.

9.4 NONLINEAR EQUATIONS

The same algorithms that solve NLP problems can be applied to solving sets of nonlinear equation problems (NEPs). Material and energy balances applied to chemical and biomolecular engineering problems often lead to NEPs. The following example is typical of such problems.

Example 9.4: Continuous Stirred Tank Reactor (CSTR)

Consider a CSTR as depicted in Figure 9.4.

Q is the volumetric flowrate (L/s), V is the reactor volume (L), and C_i is the concentration of each of the four components (gmol/L).

Also consider the following hypothetical reactions taking place in the CSTR:

$$A \xrightarrow{r_1} 2B$$

$$A \underset{r_3}{\overset{r_2}{\rightleftharpoons}} C \qquad (9.10)$$

$$B \xrightarrow{r_4} D + C$$

FIGURE 9.4 Continuous stirred tank reactor.

where

$$r_1 = k_1 C_A$$

$$r_2 = k_2 C_A^{3/2}$$

$$r_3 = k_3 C_C^2 \tag{9.11}$$

$$r_4 = k_4 C_B^2$$

k_1, k_2, k_3, and k_4 are rate constants with the proper units. Typical values are as follows:

$$k_1 = 1.5 \text{ s}^{-1}$$

$$k_2 = 0.1 \text{ L}^{1/2}/\text{gmol}^{1/2} - \text{s}$$

$$k_3 = 0.1 \text{ L/gmol} - \text{s} \tag{9.12}$$

$$k_4 = 0.5 \text{ L/gmol} - \text{s}$$

The r_i have units of gmol/L-s.

A mass balance on each of the four components leads to the following set of nonlinear equations:

$$C_A = C_{A0} + V\left(-k_1 C_A - k_2 C_A^{3/2} + k_3 C_C^2\right)/Q$$

$$C_B = C_{B0} + V\left(2k_1 C_A - k_4 C_B^2\right)/Q$$

$$C_C = C_{C0} + V\left(k_2 C_A^{3/2} - k_3 C_C^2 + k_4 C_B^2\right)/Q \tag{9.13}$$

$$C_D = C_{D0} + V\left(k_4 C_B^2\right)/Q$$

There are two ways in which to use Solver for nonlinear equations. The *direct* way is to set up the nonlinear equations as *constraints* with *no objective function*. The other way is to set up the spreadsheet to compute the *sum of squares of residuals* and use Solver to minimize this (without any constraints). The latter method is used in the following spreadsheet, where the feed consists only of component A with $C_{A0} = 1$. The volumetric flow rate is 50 gmol/s, and the reactor volume is 100 L/s. The equations are rearranged in the form $f(x) = 0$ so that the left-hand sides are *residuals* whose value at a solution is zero (within tolerance). The initial guess for all concentrations is 0.5 gmol/L.

	A	B	C	D
1	k1	1.5		
2	k2	0.1		
3	k3	0.1		
4	k4	0.5		
5	V	100		
6	Q	50		
7	Ca0	1		
8	Cb0	0		
9	Cc0	0		
10	Cd0	0		
11			Residuals	Residuals^2
12	Ca	0.5	-1.02071	1.041850288
13	Cb	0.5	2.25	5.0625
14	Cc	0.5	-0.22929	0.052573593
15	Cd	0.5	-0.25	0.0625
16			Sum of Squares	6.219423882

The Solver setup is as follows:

When Solver is invoked, the spreadsheet changes as shown below:

	A	B	C	D
1	k1	1.5		
2	k2	0.1		
3	k3	0.1		
4	k4	0.5		
5	V	100		
6	Q	50		
7	Ca0	1		
8	Cb0	0		
9	Cc0	0		
10	Cd0	0		
11			Residuals	Residuals^2
12	Ca	0.265812	2E-05	3.99284E-10
13	Cb	0.858256	8.79E-06	7.72462E-11
14	Cc	0.673331	6.87E-06	4.7148E-11
15	Cd	0.736602	2.12E-06	4.50236E-12
16			Sum of Squares	5.28181E-10

As with all nonlinear problems, it is always good practice to try several initial guesses to see if the same solution results. This is left as an exercise.

9.5 NONLINEAR REGRESSION ANALYSIS

Recall Equations 7.9 and 7.10. G is a matrix of constants for linear regression. For nonlinear regression (NLR), G is a function of the unknown parameters, and Equation 7.10 becomes a set of nonlinear equations; therefore, there are two possible approaches to solving NLR problems. One method involves treating Equation 7.10 as a set of constraints (with no objective function), and the other is to minimize the sum of squares of residuals (no constraints). The latter approach (to minimize the sum of squares of residuals) is very much the more straightforward of the two approaches.

A common NLR problem in chemical and biomolecular engineering involves finding model coefficients (parameters) for models in which the parameters occur nonlinearly. A typical problem is solved in Example 9.5.

Example 9.5: NLR in Reaction Kinetics

Consider the simple decomposition reaction of compounds A to B:

$$A \rightarrow B \tag{9.14}$$

Assuming an elementary reaction, the rate of disappearance of A is given by

$$\frac{dC_A}{dt} = -kC_A \tag{9.15}$$

where C_A is the molar concentration of A. Assuming an initial condition of

$$C_A(0) = 1 \text{ mol/L} \tag{9.16}$$

the solution to the separable differential Equation 9.15 can be obtained as follows:

1. Rearrange the differential equation to the form

$$\frac{dC_A}{C_A} = -kdt \tag{9.17}$$

2. Integrate both sides

$$\ln(C_A) = -kt + \text{constant}$$

3. From the initial condition,

$$\ln(1) = 0 = \text{constant}$$

4. So, finally, the solution is

$$C_A = e^{-kt} \tag{9.18}$$

Further assume that the rate constant, k, is a function of temperature according to the Arrhenius form:

$$k = c_1 e^{-c_2/T} \qquad (9.19)$$

where T is the absolute temperature. The overall mathematical model for this system then becomes

$$C_A = \exp(-c_1 t)\exp(-c_2/T) \qquad (9.20)$$

So, C_A is the dependent variable, t and T are the two independent variables, and c_1 and c_2 are two parameters to be determined. A typical set of data appears in Table 9.1.

To show clearly that this is an NLR problem, the derivatives of the dependent variable with respect to the unknown parameters (Equation 7.9) lead to

$$Z_1 = -t\exp(-c_2/T)\exp(-c_1 t)\exp(-c_2/T)$$
$$Z_2 = -\frac{c_1 t}{T}\exp(-c_1 t)\exp(-c_2/T) \qquad (9.21)$$

Clearly, the matrix G of Equation 7.10 involving the sum of products of the Z's is dependent on the unknown coefficients c_1 and c_2, and the equations are *nonlinear*. Because the analytical derivatives for NLR problems are often complex (such as those of Equation 9.21), it is often simpler to determine these derivatives numerically.

To solve the nonlinear equations associated with NLR, Solver can be used. However, in doing so, the statistical nature of the regression problem is ignored. It is necessary to calculate the G matrix *at the solution*. Once the parameters are known, the G matrix again becomes one of constants and can be inverted. Once the matrix G is known, all of the statistical aspects of the problem (Chapter 7) can be computed.

TABLE 9.1

Data for Kinetics NLR

Expt. No	Time, s	Temp, K	C_A
1	0.1	100	0.98
2	0.2	100	0.983
3	0.1	200	0.544
4	0.5	200	0.225
5	0.02	300	0.566
6	0.06	300	0.034

Shown below is a spreadsheet for solving this problem using Solver:

	A	B	C	D	E	F	G
1	c1=	913.583	c2=	981.0928			
2	Expt.No	Time, t	Temp, K	C_A	CaCalc	Residual^2	Residual
3	1	0.1	100	0.98	0.995002	0.000225	0.015002
4	2	0.2	100	0.983	0.990028	4.94E-05	0.007028
5	3	0.1	200	0.544	0.508342	0.0012715	-0.03566
6	4	0.2	200	0.225	0.258412	0.0011164	0.033412
7	5	0.02	300	0.566	0.499461	0.0044274	-0.06654
8	6	0.06	300	0.034	0.124596	0.0082077	0.090596
9						0.0152974	
10					se^2	0.0038243	

S_e^2 was minimized using Solver. The initial guesses for c_1 and c_2 were both 1000. The following spreadsheet segment shows the Zs calculated using finite differences (second-order correct), the requisite Z products, and the sum of Zs required for Equation 9.10. Also shown are G^{-1}, the parameter standard deviations, the optimal c values, and the t-ratios.

Z1	Z2	Z1*Z1	Z1*Z2	Z2*Z2
-5.457E-06	4.986E-05	2.978E-11	-2.721E-10	2.486E-09
-1.086E-05	9.922E-05	1.179E-10	-1.077E-09	9.844E-09
-3.765E-04	1.720E-03	1.417E-07	-6.474E-07	2.957E-06
-3.827E-04	1.748E-03	1.465E-07	-6.692E-07	3.057E-06
-3.795E-04	1.156E-03	1.440E-07	-4.387E-07	1.336E-06
-2.840E-04	8.650E-04	8.067E-08	-2.457E-07	7.482E-07
-1.439E-03	5.638E-03	5.131E-07	-2.002E-06	8.111E-06

5.131E-07	-2.002E-06	=G		
-2.002E-06	8.111E-06			

G^{-1}		StdDev	c	t-ratio
53261140	13148378	451.3191	913.6231	2.0243
13148378	3369183.5	113.5118	981.1010	8.6432

It can be seen from the t-ratios that c_1 is "borderline" well determined, while c_2 is more well determined. Since these results are based on very few data points, it is likely that the parameter behavior would improve with a much larger data set.

EXERCISES

Exercise 9.1: Linear programming.

It is required to produce one pound of an alloy that has at least 30% Pb and at least 30% Zn by mixing a number of available Pb–Zn–Sn alloys. Find the cheapest blend using the following data:

	Analysis (%)			
Available Alloy	Pb	Zn	Sn	Cost ($/lb)
1	20	20	60	6.0
2	10	40	50	6.3
3	40	50	10	7.5
4	50	30	20	8.0

Use Solver for this LP problem.

Exercise 9.2: Nonlinear regression.

A heterogeneous reaction is known to occur at a rate described by the following Langmuir–Hinshelwood expression:

$$r = \frac{k_1 P_A}{(1 + K_A P_A + K_R P_R)^2} \tag{9.22}$$

From initial rate measurements, k_1 has been determined as 0.015 gmol/s-gcat-atm at 400 K. Using the following rate data at 400 K, estimate the values of K_A and K_R:

P_A (atm)	1	0.9	0.8	0.7	0.6
P_R (atm)	0	0.1	0.2	0.3	0.4
r	3.4e-5	3.6e-5	3.7e-5	3.9e-5	4.0e-5

This is an NLR problem. Solve this by minimizing the sum of squares of residuals using Solver. After you have determined the optimal values for K_A and K_R, calculate numerically (using second-order correct formulas) the derivatives Z_1 and Z_2 at each data point, form the G matrix, calculate the parameter standard deviations, and calculate the t-ratios for each parameter.

Exercise 9.3: Nonlinear programming.

Consider the following NLP:

$$\text{minimize } x_2^2 - x_1^2$$

$$\text{subject to } x_1^2 + x_2^2 = 4$$

First, solve this problem *analytically* by solving the constraint for x_1^2 and substituting this into the objective function. Then differentiate the objective function (the only remaining variable is x_2), set the derivative to zero, and find x_2. Use the value for x_2 to find the value(s) for x_1. Next use Solver to find the solution(s). Use a starting point of [1, 1] and then [−1, −1] and see what solutions Solver finds from these starting points.

Exercise 9.4: Nonlinear equations.

The calculation of the equilibrium concentration when we have several reactions and components usually results in nonlinear algebraic equations. Consider the following three reactions involving seven components:

$$A + B \overset{x_1}{\Longleftrightarrow} C + D \qquad K_1 = \frac{C_C C_D}{C_A C_B}$$

$$B + C \overset{x_2}{\Longleftrightarrow} E + F \qquad K_2 = \frac{C_E C_F}{C_B C_C} \tag{9.23}$$

$$A + E \overset{x_3}{\Longleftrightarrow} G \qquad K_3 = \frac{C_G}{C_A C_E}$$

$$K_1 = 1, K_2 = 2, K_3 = 4$$

Here, x_1, x_2, and x_3 (the unknowns) are the *extents of reaction* at equilibrium, and the Cs are molar concentrations. Note that the extent of reaction is a number between 0 and 1. Zero indicates no production of products, while a value of 1 means that the reaction goes to completion (no reactants remain). Given the extents of reaction, the following mass balances can be written:

$$C_A = C_{A0} - x_1 C_{B0} - x_3 C_{A0}$$

$$C_B = C_{B0} - x_1 C_{B0} - x_2 C_{B0}$$

$$C_C = C_{C0} + x_1 C_{B0} - x_2 C_{B0}$$

$$C_D = C_{D0} + x_1 C_{B0} \tag{9.24}$$

$$C_E = C_{E0} + x_2 C_{B0} - x_3 C_{A0}$$

$$C_F = C_{F0} + x_2 C_{B0}$$

$$C_G = C_{G0} + x_3 C_{A0}$$

Initial conditions are $C_{A0} = C_{B0} = 1$; all others are zero. The nonlinear equations 9.23 can be expressed as

$$C_C * C_D - K_1 * C_A * C_B = 0$$

$$C_E * C_F - K_2 * C_B * C_C = 0 \tag{9.25}$$

$$C_G - K_3 * C_A * C_E = 0$$

When formulating an initial guess, make note of the following: the optimal values for the variables *must be between 0 and 1*. Use Solver for this problem.

Exercise 9.5: Solve the following linear program:

$$\text{maximize } q = 80x_1 + 100x_2$$

subject to

$$0.5x_1 + 0.5x_2 \leq 25$$

$$0.2x_1 + 0.6x_2 \leq 10$$

$$0.8x_1 + 0.4x_2 \leq 14$$

$$x_1, x_2 \geq 0$$

Exercise 9.6: Solve the following nonlinear program:

$$\text{minimize } q = (x_1 - 0.5)^2 + (x_2 - 2.5)^2$$

subject to

$$(x_1 - 2)^2 + x_2^2 \le 4$$

$$x_1 \ge 0$$

$$0 \le x_2 \le 2$$

Exercise 9.7: Solve the following nonlinear algebraic equations:

$$(x_1 - 1)^3 + x_2^2 = 0$$

$$x_1 + x_2 = 1$$

Try to find more than one solution.

Exercise 9.8: Solve the following NLR problem:

Fit the function $y = c_1 \exp(c_2/T)$ to the data shown below. After having determined the optimal values for c_1 and c_2, calculate analytically or numerically (using second order correct formulas) the derivatives Z_1 and Z_2 at each data point, form the G matrix, calculate the parameter standard deviations and calculate the t-ratios for each parameter. Comment on the significance of the two parameters c_1 and c_2.

T	y Data
100	0.63
110	0.60
120	0.57
130	0.53
140	0.51

10 Introduction to MATLAB®

10.1 INTRODUCTION

The name MATLAB® stands for Matrix Laboratory and was first published before graphical user interfaces were popular. It has evolved through many versions and is usually updated every 6 months or so. MATLAB is a popular computing environment in universities and research institutions. It is not, however, used often in industrial settings because of somewhat expensive licensing fees. Inasmuch as many students of chemical and biomolecular engineering undertake postgraduate or professional studies where MATLAB can be popular, this programming environment is introduced here with the assumption that the reader is familiar with Excel® and VBA. This brief introduction is intended only to present the rudiments of MATLAB. Full documentation is available at http://www.mathworks.com/help/techdoc/learn_matlab/bqr_2pl.html. Several of the instructions in this chapter are taken from this reference.

When MATLAB starts up (in either Windows or Macintosh environments), the MATLAB Command Window and subsidiary windows appear as follows:

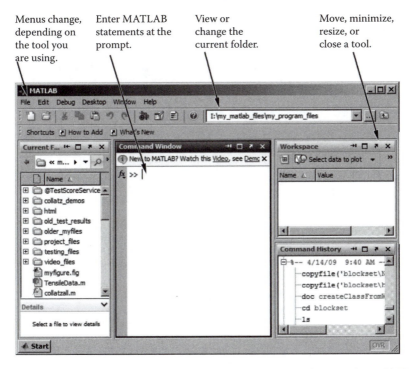

The >> icon is the command prompt. Anything entered after this is a MATLAB command. The Current Directory window displays folders and files associated with the current directory (see Path discussion below). The Workspace Window shows the names of all variables that have been created, and the Command History shows a record of the most recent commands given at the Command Prompt.

The actual arrangement of the sub-windows might be different than shown. The Command Window is somewhat akin to the Excel Spreadsheet; it is through this window that MATLAB commands are given. While many useful things can be accomplished directly in the Command Window, for present purposes, it acts as the interface to MATLAB's programming language. Most input and output are accomplished through the Command Window. For Excel/VBA users, it is convenient to think of the Command Window as comparable to the spreadsheet and MATLAB programs as similar to VBA macros. This is not a perfect metaphor but is sufficient for present purposes.

If a previously issued command is needed again, the Command History can be recalled using the ↑ key. The last command issued is shown first. Each time ↑ is pressed, the previous commands appear in reverse order of having been typed.

10.2   MATLAB BASICS

Perhaps the first thing to do when first using MATLAB is to set the *Path*. The Path is a list of directories that MATLAB searches for files. The default Path is where users usually want to store files and recover them later. A usual place for file storage might be on a thumb drive. Assume that the directory of interest is `F:\MyDocuments\MATLAB`. To put this directory into the Path and to make it the default directory, go to the File menu and click on Set Path—the following window appears:

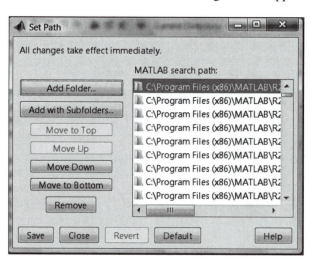

Click on `Add Folder`, `Save`, and then `Close`. Next, at the Command prompt, issue the following command (`cd` stands for change directory).

```
>> cd 'F:\My Documents\MATLAB'
```

From this point forward, any file that is saved is deposited in the selected directory, and when opening a file, this directory will be searched first.

A unique thing about MATLAB is that all variables are *matrices*. For example, the command shown in the Command Window below creates a variable named x (see that name having been added to the Workspace). Following the command, the current value of x is listed. To avoid having the value of x printed following the command, simply add a semicolon at the end of the command. It is important to note that all MATLAB identifiers are *case sensitive*.

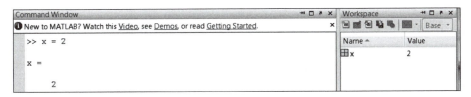

In the interest of keeping this discussion brief, emphasis is given to the *differences* between Excel/VBA and MATLAB. When the MATLAB syntax is the same as that of Excel/VBA, no explanation is given. The useful command

```
>> clc
```

clears the Command Window. It is a good idea to always begin with a blank window.

To get the feel of MATLAB, some annotated Command Window sessions are now shown.

In the following example, the command `diary` is used. This command has the following syntax:

`diary` filename

where filename is any legal MATLAB file name. Everything that appears in the Command Window following this command is recorded in the file. To terminate recording, the command

`diary off`

is entered

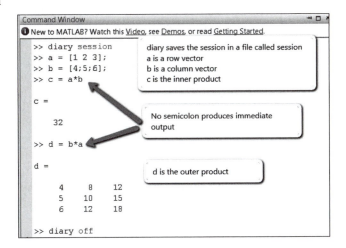

Special attention should be given to the command

```
b = [4;5;6];
```

The interior semicolons indicate "start of a new row." So, b is a column vector while a is a row vector.

Issuing the command

```
Dir
```

produces a list of the files in the current directory. Among those will be a file called session. (Note that the name session is arbitrary—any legal file name can be used.) This is a text file that can be opened within MATLAB or by any word processor. To see the file in MATLAB, go to File/Open. At the bottom of the window that appears, change the File of Type to All Files (the default is to show only MATLAB type files). Then click on the file name session and then Open. The following is displayed:

```
Editor - F:\My Documents\MATLAB\session
File  Edit  Text  Go  Tools  Debug  Desktop  Win

 1   a = [1 2 3];
 2   b = [4;5;6];
 3   c = a*b
 4
 5   c =
 6
 7          32
 8
 9   d = b*a
10
11   d =
12
13          4       8      12
14          5      10      15
15          6      12      18
16
17   diary off
```

This is an exact duplicate of the Command Window session and includes everything from the diary session command until diary off. The contents of the file are displayed in the Editor window. This window is similar to the VBA editor window in that this is where MATLAB programs are coded.

If a command (at the Command Window or in a MATLAB program) is very long, a continuation indicator is three (or more) consecutive periods ....

Table 10.1 enumerates the MATLAB operators. Most of these are the same as for Excel and VBA. A notable difference is the backslash operator \ (called left division), which is used primarily when solving sets of linear algebraic equations. Examples using this operator appear in the sequel.

The following MATLAB session shows a variety of matrix operations. It can be seen again that the semicolon is used to suppress printing when placed at the end of a command, and it also indicates the end of a row (and the beginning of a new one) in a matrix. The apostrophe is the transposition operator, and the built-in function inv takes the inverse of a matrix. If the inverse does not exist, an error message appears.

TABLE 10.1
MATLAB Operators

Operator	Meaning
+	Addition
−	Subtraction
*	Multiplication
/	Division
\	Left division
^	Power
'	Transpose
( )	Specify evaluation order
=	Assignment
>	Greater than
<	Less than
> =	Greater than or equal to
< =	Less than or equal to
= =	Equal to (logical)
~ =	Not equal to
&	Logical and
|	Logical or
~	Logical not

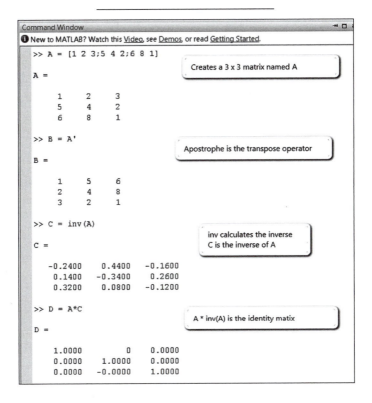

The following is a listing of a diary file where the `randn` function and the backslash operator are used:

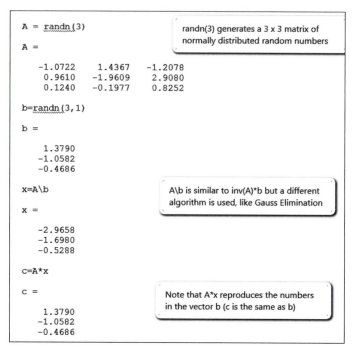

```
A = randn(3)                          randn(3) generates a 3 x 3 matrix of
                                      normally distributed random numbers
A =

   -1.0722    1.4367   -1.2078
    0.9610   -1.9609    2.9080
    0.1240   -0.1977    0.8252

b=randn(3,1)

b =

    1.3790
   -1.0582
   -0.4686

x=A\b                                 A\b is similar to inv(A)*b but a different
                                      algorithm is used, like Gauss Elimination
x =

   -2.9658
   -1.6980
   -0.5288

c=A*x

c =                                   Note that A*x reproduces the numbers
                                      in the vector b (c is the same as b)
    1.3790
   -1.0582
   -0.4686
```

The next MATLAB session shows the use of the `pinv` (pseudo-inverse function).

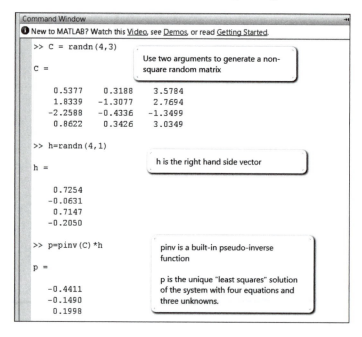

```
Command Window
❶ New to MATLAB? Watch this Video, see Demos, or read Getting Started.

>> C = randn(4,3)
                                      Use two arguments to generate a non-
C =                                   square random matrix

    0.5377    0.3188    3.5784
    1.8339   -1.3077    2.7694
   -2.2588   -0.4336   -1.3499
    0.8622    0.3426    3.0349

>> h=randn(4,1)
                                      h is the right hand side vector
h =

    0.7254
   -0.0631
    0.7147
   -0.2050

>> p=pinv(C)*h                        pinv is a built-in pseudo-inverse
                                      function
p =
                                      p is the unique "least squares" solution
   -0.4411                            of the system with four equations and
   -0.1490                            three unknowns.
    0.1998
```

Access to any element of an array is the same as in VBA. For example, using variables from the previous MATLAB session:

```
h(2)  =  -0.0631
p(3)  =  0.1998
C(2, 4)  =  0.3426
```

10.2.1 MATLAB COLON OPERATOR

The colon (:) is an important MATLAB operator. It occurs in several different contexts. The expression

```
1:10
```

produces a row vector containing the integers from 1 to 10:

```
1 2 3 4 5 6 7 8 9 10
```

To obtain nonunit spacing, specify an increment. For example,

```
100:-7:50
```

generates

```
100 93 86 79 72 65 58 51
```

and

```
0:pi/4:pi
```

produces

```
0 0.7854 1.5708 2.3562 3.1416
```

Subscript expressions involving colons refer to portions of a matrix:

```
A(1:k,j)
```

is the first k elements of the jth column of A. Therefore,

```
sum(A(1:4,4))
```

computes the sum of the fourth column assuming a 4×4 matrix. However, there is another way to perform this computation. The colon by itself refers to *all* the elements in a row or column of a matrix, and the keyword end refers to the *last* row or column. Therefore,

```
sum(A(:,end))
```

computes the sum of the elements in the last column of A.

10.2.2   MATLAB, M-FILES, AND INPUT FROM COMMAND WINDOW

While a sequence of commands in the Command Window can implement many algo-rithms, it is awkward if selection (if-then-else) or repetition (e.g., while) logic is involved. The best way to do programming in MATLAB is to construct an M-file (comparable to a VBA Macro). These are called M-files since the automatic file type is .m.

Before writing a first M-file program, input/output with the command window must be covered. For MATLAB programs with small amounts of input, the input statement is used. For output to the Command Window, the usual method involves the fprintf statement. The syntax of the input statement is

```
Variable = input('prompt')
```

For example,

```
A = input('Enter a number:')
```

Note that strings are delimited by astrophes (recall that quote marks are used in VBA).

When executed, the prompt appears in the Command Window. A number is entered and stored in the variable A.

Shown next is an M-file that reads several numbers and computes the average of the numbers. After the M-file is a Command Window session that invokes the program.

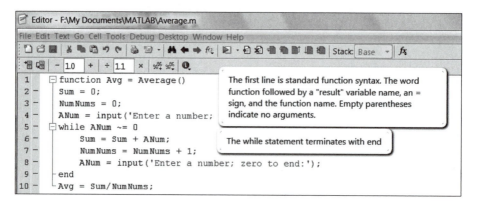

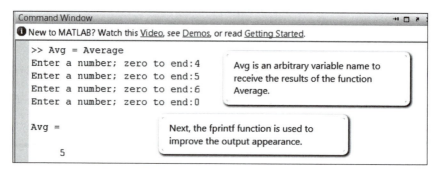

The `fprintf` function is somewhat complicated since it requires a cryptic formatting string. The format of the `fprintf` statement is

```
fprintf (<format string>, <variables>)
```

Here is an example:

```
fprintf ('The sum is%5.3f\n', Sum)
```

If the variable sum has a value of `16.12365`, the output produced by the statement is

```
The sum is 16.124
```

Note that the value of sum has been truncated to three digits after the decimal, and the last digit is rounded up. The `\n` at the end of the format string is the new line control character—any further output appears on a new line.

There are a large number of format string data type specifiers available for use with the `fprintf` function. Some of these are enumerated in Table 10.2.

Table 10.3 displays some of the available *control* characters for formatting.

Shown next is a revised version of the `Average` file. In this case, output is accomplished using the `fprintf` function. Following the program listing is the associated Command Window where the function file is called without putting the result into a variable since the result has already been output.

TABLE 10.2
Format Data Type Specifiers

Specifier	Display
`%d`	Integer/whole number
`%f`	Floating point
`%e`	Exponential
`%g`	General (shortest format possible)
`%c`	Character
`%s`	Character string

TABLE 10.3
Format Control Characters

Control Character	Description
`\n`	New line
`\t`	Tab
`''`	Two apostrophes prints one apostrophe

```
Editor - F:\My Documents\MATLAB\Average.m
File Edit Text Go Cell Tools Debug Desktop Window Help
 1    function Avg = Average()
 2        Sum = 0;
 3        NumNums = 0;
 4        ANum = input('Enter a number; zero to end:');
 5        while ANum ~= 0
 6            Sum = Sum + ANum;
 7            NumNums = NumNums + 1;
 8            ANum = input('Enter a number; zero to end:');
 9        end
10        Avg = Sum/NumNums;
11        fprintf ('The average is %5.3f\n', Avg)
```

```
Command Window
New to MATLAB? Watch this Video, see Demos, or read
>> Average;
Enter a number; zero to end:3
Enter a number; zero to end:5
Enter a number; zero to end:7
Enter a number; zero to end:91
Enter a number; zero to end:35
Enter a number; zero to end:0
The average is 28.200
```

10.3 MATLAB PROGRAMMING LANGUAGE STATEMENTS

MATLAB's programming language is similar to that of VBA. The assignment statement has already been used and is indicated by the = sign. Some of the other statements are discussed next.

10.3.1 If-Then-Else Statements

The syntax of the MATLAB If-Then-Else statement is as follows:

```
if condition1
  Statements1
else
  Statements2
end
```

Note that the words if, else, and end are all lowercase. Statements1 and Statements2 can be any other MATLAB statements. The else clause is optional.

10.3.2 LOOPING STATEMENTS (FOR, WHILE)

The syntax of the for statement is

```
for variable = initial_value:increment/decrement:final_value
  Statements
end
```

The increment/decrement is optional, and if omitted, the increment is 1. The syntax of the while statement is

```
while condition
  Statements
end
```

Sufficient MATLAB programming background is now available so that programs previously written in VBA can be demonstrated in MATLAB.

Example 10.1: MATLAB Program for Averaging Numbers

The MATLAB program listing below reimplements the one of Example 2.2. In that example, numbers were input from a spreadsheet and stored in an array, the average of the numbers calculated, and the average output to the spreadsheet. Note that the % sign is used to indicate a *comment*. Following the MATLAB program listing is a Command Window session that executes the program and inputs a set of numbers, and the result is output to the Command Window. There is great similarity between this program and that of Example 2.2 with minor syntax differences. The most significant difference is in the input/output portion. A direct comparison with the VBA program of Example 2.2 is advised.

```
1    function Average =CalcAverage2()
2
3      NumNumbers = 0;
4
5      %get the first number
6      ANumber = input('Enter a number to be averages: ');
7      %keep getting numbers until a zero (blank) is encountered
8      while ANumber ~= 0
9          NumNumbers = NumNumbers + 1;          %increment how many numbers
10         InputNumbers(NumNumbers) = ANumber; %store the number in the array
11         ANumber = input('Enter a number to be averaged: ');
12     end
13
14     Sum = 0;
15     for i = 1 : NumNumbers
16         Sum = Sum + InputNumbers(i);
17     end
18
19     if NumNumbers > 0
20         Average = Sum / NumNumbers;
21         fprintf ('Average %9.4f\n', Average)
22     else
23         fprintf ('No input numbers to average')
24     end
```

```
Command Window
❶ New to MATLAB? Watch this Video, see Demos, or read Ge
 >> CalcAverage2;
 Enter a number to be averaged: 200
 Enter a number to be averaged: 500
 Enter a number to be averaged: 666
 Enter a number to be averaged: 876
 Enter a number to be averaged: 0
 Average 2242.0000
```

10.4 MATLAB FUNCTION ARGUMENTS

All functions have this standard *function syntax*:

```
function[output1,..., outputM] = functionName(input1,...,
inputN)
```

Input arguments are passed by *value*, while output arguments are passed by *reference*. Even if an input argument is changed by the function, the altered value is not returned after the function call. (Recall the argument passing descriptions given in Chapter 2.) Typically, output arguments are not defined at the time of calling, and the values returned are set by the function. As with VBA, the argument names used in the function definition are dummy arguments; the actual arguments are those used when the function is called.

Example 10.2: Argument Passing to and from a Function

The function shown below generates two vectors using the MATLAB function linspace. This built-in MATLAB function can be handy for generating equally spaced data. It generates the number of vector elements given by the third argument. The numbers start with its first argument and are equally spaced up to the value of the second argument. Dummy input arguments are named a and b, while dummy output arguments are called p and q.

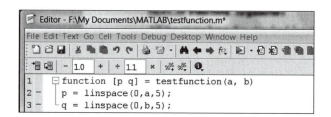

Shown next is a Command Line session that sets the first two arguments (actual argument names x and y) and then calls the testfunction. The output arguments (actual arguments f and g) are printed using the fprint command.

```
Command Window
❶ New to MATLAB? Watch this Video, see Demos, or read Get
  >> x = 2;
  >> y = 3;
  >> [f g] = testfunction(x, y);
  >> fprintf('%6.4f \t',f)
  0.0000   0.5000   1.0000   1.5000   2.0000
  >> fprintf('%6.4f \t',g)
  0.0000   0.7500   1.5000   2.2500   3.0000
```

10.5   PLOTTING IN MATLAB

MATLAB's plotting capabilities are such that professional quality graphs can be generated. The commands for producing graphs vary from the very simple to the quite complex. In this coverage, only relatively simple plot commands are discussed, but more complete discussions are readily available.

10.5.1   PLOTTING TWO FUNCTIONS ON THE SAME GRAPH

Consider the following MATLAB Command Window session:

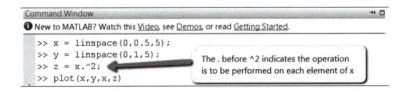

```
Command Window
❶ New to MATLAB? Watch this Video, see Demos, or read Getting Started.
  >> x = linspace(0,0.5,5);
  >> y = linspace(0,1,5);
  >> z = x.^2;          The . before ^2 indicates the operation
  >> plot(x,y,x,z)      is to be performed on each element of x
```

The graph produced by the plot command is shown below:

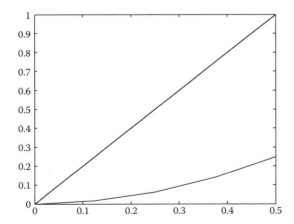

While labeling and changing the graph properties can be done under program control, it is much easier to use the plot editor. When in the plot window, go to Edit/ Axis Properties. Axis labels and a plot title can be added. Also, line types and colors can be changed along with a myriad other things. Shown below is an edited plot with some of these changes:

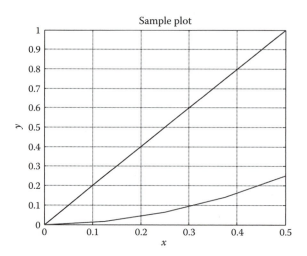

10.6 EXAMPLE MATLAB PROGRAMS

In this section, several MATLAB M-files are presented that perform operations previously visited using VBA. These include

- Solving a single nonlinear equation using `fzero`
- Solving ordinary differential equations using `ode45`
- Solving a boundary value problem using the `ode45` and the shooting method
- Nonlinear equations using `fsolve`
- Nonlinear regression using `minsearch`

Example 10.3: Solving a Single Nonlinear Equation Using `fzero`

Consider finding a zero of the function

$$f(x) = x^4 - e^{-x} + 1 \tag{10.1}$$

The syntax for the `fzero` function is

```
var = fzero('equation', init_guess)
```

where

`var`	= the final value of `x`
`equation`	= name of the function representing the function to zero
`init _ guess`	= the initial guess of `x`

A listing of the M-file for the function is as follows:

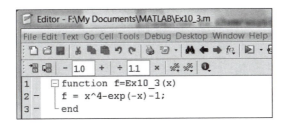

Next is a Command Window session that invokes `fzero` and finds a value of `x` that is a root of the function:

```
Command Window
❶ New to MATLAB? Watch this Video, see Demos, or read G

>> x = fzero('Ex10_3', 2)

x =

      1.0761
```

The function `fzero` uses a combination of the methods discussed in Chapter 1. Depending on the initial guess, it might find different roots than the one shown.

Example 10.4: Solving Ordinary Differential Equations Using `ode45`

The built-in MATLAB function `ode45` uses a Runge–Kutta method and a *variable time step*. Based on how rapidly the solution functions are changing, the time step is altered to improve accuracy. The user need not be aware of the details of the algorithm, but when it is necessary to know the number of time steps, it can be useful to call the function `length`, which is illustrated in the example problem below.

Recall the problem of Example 5.5. Suppose the following chemical reactions take place in a continuous stirred tank reactor (CSTR):

$$A \underset{k_2}{\overset{k_1}{\Longleftrightarrow}} B \underset{k_4}{\overset{k_3}{\Longleftrightarrow}} C \tag{10.2}$$

where the rate constants are as follows:

$$k_1 = 1 \text{ min}^{-1},\ k_2 = 0 \text{ min}^{-1},\ k_3 = 2 \text{ min}^{-1},\ k_4 = 3 \text{ min}^{-1}$$

The initial charge to the reactor is all A, so the initial conditions are (in mol/L)

$$C_{A_0} = 1 \quad C_{B_0} = 0 \quad C_{C_0} = 0$$

An unsteady-state mass balance on each component leads to the following set of ODEs:

$$\frac{dC_A}{dt} = -k_1 C_A + k_2 C_B$$

$$\frac{dC_B}{dt} = k_1 C_A - k_2 C_B - k_3 C_B + k_4 C_C \qquad (10.3)$$

$$\frac{dC_C}{dt} = k_3 C_B - k_4 C_C$$

The syntax of the ode45 function is

```
[t, y] = ode45(@rhs_function, tspan, initial_conditions)
```

where

t	=	the independent variable vector.
y	=	the dependent variable matrix (first column is the first dependent variable, second column is the second, etc.).
rhs _ function	=	an M-file function defining the right-hand sides of first-order ODEs. The @ sign signifies this as a function name.
tspan	=	a vector of initial and final values of t.
initial _ conditions	=	a vector of initial conditions.

The following is an M-file listing of a function called chemrxsys. Within the function, the rate constants are fixed, and the right-hand-side functions are identified.

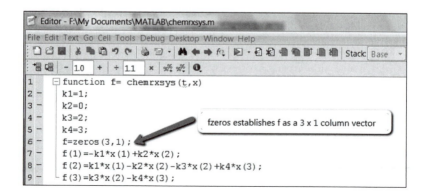

```
Editor - F:\My Documents\MATLAB\chemrxsys.m
File Edit Text Go Cell Tools Debug Desktop Window Help
                                                        Stack: Base
    - 1.0  +  ÷ 11   ×
1    function f= chemrxsys(t,x)
2      k1=1;
3      k2=0;
4      k3=2;
5      k4=3;
6      f=zeros(3,1);
7      f(1)=-k1*x(1)+k2*x(2);
8      f(2)=k1*x(1)-k2*x(2)-k3*x(2)+k4*x(3);
9      f(3)=k3*x(2)-k4*x(3);
```

fzeros establishes f as a 3 x 1 column vector

Shown next is a Command Window session in which `ode45` is called to solve this problem. Also, a graph is produced (the graph shown was enhanced by editing it).

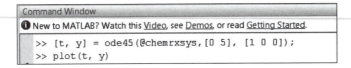

```
>> [t, y] = ode45(@chemrxsys,[0 5], [1 0 0]);
>> plot(t, y)
```

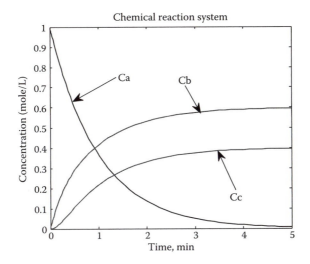

When comparing this graph to that of Example 5.5, the two plots are essentially identical.

Example 10.5: Solving a Boundary Value Problem Using `ode45` and the Shooting Method

Recall the problem of Example 6.2 involving heat conduction in a rod. The requisite ODEs and boundary conditions are

$$\frac{dT}{dx} = F$$

$$\frac{dF}{dx} = \frac{4h}{Dk}(T - T_a)$$

$$T(0) = 100$$ (10.4)

$$F(0) = ?$$

$$T(1) = 0$$

Shown below is a MATLAB script (no function heading) file that prompts the user for two initial guesses for $F(0)$. These guesses are used by the *secant method* to converge the right and boundary conditions, which is $T(1) = 0$.

```
File Edit Text Go Cell Tools Debug Desktop Window Help

                                                              Stack: Base

       - 1.0  +  ÷ 1.1  ×

 1     Fone = input('\n First guess for F(0):');
 2     Ftwo = input('\n Second guess for F(0):');
 3     xspan=[0 1];
 4     T = zeros(2,1);
 5     %T(:,1) = temperature
 6     %T(:,2) = derivative of temperature (F)
 7
 8     %get the first two function values for the secant method
 9     init_T1=[100 Fone];
10     [x,T]=ode45('RodConduction', xspan, init_T1);
11     TLone = T(length(T),1);
12     init_T2 = [100 Ftwo];
13     [x,T]=ode45('RodConduction', xspan, init_T2);
14     TLtwo = T(length(T),1);
15     %apply the secant method to the right hand boundary condition
16     while abs(TLtwo) > .001
17         Fnew = Ftwo - (((Ftwo-Fone)*(TLtwo))/((TLtwo)-(TLone)));
18         init_T3 = [100 Fnew];
19         [x,T]=ode45('RodConduction', xspan, init_T3);
20         TLnew = T(length(T),1);
21         Fone = Ftwo;
22         Ftwo = Fnew;
23         TLone = TLtwo;
24         TLtwo = TLnew;
25     end
26
27     plot(x,T(:,1));
```

The function `RodConduction`, which defines the two right-hand-side functions for the two ODEs, appears below:

```
File Edit Text Go Cell Tools Debug Desktop Window Help

       - 1.0  +  ÷ 1.1  ×

1      function RHS = RodConduction(x, T)
2      % T(1) is the temperature T
3      % T(2) is the derivative of T
4      c = 12.82;  %c = 4h/Dk
5      Ta = 25;    %Ta is the air temperature
6      RHS = zeros(2,1);
7      RHS(1) = T(2);
8      RHS(2) = c*(T(1) - Ta);
```

The plot generated by the script (with initial guesses of −100 and −150, respectively) is shown below:

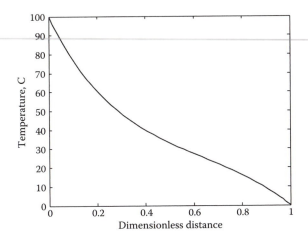

This plot is essentially identical to that of Example 6.2.

Example 10.6: Nonlinear Equations Using `fsolve`

Recall the problem of Example 9.4 involving a CSTR. The appropriate equations and data are as follows:

$$k_1 = 1.5\ s^{-1}$$
$$k_2 = 0.1\ L^{1/2}/gmol^{1/2} - s$$
$$k_3 = 0.1\ L/gmol - s \qquad (10.5)$$
$$k_4 = 0.5\ L/gmol - s$$
$$Q = 50\ gmol/s$$
$$V = 100\ L/s$$

$$C_A = C_{A0} + V\left(-k_1 C_A - k_2 C_A^{3/2} + k_3 C_C^2\right)/Q$$
$$C_B = C_{B0} + V\left(2k_1 C_A - k_4 C_B^2\right)/Q$$
$$C_C = C_{C0} + V\left(k_2 C_A^{3/2} - k_3 C_C^2 + k_4 C_B^2\right)/Q \qquad (10.6)$$
$$C_D = C_{D0} + V\left(k_4 C_B^2\right)/Q$$

The MATLAB function `fsolve` is used to solve sets of nonlinear equations. The syntax for `fsolve` is as follows:

```
x = fsolve(func, x0)
```

where

    x   = a vector of unknowns
    func = a function M-file that evaluates the right-hand side of $f(x) = 0$
    x(0) = a vector of initial guesses for x

The following is a listing of a function CSTR, which codes Equation 10.6 in the form $f(x) = 0$:

```
File Edit Text Go Cell Tools Debug Desktop Window Help

                                                    Stack: Base    fx
   -  1.0  +  ÷ 1.1  ×

1      function f = CSTR( C )
2        k1 = 1.5;
3        k2 = 0.1;
4        k3 = 0.1;
5        k4 = 0.5;
6        Q = 50;
7        V = 100;
8        CInit = [1; 0; 0; 0];
9        f(1) = -C(1) + CInit(1) + V*(-k1*C(1) - k2*C(1)^1.5 + k3*C(3)^2)/Q;
10       f(2) = -C(2) + CInit(2) + V*(2*k1*C(1) - k4*C(2)^2)/Q;
11       f(3) = -C(3) + CInit(3) + V*(k2*C(1)^1.5 - k3*C(3)^2 + k4*C(2)^2)/Q;
12       f(4) = -C(4) + CInit(4) + V*(k4*C(2)^2)/Q;
13     end
```

A MATLAB Command Window session where the initial guess for the concentrations is given and `fsolve` is called appears below. The solution vector for the concentrations is essentially identical to that of Example 9.4.

```
File Edit Debug Parallel Desktop Window Help

                                    C:\Users\Victor Law\Desktop

Shortcuts  How to Add  What's New

 New to MATLAB? Watch this Video, see Demos, or read Getting Started.

 >> C0 = [0.5;0.5;0.5;0.5];
 >> C = fsolve('CSTR', C0)
 Optimization terminated: first-order optimality is less than options.TolFun.

 C =

     0.2658
     0.8583
     0.6734
     0.7366
```

Example 10.7: Nonlinear Regression Using `minsearch`

The built-in function `minsearch` is based on a rather unsophisticated algorithm. There are more robust unconstrained minimization functions available in some of the MATLAB Toolboxes, but unfortunately, these are not standard. For simple problems, `minsearch` often works well enough. It is used here to minimize the sum of squares between fictitious data (program generated data) and a function in which the regression coefficients appear nonlinearly.

The syntax of `minsearch` is as follows:

```
params = fminsearch(@function, initial_guess,[], xdata,
ydata)
```

where

`params`	=	The regression coefficients
`function`	=	The name of the function that calculates the sum of squares
`initial_guess`	=	Vector of initial guesses for coefficients
`[]`	=	An "empty" argument that is not needed for present purposes
`xdata, ydata`	=	Vectors holding the experimental data

The specific nonlinear regression problem to be considered is to find the coefficients, $c(1)$ and $c(2)$, in the function of Equation 10.7.

$$ycalc = c(1)e^{c(2)x} \tag{10.7}$$

A MATLAB script file saved as NIRMAIN.M that generates data using $c(1) = 2$ and $c(2) = 0.5$, adds Gaussian random noise to these data (to make things a bit more realistic), calls `fminsearch`, and prints results is shown below:

```
File Edit Text Go Cell Tools Debug Desktop Window Help

 1    x=1:10;
 2    y=2*exp(-0.5*x);              Generates "perfect" data
 3    y = y + 0.01*randn(1,10);     Adds random "noise" to the data
 4    coeffs=fminsearch(@NLRegress,[1 1],[],x,y);
 5    yfit=coeffs(1)*exp(-coeffs(2)*x);
 6    fprintf('optimal coefficients: ');
 7    coeffs
 8    fprintf('   x       y       yfit \n')
 9    for i = 1:10
10        fprintf('%6.3f %6.3f %6.3f  \n', x(i), y(i), yfit(i))
11    end
```

Shown next is a listing of the function NLRegress, which provides `fminsearch` with the objective function to minimize. In this case, it is the sum of squares of residuals between data and calculated values.

```
File Edit Text Go Cell Tools Debug Desktop Window Help

1         function f=NLRegress(c,X,Y)
2             a = c(1);
3             b = c(2);
4
5             YCalc = a * exp(-b*X);
6             Resid = YCalc - Y;
7             SSQ = Resid.^2;
8
9             f = sum(SSQ);
```

Finally, the following shows a MATLAB Command Window session that calls the script file. The results are displayed in the Command Window.

```
File Edit Debug Parallel Desktop Wind

Shortcuts  How to Add   What's Ne
New to MATLAB? Watch this Video, se

>> NLRMain
optimal coefficients:
coeffs =

    2.0174    0.4989

    x        y       yfit
 1.000    1.227    1.225
 2.000    0.739    0.744
 3.000    0.448    0.452
 4.000    0.287    0.274
 5.000    0.156    0.166
 6.000    0.107    0.101
 7.000    0.069    0.061
 8.000    0.034    0.037
 9.000    0.024    0.023
10.000    0.002    0.014
```

The coefficients used to generate the data were 2 and 0.5, respectively. The values of 2.0174 and 0.4989 are optimal for the data with random noise added.

Note: Each time this program is run, the results will be slightly different. This is because a different set of random numbers is generated on each run.

10.7   CLOSING COMMENT REGARDING MATLAB

As with the coverage of VBA in this text, this chapter has only touched the "tip of the iceberg" with respect to MATLAB. It is intended that with the background of the introductory material present here, a student can explore the vastness of available MATLAB features and functions. For example, the nonlinear regression example (Example 10.7) used the function minsearch, which is not a highly robust minimization algorithm. Another MATLAB function that is particularly suited to nonlinear regression is nlfit and its companion nlparci, which provides confidence intervals for each parameter. A MATLAB add-on Optimization Toolbox provides several algorithms for nonlinear programming. There are many other Toolboxes available for specialized areas. If one searches diligently, a MATLAB function set can be found for any of a vast number of application areas.

EXERCISES

Exercise 10.1: Solve the problem described in Exercise 5.1 using MATLAB.

Exercise 10.2: Solve the problem described in Exercise 5.4 using MATLAB.

Exercise 10.3: Solve the problem described in Exercise 5.7 using MATLAB.

Exercise 10.4: Solve the problem described in Exercise 5.10 using MATLAB.

Exercise 10.5: Solve the problem described in Exercise 5.11 using MATLAB.

Exercise 10.6: Solve the problem described in Exercise 6.1 using MATLAB. Implement the secant method, as in Example 10.5, to converge the right-hand boundary condition. Use `ode45` to solve the ODEs.

Exercise 10.7: Solve the problem described in Exercise 6.2 using MATLAB. Implement the secant method, as in Example 10.5, to converge the right-hand boundary condition. Use `ode45` to solve the ODEs.

Exercise 10.8: Solve the problem described in Exercise 6.4 using MATLAB. Implement the secant method, as in Example 10.5, to converge the right-hand boundary condition. Use `ode45` to solve the ODEs.

Exercise 10.9: Solve the problem described in Exercise 6.5 using MATLAB. Implement the secant method, as in Example 10.5, to converge the right-hand boundary condition. Ignore part b and use `ode45` to solve the ODEs.

Exercise 10.10: Solve the problem described in Exercise 6.6 using MATLAB. Implement the secant method, as in Example 10.5, to converge the right-hand boundary condition. Use `ode45` to solve the ODEs.

Exercise 10.11: Solve the problem described in Exercise 9.2 using MATLAB. Use the `fminsearch` function to minimize the sum of squares of residuals.

Exercise 10.12: Solve the problem described in Exercise 9.4 using MATLAB. Use the `fsolve` function to minimize the sum of squares of residuals.

Exercise 10.13: Solve the problem described in Exercise 9.7 using MATLAB. Use the `fsolve` function to minimize the sum of squares of residuals.

Exercise 10.14: Solve the problem described in Exercise 9.8 using MATLAB. Use the `fminsearch` function to minimize the sum of squares of residuals.

Appendix: Additional Features of VBA

A.1 INTRODUCTION

VBA is a mega system of programming language and objects. It is probably impossible for any one person to be familiar with all of the documented features of VBA. However, the object-oriented nature of the system makes it extensible both by "official" Microsoft® documented items as well as those added by third party developers and individual programmers. In this appendix, a few additional features of VBA are presented that might be useful to chemical and biomolecular engineering students and practitioners. The following items are covered:

1. How to call both built-in functions and Add-In functions from VBA Macros
2. How to include user-defined functions as Add-Ins that can be accessed by other VBA subs and functions
3. How to return arrays from functions, which can in turn be included as Add-Ins

Warning: The author is not a VBA expert. The methods presented in this appendix are ones that have been found to work. No claim is made for their uniqueness or efficiency. When viewed by a true VBA "guru," these techniques might be considered naïve. Engineers often settle for things that work as opposed to ones that are perfect.

A.2 CALLING EXCEL® BUILT-IN FUNCTIONS IN VBA MACROS

To use Excel functions for which there is no VBA counterpart (e.g., ATAN2), the `Application` object can be used as shown in Chapter 2. Given the code

```
Sub testExcelFunctionCall()
Dim Rads As Double
  Rads = Application.Atan2(0, 1)
End Sub
```

the variable `Rads` is assigned the value $\pi/2$.

Note that using the `Application` object does not work with functions for which there is a VBA counterpart (even if the names are different). For example, the code shown below produces the error message that appears after the code:

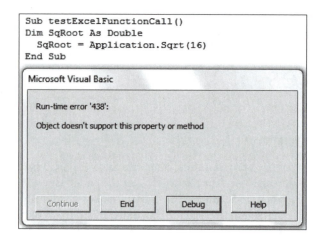

```
Sub testExcelFunctionCall()
Dim SqRoot As Double
  SqRoot = Application.Sqrt(16)
End Sub
```

A.3 CALLING EXCEL® ADD-IN FUNCTIONS IN VBA MACROS

Consider the following VBA sub:

```
Option Base 1
Option Explicit
Sub TestMatrixXLAFunctionCall()
Dim A() As Variant
Dim Ainv() As Variant
Dim b() As Variant
Dim bb() As Variant
Dim x() As Variant

ReDim A(3, 3), b(3), x(3), bb(3), Ainv(3, 3)

  A = Application.Run("MatRnd", 3)
  b = Application.Run("MatRnd", 3, 1)
  Ainv = Application.Run("Mat_Pseudoinv", A)
  x = Application.MMult(Ainv, b)
  bb = Application.MMult(A, x)

End Sub
```

Since the sub involves matrices and vectors, the usual `Option Base 1` is used. Five variables are declared as dynamic arrays whose elements are of type `Variant`. This data type has been avoided in purely numerical computations but is necessary here to allow assignment to array variables. The statement

```
A = Application.Run("MatRnd", 3)
```

executes the `MatRnd` function from the `Matrix.XLA` add-in and produces a 3×3 matrix of random integers (the default range is −10 to 10), and this matrix is assigned to the variable A. The next statement creates a 3×1 vector (b) of random integers.

The statement

```
Ainv = Application.Run("Mat_Pseuoinv", A)
```

calls the `Matrix.XLA` add-in function to calculate the pseudoinverse of A, which is then stored in the variable `Ainv`. The next statement uses the Excel function MMULT to produce the solution to the system $Ax = b$. The last statement again uses MMULT; in this case, the variable bb should have the same values as the original right-hand side, b. To see these results, it is best to run the Macro in Debug mode and use the Set Watch feature to display the values stored in each variable. This is the first use of the Variant data type in this text. See the next section for more details on this type.

Shown below is another VBA sub that performs the same operations as the last example.

```
Option Base 1
Option Explicit
Sub TestMatrixXLACalls2()
Dim A, Ainv, b, bb, x

  A = Application.Run("MatRnd", 3)
  b = Application.Run("MatRnd", 3, 1)
  Ainv = Application.Run("Mat_Pseudoinv", A)
  x = Application.MMult(Ainv, b)
  bb = Application.MMult(A, x)

End Sub
```

Here, an "anonymous" Dim statement is used for all variables (no data type is indicated). The structure and data type of these variables are established when they are assigned something. For example, in the case of the variable A, the assignment statement stores a 3 × 3 matrix of random integers. Although this example is more compact than the previous one, it is not as explicit. From an engineering perspective, since they both work the same, there is no reason to prefer one over the other.

A.4 VARIANT DATA TYPE

The following is an excerpt from "The Power of Variants" (http://www.tushar-mehta .com/publish_train/book_vba/08_variants.htm#_ftn1):

A variable declared as type *Variant* can contain any type of data. Unlike a variable that declared on a specific type, say, String or Integer, which can only contain a text string or a specific range of integers respectively, a variant can contain any data—text or an integer value or a real, i.e., a floating point, value. It can even behave like an array or refer to an object, either built into Excel or a user defined type. Essentially, there are almost no rules on what a developer can do with a variant. For example, with `aVar` declared as a variant each of the assignment statements is legitimate.

```
Dim aVar as Variant
aVar = "a"
aVar = 1
aVar = Array (1,22,333)
set aVar = ActiveSheet
aVar = 3.1415927
```

The common wisdom is that one should stay away from variants. By and large, that is true. If one knows the data type of a variable it is best to declare it correctly. There are many benefits to doing so, the most significant being that the VBA compiler can ensure data and program integrity. With a variant one could accidentally assign a text string to what might be intended to be a number. Essentially, the developer gets the flexibility of a variable that can take any type of data together with the responsibility of ensuring proper data type use. That's a steep burden and one best avoided whenever possible. Consequently, in those cases where the data type is pre-determined and will not change, it is indeed best to declare the variable of the particular type.

However, *there are many instances where a variant allows one to do things that otherwise would be impossible.* The power of a variant comes from the fact that it is a simple data element and yet can contain *any*—and that means *any*—type of data. It can be a string or a Boolean or an integer or a real number. Hence, when the data type returned by a function can vary, one is obligated to use a variant for the returned value.

As we will see in a later section of this chapter, the ability to create an array in a variant makes it possible to create functions that would otherwise be impossible. For example, a function can return either an error condition or an array of values. It also allows a developer to write a User Defined Function (UDF) that returns multiple values in a single call to the function. A variant is also one way to pass an array as a 'by value' argument to a procedure. One Excel-specific reason to use a variant is that it provides a very efficient way to exchange information between Excel and VBA.

Finally, in the advanced section of the chapter, we will use the variant data type to create *an array of arrays.* This makes it possible to create, and work with, data structures that would otherwise be impossible. It also allows one to operate on an entire row of an array.

In the hands of a creative—and defensive—developer, the power of a variant can be nearly limitless.

With this background, it can be seen why the Variant data type was used in the prior example when it was desired to assign an entire array to a variable.

A.5 VBA FUNCTION THAT RETURNS AN ARRAY

Consider the following VBA Function Macro.

```
Option Base 1
Option Explicit
Public Function Trapezoid(x, f) As Variant
Dim XX, FF, IntTrap
XX = x
FF = f
IntTrap = f  'This defines IntTrap as an Object whose elements are accessed as an array.

Dim i As Long
Dim Npts As Long
Npts = UBound(XX)

IntTrap(1, 1) = 0#
For i = 2 To Npts
  IntTrap(i, 1) = IntTrap(i - 1, 1) + 0.5 * (FF(i, 1) + FF(i - 1, 1)) * (XX(i, 1) - XX(i - 1, 1))
Next i
Trapezoid = IntTrap
End Function
```

Before going into any detail about this code, it is important to see how one enters a stand-alone function into the VBA system. Recall that when writing a Sub Macro, the Macros option is chosen from the Developer tab, in which case the VBA editor appears with a blank Sub. It is not possible to change the word Sub to Function and proceed as usual. Instead, the Visual Basic® option must be chosen from the Developer tab, which gives a blank screen in the VBA editor. Look for the name of the associated VBA project (something like Book 2, for example); right click on the project name and select Insert/Module. This gives a blank editor page where the code for the function is to be entered. When the associated Excel Worksheet is saved, the function Macro is saved with it (the Worksheet must be saved as a macro-enabled one). Another oddity of function Macros occurs when they are to be edited. The function name does not appear when visiting Developer/Macros. However, if the name of the function is typed on the appropriate line, then the Edit button activates and the Macro can be edited as though it were a Sub Macro.

Referring now to the code for the function Trapezoid shown above, there are several things to note:

1. The function has two arguments called x and f, which are arrays of independent variable and associated function value whose definite integral is required. The function uses the trapezoidal rule to calculate the integral, which is returned to the calling spreadsheet.
2. The Public declaration might not be required, but it guarantees that anyone can use the function without permission.
3. The Variant type of the function allows an *array* to be returned from the function.
4. The anonymous Dim statement allows *anything* to be assigned to the associated variables.

5. The three assignment statements create Objects whose elements can be accessed with subscripts. They are actually accessed as two-dimensional arrays with a second subscript of 1—that is, they are addressed as Npts × 1 arrays instead of vectors of length Npts. The items to the right of the = sign are function arguments.
6. The variable Npts takes on the upper bound of subscripts of XX, whose length is that of the input arrays.
7. The initialization IntTrap(1, 1) = 0# sets the first value of the integral to floating point zero (that is what the #means).
8. The For loop computes the integral at each x-value using the trapezoidal rule.
9. The final assignment Trapezoid = IntTrap returns the array of integral values in the function name.

Shown next is a spreadsheet in which the x and y columns contain an independent variable varying between 0 and 1 and associated function values for the simple function $e^x \sin x$. The third column was selected and the text = trapezoid(was typed. At this point, the data values for x were selected followed by a comma, then the selection of the second column of values (for f), and finally a close parenthesis.

	A	B	C	D	E
1	x	f	Int[f(x)]		
2	0	0	=trapezoid(A2:A12,B2:B12)		
3	0.1	0.110333			
4	0.2	0.242655			
5	0.3	0.398911			
6	0.4	0.580944			
7	0.5	0.790439			
8	0.6	1.028846			
9	0.7	1.297295			
10	0.8	1.596505			
11	0.9	1.926673			
12	1	2.287355			

Note that the cell ranges for x and f appear within the parentheses. To view the entire column of results, it is necessary to strike Shift/Ctrl/Enter. The results appear in the spreadsheet shown below.

	A	B	C
1	x	f	Int[f(x)]
2	0	0	0.00000
3	0.1	0.110333	0.00552
4	0.2	0.242655	0.02317
5	0.3	0.398911	0.05524
6	0.4	0.580944	0.10424
7	0.5	0.790439	0.17281
8	0.6	1.028846	0.26377
9	0.7	1.297295	0.38008
10	0.8	1.596505	0.52477
11	0.9	1.926673	0.70093
12	1	2.287355	0.91163

A.6 CREATING EXCEL® ADD-INS

Suppose that the function for calculating an integral using the trapezoidal rule is to be used in more than one spreadsheet and is to be used frequently. It would be useful to have this function available as an Excel Add-In (like the Matrix.xla functions). To do this, once the function has been thoroughly tested, simply save the spreadsheet as an Excel Add-In. When saving, change the file type accordingly. The file is saved in a special directory reserved for Add-Ins (this can be overridden if desired). To activate the Add-In, choose the File/Options/Add-Ins menu. At the bottom of the screen is an option to manage Add-Ins—choose Go. This brings up a screen with Add-In names and a check box in front of each. Be sure the box is checked for the macro to be activated. If the name of the Add-In does not appear, choose Browse to locate the file for the Add-In.

Shown below is a spreadsheet in which the Trapezoid Add-In is used. The column where results are to appear is selected and then Formulas/Insert Function. Choose User Defined from the "Or Select a Category" menu. Find and select Trapezoid and click OK.

Note that the user is automatically prompted to select the range for x and f. This feature is added when the Add-In is created and activated—no special programming is required.

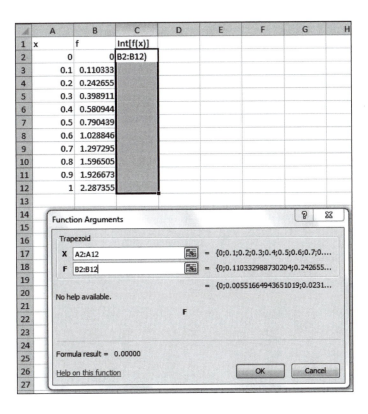

After selecting the range for x and f, hit Shift/Ctrl/Enter to display the results as follows:

	A	B	C
1	x	f	Int[f(x)]
2	0	0	0.00000
3	0.1	0.110333	0.00552
4	0.2	0.242655	0.02317
5	0.3	0.398911	0.05524
6	0.4	0.580944	0.10424
7	0.5	0.790439	0.17281
8	0.6	1.028846	0.26377
9	0.7	1.297295	0.38008
10	0.8	1.596505	0.52477
11	0.9	1.926673	0.70093
12	1	2.287355	0.91163

Index

Seeking the Perfect Game

Recent Titles in Contributions to the Study of Popular Culture

Seeking the
PERFECT GAME

Baseball in American Literature

CORDELIA CANDELARIA

Contributions to the Study of Popular Culture, Number 24

GREENWOOD PRESS

New York • Westport, Connecticut • London

Library of Congress Cataloging-in-Publication Data

Candelaria, Cordelia.
    Seeking the perfect game : baseball in American literature /
Cordelia Candelaria.
        p.   cm.—(Contributions to the study of popular culture,
    ISSN 0198–9871 ; no. 24)
    Bibliography: p.
    Includes index.
    ISBN 0–313–25465–6 (lib. bdg. : alk. paper)
    1. American literature—History and criticism.   2. Baseball
stories, American—History and criticism.   3. Baseball in
literature.   I. Title.   II. Series.
PS169.B36C36   1989
813.009′355–dc20   89–2157

British Library Cataloguing in Publication Data is available.

Library of Congress Catalog Card Number: 89–2157
ISBN: 0–313–25465–6
ISSN: 0198–9871

First published in 1989

Greenwood Press, Inc.
88 Post Road West, Westport, Connecticut   06881

Printed in the United States of America

The paper used in this book complies with the
Permanent Paper Standard issued by the National
Information Standards Organization (Z39.48–1984).

10   9   8   7   6   5   4   3   2   1

For Fidel and Clifford,
ideal teammates, coaches, fans,
with infinite love

CONTENTS

INTRODUCTION

At least since "Casey at the Bat" (1888), American writers simply have not been able to leave baseball alone. They have written about all aspects of the game and its history, chronicled the lives of every significant participant in the sport, have examined its relations to myriad other phenomena from agrarianism to Zen, and they have employed every journalistic and literary genre in seeking to capture its essence and meaning. What explains this extraordinarily wide, profoundly deep interest in what Virginia Woolf described as America's equivalent of society (Woolf, 118) and George Bernard Shaw called "America's tragedy"? More recently, what led *Esquire Magazine* to query solemnly on one of its covers, is "BASEBALL THE LAST THING LEFT TO BELIEVE IN?"

One answer is that because baseball exists in the fluid public domain, its definition can only be dialectical, for it derives from and depends on more than a single, unitary identity. That is to say that baseball—at once a folk game, professional sport, business, entertainment, occupation, gambling activity, and so on—in its existential state of becoming, has been multifold, dynamic, contradictory, literal and measurable, as well as imaginative, poetic, and intangible. Thus, the thousands of written treatments of the game, both those actually set on the field and those caught in the playfields of the imagination, evince the unwavering commitment of hundreds of writers "to learn baseball, the rules and realities of the game" in order to "know the heart and mind of America" (Barzun, 55).

Another answer is that baseball has been written about, as it has been played and watched, to provide both recreation and re-creation. Providing recreation, the fiction about the game, like the game itself, fulfills society's need and desire for the ludic and agonistic exaltation identified by philosophers such as Huizinga and Caillois who, for the most part, see play as separate from "real life." Later philosophers have gone further by refusing to limit play to an isolated arena of impractical, gratuitous activity. By asserting that play is "a basic existential phenomenon" with its own purposiveness, Jacques Ehrmann, for example, grounds play firmly within, not apart from, reality, just as Robert J. Higgs believes that like literature it "invariably reflects the *Zeitgeist* of a society" (Ehrmann, 19; Higgs, 7). These basic and contradictory impulses within the primal center of baseball have attracted numerous chroniclers seeking to present their understanding of its complex, yet lovely, simplicities.

Accordingly, as with any play activity, but especially in one permanently enmeshed in a society's folklore and culture, re-creation is indispensable to baseball. The writers discussed in this study well know that the rules and object of the game, coupled with its rich social texture, make it especially serviceable as literary subject, metaphor, and theme. Indeed, many writers have employed the game as an analogy for life, as a thematic threshold to a clearer perception of American culture, as an allegory of Yankee values, and in other similarly emblematic ways. Depicting the game as both a lamp and a mirror of society, writers have shown baseball to be an apt, often elaborate metonym of the vast, multitudinous United States. To describe, classify, and analyze a literature so varied and vast in one book would be as difficult as negotiating an expansion league with Japan. Clearly, the multifarious nature of baseball fiction makes it elusive of neat conceptual categories and, in fact, its multilayered literary texture makes it richly amenable to multiple interpretations. Still, the more elusive the pitch, the more certain must be the batter's eye and swing.

Hence, *Seeking the Perfect Game: Baseball in American Literature* endeavors, first, to survey the baseball imagery and metaphor found in U.S. American literature and, second, to explain how this figurative treatment of the sport creates its own framework of the imagination, its own fictive universe. From this twofold investigation emerges the book's thesis: baseball fiction discloses a movement from the allegory and romanticism of its earliest forms to the realism and solipsism of its contemporary renderings. Stylistically this movement reveals a shift from linear, mostly one-dimensional narratives to those presented multidimensionally in the fragmented chronology usually derived from ironic, autotelic viewpoints. As a subgenre of American literature, baseball fiction has continually progressed to increasingly complex levels of literary abstraction, a progression that itself mirrors baseball's metamorphosis from the primitive fact of ritual to the stylized realm of (meta)fiction, where symbols are beheld at the highest level of refinement of their cultural origins.

This study thus begins in chapter one with a look at origins, at the aboriginal sources of "America's national pastime," a description put forth by the founding barons of baseball in their singleminded concern with ticket sales and the sport's institutionalization into an indispensable profitmaking industry (Seymour, *Golden Age*, 19–20). Indeed, "by 1869 the promotional myth that baseball was the 'national game' had rooted itself into the culture" (Voigt, *From Gentleman's Sport*, 4). The first chapter also summarizes the early folkloric and literary treatments of the game before moving on to chapter two and analysis of its emergence as a romantic metaphor in juvenile and pulp fiction. Chapter two also discusses how Ring Lardner's trenchant irony saves baseball from the romantic escapism of juvenalia. The metaphor undergoes further transformations—discussed in chapter three—in its appearance as an allusion, as opposed to a controlling metaphor, in the fiction of both major and minor writers from Mark Twain to the present. Baseball's mythic dimension, running through all the literature like the hidden reality of the Negro leagues, comprises chapter four. It focuses on the treatment of myth in the first baseball fiction by a major contemporary author, Bernard Malamud's *The Natural*, and also on the lesser-known *The Rio Loja Ringmaster* by Lamar Herrin, a novel built around pre-Columbian myth. Chapter five examines the metonymic properties of the game by discussing Mark Harris's tetralogy about pitcher-writer Henry Wiggen, along with two fictive "resurrections" of Babe Ruth by writers John Alexander Graham and Jay Neugeboren. The final chapter presents the brilliant culmination of the multifold development of the baseball figure found in Robert Coover's *The Universal Baseball Association, Inc., J. Henry Waugh, Prop.* and Philip Roth's *The Great American Novel*.

Because of the impossibility of providing comprehensiveness in a study of this length, *Seeking the Perfect Game* focuses primarily on novelistic fiction. Moreover, just as baseball fiction spans genres, it also spans over a century of U.S. American literature, a fact that places the baseball metaphor among such quintessential American themes as the wilderness and frontier, the new Eden, technology or the "machine in the Garden," the "holy grail" of materialism and mercantile success, the idea of The Great American Novel, among others.

One of those other themes is the crucial treatment of the issues of racism and, to some extent, sexism. Mirroring the pervasive ethnocentrism and male chauvinism of American society, baseball fiction is, with few notable exceptions, largely about white men participating in a closed activity of the dominant—that is, white, male, Christian, and capitalist—culture. Moreover, because that culture deliberately strove to remain closed for well over fifty years, the institution of baseball, and by extension fiction about it, suffers from the same part equals whole mythology that typifies recorded human history in which a fraction of experience is equated with all of experience. That conception views patriarchy as the whole of civilization not, more correctly, as only part of it. A part of dominant, culture ballplaying was perceived (and, as the 1987 Dodgers' Al Campanis controversy again proved, still is [Jackson, 41]) as being the whole of baseball. In fact, black and other nonwhite ballplayers, both professional and amateur, existed alongside those of the dominant society, but in second-class (or worse) citizenship. In the words of baseball historian David Q. Voigt:

As Jim Crowism was applied to the social institutions in American life, it affected both big-league and minor-league baseball...Negro players were not only refused the same accommodations as white players, but were also ridiculed by whites.... Today, apologists excuse such conduct on the grounds that men of the era were "prisoners of their times."...This rationalization, however, persuades the more objective individual to view major-league baseball as a reflector of the emotions and the values of a culture and to question the claim that the game builds character by its "intrinsic" values of "Americanism" or "gentlemanly sportsmanship." (Voigt, *From Gentleman's Sport*, 278–79)

Consequently, *Seeking the Perfect Game* reflects the bias of fiction that itself reflects the culture's basic exclusivity, even though I am keenly aware of the inadequacy of such an approach in acknowledging the major contributions of Black, Latino, Caribbean, and other excluded participants, including women, filling out a truer, whole picture of the baseball universe.

One final prefatory note: in writing this book the difficulty of targeting its audience was never far from thought. It frequently occurred to me that at one end of the spectrum of prospective readers are those baseball aficionados so knowledgeable about the sport, its history, and its essential sabermetrics that they would likely consider baseball fiction and, certainly, literary criticism about that fiction as insignificant dilettantism or esoterica. On the other end are those readers who perceive baseball—and, indeed, all professional sports—as the banal phenomena of mere popular culture, better fit for sociological investigation than for serious literary study. Not altogether at ease about addressing either extreme, reasonable though they might be, I decided instead to write for two special readers of my own: Holly Wiggen and Roberto Clemente, paragons of talent, good taste, compassion, and, not unimportantly, muteness. They will understand that the book's unavoidable omissions, frequently synoptic analyses, and very modest coverage of baseball qua sport have been dictated by limitations of space—both on these pages and within the author's head. And they will be sympathetic.

REFERENCES

(Unless otherwise noted, place of publication is New York. Full identification is given only for the first citation; subsequent citations in later chapters will consist only of author/title/date. For complete listing please refer to the Bibliography following chapter six.)

Jacques Barzun, *God's Country and Mine* (Vintage, 1954).
Roger Caillois, *Man, Play and Games*, Meyer Barash, trans. (1958; reprint, Free Press of Glencoe, 1961).

Jacques Ehrmann, *Game, Play, Literature* (New Haven, CT: *Yale French Studies* 41, 1955).

Robert J. Higgs, *Laurel & Thorn: The Athlete in American Literature* (Lexington: University of Kentucky Press, 1981).

Johan Huizinga, *Homo Ludens: A Study of the Play Element in Culture* (1950; reprint, Boston: Beacon Hill Press, 1955).

Reggie Jackson, "We Have a Serious Problem that Isn't Going Away," *Sports Illustrated* (May 11, 1987), 40–48.

Harold Seymour, *Baseball: The Golden Age* (Oxford University Press, 1971).

David Quentin Voigt, *American Baseball: From Gentleman's Sport to the Commissioner System* (Norman: University of Oklahoma Press, 1966).

Virginia Woolf, "American Fiction," *Collected Essays*, Vol. 2 (Harcourt, Brace and World, 1967).

1

BASEBALL FROM RITUAL TO FICTION

> It seems to me that next to *Homo Faber*, and perhaps on the same level
> as *Homo Sapiens*, Homo Ludens, Man the Player, deserves a place in our
> nomenclature.
> Johan Huizinga, *Homo Ludens: A Study of the Play Element in Culture*

GENESIS

Like the ancient belief in the earth's flatness, the modern educated mind
finds harmlessly quaint the belief in the Doubleday myth of baseball's origin.
Supposedly invented by Abner Doubleday, a young man who later became
a Civil War hero, the game's apocryphal genesis on U.S. soil was seized as
an effective marketing tool by many entrepreneurs, most notably Albert G.
Spalding, millionaire sporting goods manufacturer, who was "unscrupulous
in his chauvinistic determination to 'prove' the American origin of the game"
(Voigt, *From Gentleman's Sport*, 5). Accordingly, just as many "round earth"
partisans were once tortured and killed for their heresies, the quaint Doubleday myth of baseball's miraculous creation is not so harmless when considered in the full context of similar deliberate lies and distortions throughout
U.S. history. Usually promulgated by businessmen and politicians in cozy
collusion, such lies undergird the official accounts of, for example, the U.S.
War with Mexico, the Teapot Dome Scandal, Watergate, U.S. illegalities in
Vietnam and Central America including the Iran/Contra Scandal—to name a
handful that come readily to mind. These and other profitmaking political
corruptions pervade baseball fiction, as writers of diverse backgrounds and
ideologies have sought to come to terms with the place of both army and
industry within American social history.

In fact, as many baseball devotées have long known (Voigt, 6) and as scholars like Robert W. Henderson and H. J. Massingham have definitively shown, the game did not have a specific instant of invention like, for example, basketball. Its development has been evolutionary with its true origins traceable to ancient ball-and-stick rituals. In *Ball, Bat and Bishop* Henderson categorically states and carefully demonstrates that "all modern games played with bat and ball descend from one common source: an ancient fertility rite observed by Priest-Kings in the Egypt of the Pyramids" (4). Despite this and other evidence, the most current edition of the official *Baseball Encyclopedia* continues to evade the subject in its section on the sport's development: "Before [1846] the origin is vague. Credit has been given to the Egyptians and every succeeding culture" (11). To the contrary, the origin and development is anything but vague—if, that is, the student of the game is prepared to explore and probe disinterested sources other than those pushed by the baseball entrepreneurs and their sycophants.

The ancient fertility rite singled out by Henderson as seminal was performed as mock combat between two opposing sides, with the god Osiris at the center of the ritual battle: "At first Osiris was a nature god and embodied the spirit of vegetation which dies with the harvest to be reborn when the grain sprouts. Afterwards he was worshipped...as god of the dead" (*New Larousse*, 16). The extraordinary importance of Osiris—and his counterparts the world over—can be measured in one way by noting the many civilizing attributes he was assigned over time, including the start of agriculture, the organization of religious worship and social law, the creation of music, and development of villages and towns (*New Larousse*, 16). Consequently, the two sides in the ritual conflict paying homage to Osiris presumably symbolized such fundamental dualisms as winter and summer, life and death, day and night, good and evil, and so on. As later chapters, particularly chapter four, of *Seeking the Perfect Game* show, this primeval dichotomy has been vigorously exploited as archetypal metaphor by many writers sensitive to the literary possibilities of baseball.

An exhaustive survey of the world's ancient myths and rituals, and the anthropological scholarship about them, indicates that agricultural fecundity was the primary impetus for these prehistoric rites, which were under the supervision of an elite priesthood (Henderson, 3–8, 31). With time, the ceremonies expanded to include nonreligious participants—a secularization typical of the changes slowly wrought by modernity. Researchers believe that out of these and similar rites of fertility and mock-warfare emerged the concept of opposing teams *pray*ing for the goal of fruitful harvests (Henderson, 15). The rites may be seen as a primal ur-baseball, a "*play*ing" between teams.

It appears that the ritual significance of the ball evolved out of the use of a good-luck effigy, possibly of a king or god, which became streamlined over time into use of only the head, for centuries regarded as the most potent part of the human anatomy (Henderson, 16–18; Frazer, 343–44). The object of the rite was to safeguard the god-effigy-head safely home to his temple. Noting that many writers believe the ball originally symbolized the sun, Henderson nonetheless favors the other interpretation because of the predominance of head cults throughout the primitive world, including Mesoamerica, which had similar ball-and-stick rites and games. Nevertheless, whether an emblem of the sun or the mummified head of a generic Osiris, he agrees that the ball represented "the idea of fertility, the life-giving principle" (19), an essential aspect of popular belief and custom that even worked its way into early and medieval Christian rituals (36–38).

A similar difference of interpretation of symbols occurs with the primal significance of the bat. Although it is usually related to the procreant shape and energy implied by its obvious phallic form, its prototype was more likely a simple farming tool, a threshing flail or shepherd's staff (Henderson, 22–25). Since both flail and staff could also be used as weapons, their war-game utility was also central to the ritual stick's eventual evolvement into, for example, the polo mallet, tennis racket, and baseball bat (Henderson, 26).

These prehistoric seeds producing baseball's roots relate directly to the subject of *Seeking the Perfect Game* because most of the writers discussed in this study recognize the ways in which baseball is embedded in the country's oldest folk customs and rituals, as well as in its cultural values and practices. The beliefs and habits of any given society are usually modern adaptations of ancient activities. As anthropology has proven, the interconnection between a people and its recreational habits reveals a great deal about that society's inner life and total personality. Hence, recognizing the culture-reflecting, symbolic possibilities of the game, baseball fiction writers have consistently underscored, sometimes in lyric pastoral and sometimes in ironic parody, its mythic place in American society.

Jumping ahead with base-stealer speed from prehistory to the earliest recorded mention of baseball in English, we find in 1700 a Puritan minister in England objecting to "baseball," dancing, and playing "many other sports on the Lord's Day" (Henderson, 132). Other eighteenth-century references to "base-ball," as it was usually spelled, describe a bat-and-ball game played by running to bases and home. While these references would not alone disprove the Doubleday story of baseball's American origin, in their sociohistorical context they clearly support the notion of a gradual development over time of the specific objectives that precede the emergence of distinct games with fixed rules.

To be sure, baseball's emergence from two popular English games, stoolball and rounders, is easily shown through an examination of the older games' principal features. Traditionally played at Easter, stoolball required its players, usually mixed groups of young men and women, to stand before an overturned three-legged stool and to bat or catch a ball until s/he was replaced by someone else, the object being to amass the most hits. The association of stoolball with Christian springtime celebrations leads Henderson to conclude, after detailing many examples, that "the Easter stoolball festivities had a direct association with the ancient pagan rites, with connotations of human fertility" (71). Interestingly, the game's mention in Governor William Bradford's early American history, *Of Plymouth Plantation* (1630), is the first reference to recreational playing in the United States, and it was a reprimand to young men for the sin of playing while the older pilgrims worked.

Rounders is the common name that eventually emerged, *and is still used*, in England for the game of stoolball. Moreover, it is generally agreed that rounders is a very near relation to American baseball. The *Oxford English Dictionary* describes it as a "game, played with bat and ball between two sides, in which each player endeavors to hit and send the ball away as far as he can, and to run to a base or right round the course without being struck by the fielded ball" (2:830). Henderson reports that although this form of the game was often called "base-ball" in late eighteenth-century England, by 1829 the term "rounders" was in widespread use (145). Even though the most current *Baseball Encyclopedia* dismisses as "vague" the early origins of the game, it does not hesitate to acknowledge that "in America...the English game of 'rounders' finally evolved into baseball" (11).

AMERICAN DEVELOPMENT

Ball, Bat and Bishop contains over a dozen eighteenth-century references to the game in America, where it appears to have shifted at an early period from the coeducational stoolball activity recorded in England. For many generations it seems to have been a pastime largely restricted to boys, for most of the published accounts of it depict boys at play and also address children readers. And yet the first extant record of an actual baseball game on American soil appears in an adult's diary, in the April 7, 1778, journal entry of one of George Washington's revolutionaries (Henderson, 136). Town ball, a later form of the still casual, unorganized game almost always played by adults, is credited by some baseball scholars as the folk precursor closest to the organized sport (Voigt, xxvi).

According to *The Baseball Encyclopedia*, the very "first seeds that led to organized baseball in the United States were planted on the Elysian Fields in Hoboken, New Jersey, on June 19, 1846" (11). At that time, two amateur teams played the game following the rules written by a committee headed by that game's umpire, Alexander J. Cartwright, himself an amateur athlete and land surveyor. This contest between the New York Knickerbocker Base Ball Club and the New York Club constituted the leisure-time exercise of two groups of refined gentlemen wealthy enough to support membership in expensive private clubs. This point bears stressing in light of Albert G. Spalding's relentless insistence on the game's invention by Doubleday on a cow pasture in rural Cooperstown, New York. Clearly, the exclusive elegance of these clubs was as unlike the Cooperstown story as a bunt is from a homer. Although they played the game in much the same manner as its unorganized common form, the fact that the Knickerbockers adhered to written rules, one of which required keeping a written record of all game scores, probably accounts for the survival of the Cartwright committee version. And certainly one of baseball's fascinations to the literary sensibility over the years derives from its requisite bookkeeping—the game-chronicling that produces its own history, biography, and even its own historiography.

The 1846 Elysian Field game contained modern baseball's fundamental outline and many of its key essentials. For example, play occurred on a diamond with a pitcher centrally located on the field, three strikes led to an out, and the running player could be tagged out on base; but, unlike rounders and cricket, "in no instance is a ball to be thrown at" the runner (Henderson, 165). Nevertheless, many of the Knickerbocker Club regulations changed dramatically as the game metamorphosed into a professional, money-making sport. For instance, in the Knickerbocker rules the object of the game was to make twenty-one runs regardless of the innings played, and the size of the teams varied depending on the availability of players. Still, the efficiency (and arrogance) of the club in promoting its rules and social style, along with the game's growing popularity, soon let to vigorous, if gentlemanly, com-petition—a by-product that disturbed many of the original participants who felt the raison d'etre of gentlemen's games to be sport, not competition. Eventually, Knickerbocker play led to the formation of a formal organization of players, the National Association of Base Ball Players, which modified the Cartwright rules in 1859 (Voigt, 9).

In describing this period of baseball in America as "gentlemanly," it should be further stressed that the early associations sought to model themselves after the aristocratic private clubs of England, where games were played more for exercise and social camaraderie than for victory. The cultural differences between the traditionally classist "mother" country and her frontier colony quickly surfaced in the democratic fields of a nation lacking not only the titles but also the social graces that accompany demarcated lines of class. Thus, the "more dominant American values" of "fierce competition and creep-ing commercialism" soon "disrupted the climate of [English] sportsmanship" as box office revenues, newspaper publicity, and "winning" became the cen-tral goals of the game (Voigt, 10). By the late 1860s, it seemed that "many baseball men would willingly trade the elusive claim of gentleman for the certainty of hard cash" (Voigt, 13).

True to its multilinear origin and growth, then, the professionalization of baseball into a salaried "sport" evolved from multiple sources too numerous and cobwebbed to chronicle effectively here. Suffice it to observe that by 1869, fresh from the Civil War's stimulating effect on baseball's national growth, Henry Chadwick, the highly influential sportswriter and staunch sup-porter of baseball, was forced to admit that professional baseball was, indeed, only another American business (quoted in Voigt, 21). While not widespread among the public or even among most players, this attitude was held by those entrepreneurs—both above-board businessmen and under-the-table racketeers—with commercial interests in the sport. As nearly all the baseball fictionists discussed in this study attest, that entrepreneurial attitude, which derived from America's fundamental mercantile ethos, was the yeast that leavened the Black Sox scandal of 1919.

Henderson points out that "athletic games" have "always" produced two kinds of followers, the ones "interested in the game for the game's sake" and those primarily interested in "methods of exploiting the skill of others, the gamblers and the professionals" (168). For example, as Eliot Asinof's *Eight Men Out: The Black Sox and the 1919 World Series* systematically demonstrates, in late 1919 and throughout the 1920 season, those partisans who had considered the sport a pristine pastime were dismayed to learn that the 1919 World Series between the American League Chicago White Sox and the National League Cincinnati Reds had been fixed, and that an elaborate cover-up of the fix was taking place (Asinof, 121ff). Investigation finally exposed bigtime gambler Arnold Rothstein, in collusion with assorted other gamblers, businessmen, and eight members of the White Sox team, all in a huge scam. The gamblers paid the White Sox players to throw the game and assure Cincinnati's victory (Asinof, 3–38; Seymour, *Golden Age*, 294–310; Voigt, 98). Although the originators of the scheme, and its biggest profit-makers, were definitely not the ballplayers, their faked play accounts for the "Black Sox" label by which the 1919 Series is remembered. In fact, the players ultimately lost the most—jobs, careers, reputations, and self-worth—while Rothstein, his cohorts, and interlocked corporate owners continued to gain munificently and with impunity from baseball. Baseball historians today generally agree that owners like Albert Spalding, Ban Johnson, and Arthur Soden actually facilitated the fix (Asinof, 49; Seymour, 32–34).

The Black Sox scandal, a stain on sportsmanship, exists in baseball mythology as a fall from grace. It has naturally found its way into U.S. fiction, where it usually appears as an emblem of the inevitable consequence of unchecked commerce and materialism in a secular society. For example, in *The Great Gatsby* (1925) the narrator is "staggered" by the very "idea" that "one man" could even "start to play with the faith of fifty million people" as he remembers "that the World Series had been fixed in 1919 . . . the end of some inevitable chain" (55). Sixty years and dozens of fictional treatments later, *The Celebrant*, a baseball novel by Eric Rolfe Greenberg, details exactly how that "inevitable chain" developed link by link in the money-grubbing hypocrisy and rapacity of free enterprise. Hence, like the Doubleday myth of baseball's genesis, the Black Sox motif recurs throughout baseball fiction as a reminder of the frail, imperfect human context that produced the sport, and that looks to play, myth, and art for the models of perfection humanity requires to buoy itself above its frailty and imperfection.

BASEBALL FICTION: DEFINITIONS

A capsule summary of the history of baseball just presented starts with the earliest known, prehistoric fertility rituals performed as mock combat between opposing sides. These rites were adapted into Christian springtime and Easter folk celebrations, which later became associated with specific ball-and-stick games, and subsequently evolved into the fourteenth-century stoolball. Secularized into an array of nonreligious games, stoolball was the direct precursor of such games as feeders, poisoned ball, and English rounders, the immediate antecedent of American baseball, which itself underwent an active period of amateur organization and formalization between 1840 and 1850. Despite the complexity of the game's transformation from a folk game into a professional sport, its commercialization into a profit-making industry recognizable to contemporary minds was completed at least by 1869.

To highlight the natural correspondences existing between baseball and literature, the foregoing summary may be paraphrased in an even tighter condensation especially appropriate to *Seeking the Perfect Game*'s critical method. Baseball first evolved from its origin in ancient religious ritual to an infancy as folk play associated with Christian practices in spring, which were then followed by a childhood markedly divided into a variety of informal folk ball games eventually leading to an adolescence consisting of such distinct forms as rounders and town ball. Its rite of passage as an organized sport in the United States was, almost from the start, inseparable from the commerce and corruption around it, as exemplified in the Doubleday-Cooperstown lie and cover-up about baseball's genesis, and in the more criminal but analogous Black Sox scandal.

Both summaries suggest the remarkable richness, density, and texture of baseball's growth and development, and both hint at the reason for its seduction of American writers even during its very nascence as an organized sport in the nineteenth century. There are nine important characteristics of baseball that not only help explain that seduction but also serve as a taxonomy of the literary motifs recurrent in baseball fiction.

1. Its genesis in aboriginal myth and ritual give it a depth of cultural substance similar to that found in other mythologies and human customs.

2. Its slow development over time within the active, changing lives of numerous folk societies places it inside a dynamic, rich matrix of grassroots sociohistory.

3. The preceding traits suggest that, in its antecedent folk form, the game satisfied such basic restorative human needs as physical play, relaxation, and fun.

4. Similarly, these traits indicate that baseball's antecedent folk forms also satisfied such basic agonistic human drives as rivalry, physical competition, and controlled outlets for the exercise of physical force.

5. Its emergence in the United States as a formal leisure activity for gentlemen provides historical precedent for the racial and gender exclusivity and social biases of the

modern organized sport. Racism and sexism pervade society regardless of class, of course, but baseball's particular social biases correspond, at least symbolically, to its original overt classism.

6. Organized baseball's false genesis legend, which was actively contrived by its earliest commercial promoters, plants a kernel of fiction into the very history and cultural essence of the sport.

7. Its inextricable involvement, both overt and covert, with commerce and industry early in its beginnings as an organized sport may today be viewed as foreshadowing the 1919 "fall from grace" caused by the Black Sox Scandal that forever after tainted the now post-lapsarian social phenomenon.

8. As a consequence of its deeply ingrained place within the country's history, culture, and lore, baseball is such an important political symbol that it has traditionally required the President of the United States to start each box office season.

9. In its essential and fundamental form, organized baseball contains an array of inherently symbolic elements (for example, the symmetry of its field game and internal mathematics, its required record keeping and box scores, its fictional origin, its symbolic numbers, and so on) that have quite naturally conduced extensive literary treatment.

These important characteristics, most of which are specific to baseball, will be recalled again and again in the course of this book's discussions of both individual titles and thematic clusters, for together they shape the contours of baseball fiction and help define its constituent elements. They also confirm a general idea recurring throughout baseball fiction: "sport in American life has always been contradictory. It liberates in play, but it binds its players to the most strenuous work. Such irony leads to even greater ironies" (Messenger, 2).

In addition, these characteristics also bring to the fore the underlying thesis of *Seeking the Perfect Game*: baseball originated in the primitive, informal fact of ritual and folk play and has evolved to the most formal refinements of those origins in regulated industry and stylized literature. Furthermore, just as the game has evolved from ritual to art and literature, baseball fiction itself, as a subgenre of American literature, discloses a parallel stylistic evolution. The movement within baseball fiction is from flat, one-dimensional stories sentimentalizing baseball as a symbol of the national spirit to multi-dimensional metafictions in which baseball is encapsulated as a fiction-within-a-fiction.

Before illustrating the progression from ritual to fiction through the consideration of specific texts, concise explanations of the subject and title of this book are in order. *Baseball fiction comprises both poetry and prose literature that derives its primary subject and controlling metaphors from the sport of baseball and/or from baseball's history and its sociocultural ambience*. The richest literature about baseball is most often that which dynamically integrates both these features and most of the nine specifications enumerated above. *Through its title this book also applies the elusive (and controversial) ideal of the sport's concept of the "perfect game" to the literary and philosophical search for truth about America's fundamental soul and substance.* In baseball a perfect game refers to a no-hitter, in which no batter from the opposing team reaches first base; it is a full game in which a pitcher does not allow any opposing players to reach first base safely, but retires the opponent with twenty-seven consecutive outs. Chapter six fully discusses baseball's "perfect game" and its salience to the fictive search for the "Perfect Game." At this point, it is enough to note that the perfect game serves as an apt analogy for the persistent strivings toward the ideal that characterize the human artistic record.

LOCAL COLOR AND EARLY BASEBALL FICTION

"Where the sportswriter is required by the nature of his profession to foster his accounts from kernels of truth and to anchor his imagination to actual events," comments Tristram Coffin in his classic, *The Old Ball Game: Baseball in Folklore and Fiction*, "the pulp writers have no such restrictions . . . [they] conceive, artificially, as it were, whatever they will" (136). Except for sundry early poems, usually of a sternly didactic nature, the earliest baseball fiction was of the pulp variety, mostly intended for children. Attaining extraordinary popularity in the postbellum nineteenth century, these early writings were almost interchangeable in plot, character, and theme. Not handicapped in the market by belletristic standards, works like the Frank Merriwell baseball series by Gilbert Patten, which sold over one hundred million copies, could be repeated endlessly through cosmetic disguises of formulaic plots.

The tradition of boy's sports fiction girds "every work of baseball fiction, no matter how metaphysical" (Knisley, 74). And girding that particular tradition of juvenile literature is the local color movement that dominated popular writing in the United States in the 1880s. Marked by the detailed foregrounding of the peculiar speech, landscape, customs, and values of a particular region, local-color fiction places its narrative emphasis, as Hamlin Garland noted, on locale-specific qualities that starkly differentiate "this particular place" from "everything around it." Garland also stressed the absolute "requirement" that writers first understand a locale thoroughly before writing about it to avoid damaging the region by superficial treatment (Garland, 64).

The late nineteenth-century local colorists apparently got their creative impetus from the coincidence of a massive growth in publishing markets with three important results of the Civil War (Simpson, 1–3). First, the major U.S. population shift westward led a few writers like Bret Harte and Joaquin Miller to inform Eastern readers of the new frontier and its people. A second stimulus was the great European migration to the U.S. that began after the War and led many—like Kate Chopin, Owen Wister, and, later, O. E. Rölvaag—to recognize and seek to record the country's ethnic heterogeneity for the first time. Suspicion of traditional New England leadership for the country's values and taste, promulgated by the post-War South and encouraged by the Northeast's own arrogant parochialism, provided another important spur to the local-color movement (Simpson, 3–8). In short, this literary movement "assume[d] importance at a time when the integrity of diverse subcultures within the nation was ... of ... public concern, and when a substantial group of general periodicals offered a sizable steady market for the short story" (Simpson, 7).

Although some scholars have mentioned in passing the link between local colorism and sports fiction (for example, Coffin, 55; Messenger, 109), to my knowledge no one has systematically developed a literary critical case for the connection. Quite arguably, early baseball fiction qualifies as local color both temporally, as a phenomenon of the post–Civil War "gilded age," and also in definitional details. These details include the careful foregrounding in baseball fiction of the particular locus and life of baseball's manifold sub-culture. From its sketches of the ballpark, diamond, and dug-out to its por-trayal of athleticism, team transience, masculine competitiveness, and regional dialects, early baseball fiction presented the particularity of the sport and its participants with the same attention to ambience, if not skill, as that found in Harte's "The Luck of Roaring Camp" (1870) and Chopin's *Bayou Folk* (1894). Ring Lardner, for example, like local colorists before him, "presented the athlete ... and his environment as representative of society," and was quite as masterful as Harte or Joel Chandler Harris in "creating vernacular speech" (Messenger, 109).

Moreover, just as local color writing that does not transcend its setting to engage the entire literary sensibility often remains forgotten on the eternal bench of faded memory so, too, has most early baseball fiction failed to last beyond its *fin de siecle* season as popular literature. For instance, Eggleston's sentimental *The Hoosier School-Master* and Longstreet's sketchy *Georgia Scenes*, two examples of local color writing, are today primarily of historical interest only to the literary scholar. Yet, Twain's *Adventures of Huckleberry Finn* and Wharton's *Ethan Frome*, novels firmly rooted in minutely specific locales but both simultaneously and primarily concerned with the depths and nuances of the human condition, are still remembered and read in and outside the classroom. Similarly, the scores of Beadle's dime baseball novels, juvenile baseball books like the famous Frank Merriwell series, serialized newspaper baseball stories, and many other pulp baseball fictions of the early period remain with us largely as historical trivia. They are of interest primarily for what they reveal about the institutionalization of homogenized "American" values peddled by such dominant-culture males as, for example, Horatio Alger, Henry Chadwick, Albert G. Spalding, and not because of any lasting literary merit.

Along with the juvenile pulp literature, then, baseball fiction owes its emer-gent vigor to the local color movement of the late 1800s when America was taking interest in its polymorphic self. As such, baseball fiction is directly part of the dominant thrust of America's modern literature.

REFERENCES

Eliot Asinof, *Eight Men Out: The Black Sox and the 1919 World Series* (Holt, Rinehart, & Winston, 1963).

The Baseball Encyclopedia, 6th ed. Joseph L. Reichler, ed. (Macmillan, 1985).

Tristram Potter Coffin, *The Old Ball Game: Baseball in Folklore and Fiction* (Herder and Herder, 1971).

F. Scott Fitzgerald, *The Great Gatsby* (Charles Scribner's Sons, 1925).

Sir James George Frazer, *The Golden Bough* (1922, reprint Collier, 1963).

Hamlin Garland, *Crumbling Idols*, 1894.

Robert W. Henderson, *Ball, Bat and Bishop: The Origin of Ball Games* (1947, reprint, Detroit: Gale Research, 1974).

Huizinga, *Homo Ludens*, 1950.

Patrick Allen Knisley, "The Interior Diamond: Baseball in Twentieth Century American Poetry and Fiction" (Ph.D. diss., University of Colorado at Boulder, 1978).

H. J. Massingham, "Origins of Ball Games," in *The Heritage of Man* (London, 1929).

Christian K. Messenger, *Sport and the Spirit of Play in American Fiction: Hawthorne to Faulkner* (Columbia University Press, 1981).

New Larousse Encyclopedia of Mythology (1959; reprint London: Hamlyn, 1968).

Harold Seymour, *Baseball: The Golden Age.* (Oxford, 1971).

Claude M. Simpson, ed., *The Local Colorists: American Short Stories 1857–1900.* (Harper & Brothers, 1960).

David Quentin Voigt, *American Baseball: From Gentleman's Sport to the Commissioner System*, 1966.

2

EMERGENCE OF A METAPHOR

Whoever wants to know the heart and mind of America had better learn baseball, the rules and realities of the game.

Jacques Barzun, *God's Country and Mine*

HEROES

Long before baseball stars became what promoter-owner Bill Veeck, Jr., calls today's "good gray ball players, playing a good gray game and reading the good gray *Wall Street Journal*," they were heroes in the classic American grain (Seymour, *Golden Age*, 75). The kind of heroes that Thomas Wolfe movingly describes as "strong, simple full of earth and sun" (1956, 723), and that even so disingenuous a baseball reporter as veteran Fred Lieb can still recall with goosebump excitement as he does the "truly great" shortstop, "Flying Dutchman" Honus Wagner, whom he believes stands in "a class by himself" (48). Writer Charles Merz summarizes this admiring view of athletes as vicarious self-projection, by observing that the "coming of a hero" brings both ephemeral joy and "solid satisfaction." This results from the fan's active self-identification with the "champion" who breaks through "the humdrum routing of ordered living" in an uplifting, though admittedly "fleeting bit of glory" (quoted in Seymour, 96).

Apart from baseball, volumes have been written from a variety of perspectives about heroes, heroics, and the nature of heroism. The classical view conceived of heroes as quasi-divine men whose deeds of valor and wisdom led to fame and public worship. John Dryden in *Absalom and Achitophel* (1681) saw this view as finding the "Etherial Root" of great "Heroes" to be "Jove himself." This position was immortalized by Thomas Carlyle in his famous lectures *On Heroes, Hero-worship and the Heroic in History* (1840) where, with smug patriarchal assurance, he asserted that "the history of what man has accomplished in this world is at bottom the History of . . . Great Men." Finally, altering some of the skin but not the meat of Carlyle's argument, Walt Whitman's (1819–92) conception of the American hero emerged from a fusion of Carlyle's "Great Men" with his own romanticized notion of the transcendental "common man" of democratic societies.

These and countless other similar "hero-archical" views of supposedly peerless human icons lie at the heart of the twentieth-century apotheosis of baseball players and other sports celebrities. Ironically, the deification of ballplayers coincides almost exactly with the rapid growth of modern journalism in the last 150 years. The irony lies in the fact that despite the press's exposure of the frailties and corruptions of public figures through mountains of information about them, the general public has insisted upon worshipping them—professional athletes in particular. For example, the earliest issues of *The Sporting News*, between 1886 and 1921, contained frequent news stories about such topical and nonheroic baseball issues as salary disputes, unionization, gambling, and excessive drinking. For over a century, the daily newspapers have published sufficient information to challenge effectively the "Etherial Root of Jove" view of baseball heroes. That this common wisdom has consistently been overshadowed by the pedestal the public continues to build for its heroes reflects as much on the fans as on the players they canonize—a thought central to "Casey at the Bat," discussed below.

For the purposes of this study, the analyses of heroes in Robert J. Higgs's *Laurel & Thorn: The Athlete in American Literature* and Ralph Andreano's *No Joy in Mudville: The Dilemma of Major League Baseball* are particularly useful. Higgs divides the fictional athletes in American literature into three types—the widely used Apollonian and Dionysian categories and the less common, Adonic, based on the Greek myth of Adonis. He employs the terms

to indicate conformity to some fixed code, that is, to the artificial in the case of Apollonian, or the degree of revolt against society and the indulgence of the body or the return to nature in the case of the Dionysian. Each type always suggests its opposite and each always raises the question of emphasis on body or self, nature or art. (9)

Emphasizing the potential tyranny of the Apollonian model, he exemplifies this group of fictional athletes with references to "the busher, the sporting gentlemen, the apotheosized WASP, the booster alumnus, the muscular Christian, and the brave new man." Higgs finds that each of these types "attempts to embody or uphold some concept or code of the unity of body and self" (9).

Oppositely, the

Dionysian type . . . is the true "natural" who has accepted his body as himself and feels no need to conform to . . . order of any sort. He is, in fact, narcissistic in that he worships his own body as an end in itself. (10)

Stressing the self-injuring narcissism inherent in this model, Higgs points to "the familiar babe, bum, or beast" for his examples. These, "like the Apollonian," are "invariably guilty of hubris," but they are not similarly self-deceived, for, importantly, they have "no self to deceive" (10).

Yet another type of Dionysian figure according to Higgs, but one muted by reason and a measure of self-awareness, is the Adonic. He is

a rebel who reminds us of greater "beyonds." . . . He does not kill himself [through sensual self-indulgence], nor does he uncritically conform. . . . He [represents] . . . a sort of middle way between . . . animal indulgence on the one hand and authorized forms of behavior on the other. . . . Hence he lives in a world of tension, pain, struggle, and hope. (8–11)

True to the metaphysical synthesis implied by this model, characters classified as Adonic are more mature and psychologically whole than their Apollonian or Dionysian counterparts. In this group Higgs places the "folk hero, the fisher king, the scapegoat, the absurd athlete, and the 'secret' Christian" (11).

The utility of the Higgs taxonomy resides more in its broad triad of conceptual categories than in a meticulous application of its subordinate groupings to particular texts. Therefore, in discussing here the literary characterizations of ballplayers and their fictive peers, I rely on Higgs's classification of Apollonian, Dionysian, or Adonic traits to amplify, deepen, and clarify my own understanding of character and meaning in each text.

In *No Joy in Mudville: The Dilemma of Major League Baseball*, Andreano presents his view of baseball heroes in terms of what he calls the "Folk Hero Factor":

The key to understanding the historic role of baseball, most especially professional major league baseball, in American society is . . . the Folk Hero Factor. This is, that baseball—its players, its institutions, and its success—is best explained by the appeal it makes, consciously or indirectly to the American need for symbolic characters, myths, and legends. (4)

While not suggesting that such appeals are unique to American culture, Andreano does suggest that the link between "heroic value" and "strength and endurance" as opposed to, say, compassion and tolerance, may in fact be specific to a democratic, capitalistic nation rooted in the myths of rugged individualism and staunch self-reliance.

Moreover, the discrepancy between the myths about baseball and the reality of the baseball profit-making industry has historically been ignored by both promoters and fans. The fans, in particular, prefer to consume the sport spiced by the distortion of image-making and hero-worshipping—just as they like their hotdogs smothered in relish. Angry labor disputes and fixed games, for instance, have somehow managed to coexist with the sport's image of healthy, patriotic innocence. On the other hand, however, most baseball fiction writers have looked at the organized sport honestly and objectively and have seen in the gap between reality and appearance a fertile valley of literary possibilities. Indeed, this fundamental discrepancy, considered within the context of the Folk Hero Factor, as well as within Higgs's related taxonomy of heroism, dominates most of the baseball fiction examined below.

FALLEN CASEY—OR, THE TRAGEDY OF HUMPTY DUMPTY

In Aristotelian terms, "the tragedy of Humpty Dumpty" refers to one possible interpretation of the permanent and irreparable destruction of a well-known nursery hero. This pert, though not altogether frivolous interpretation, derives from William DeWolf Hopper, the person most responsible for the lasting popularity of "Casey at the Bat," first published in 1888 (Murdock, 3–6). A professional actor, Hopper saw in the mock-heroic baseball poem "a drama as stark as Aristophanes," one offering insight into U.S. society and some of its most accepted habits and values (Gardner, 5). For this reason, he introduced it into his stage readings, where enthusiasm for it was so keen it quickly became part of his standard repertoire, which in turn led to ever wider dissemination. Hence, more than its author, Ernest Lawrence Thayer (who first published it under the pseudonym Phin), more than its appearance in the *San Francisco Examiner*, and more than the later controversy that developed over its authorship, Hopper above all is credited for "Casey's" lasting place in American literature.

Much of the poem's appeal to Hopper derived from the actor's belief that "there is no more completely satisfactory drama in literature than the fall of Humpty Dumpty" (Gardner, 3–4). This opinion reflects the actor's strong aversion to the popular theater and commercially successful pulp fiction of his day, in which the sentimental renderings of romantic heroes living happily ever after taxed the patience of mature sensibilities. For Hopper, "Casey" succeeded as drama "because it rendered, for once, the absolute failure of a superhuman hero" (Knisley, 20). Indeed, most of us can easily remember the ultimate, the fateful lines that tell of that failure:

> And somewhere men are laughing, and somewhere children shout;
> But there is no joy in Mudville—mighty Casey has struck out. (Thayer, 23)

Not wanting to destroy the mock-heroic tone of Ernest Lawrence Thayer's ballad, or its effectiveness as light verse, it is worthwhile to note that these simple lines encapsulate three important themes: the fall of a folk hero, its effect on his community, and the meaning of both to the universe of humans.

To begin with, the fall of this Apollonian folk hero is the very heart of the ballad's plot and suspense. Right after opening with the "outlook wasn't brilliant for the Mudville nine that day," Thayer introduces Casey as the only possible hope for victory, but then devotes many of the remaining thirteen stanzas to preparing his audience for the exact opposite—Casey's fall. In the process, we discover the protagonist's personal qualities—for example, his singleminded attempt, in Higgs's words describing the Apollonian figure, to "conform to some stereotyped conception of [athletic] completeness," and to give the "illusion of maturity" even though he is "unrealistic" and "immature" (Higgs, 9).

Although Higgs does not mention the poem in his book, it would seem that he had stanzas six through ten in mind when he wrote his description of the Apollonian type. Those stanzas depict the brashly confident Casey deliberately and foolishly "ignoring" hittable pitches as he seeks to perform the expected role of hero to the "rising tumult" of the "benches, black with people." But the flaw in both his conventional Apollonian pose and his playacting is exposed when reality, in the form of the strikeout, shatters the prefabricated "style" just as "the air is shattered by" his fruitless swing.

Thayer sets up this baseball Humpty Dumpty's fall by employing psychological irony and by prolonging plot suspense. The irony stems from the Mudville fans' misreading of the basic human failings that flaw their "mighty" hero's character like a crack in the wood of a Louisville Slugger. Where arrogance "sneer[s]" in "haughty grandeur," the fans see jaunty "ease in Casey's manner"; where the blood of "scorn" and "hate" pulses, they see "pride in Casey's bearing . . . as he lightly doffed his hat"; where sham shines on "great Casey's visage," they see "a smile of Christian charity"—in sum, they either fail to see or simply ignore the hubris of his cocky "defiance." Thus when the "maddened thousands" cry out at the umpire, "Fraud!" they unwittingly describe Casey, and even when the diamond "echo[es]" the cry "fraud" back to them again, they are "silenced" in "awe" by a "scornful look from Casey."

This exchange between worshippers and their doomed idol brings us to the second point mentioned above, the effect a hero's fall has on his community. That Thayer intended his poem to be understood as being as much about the Mudville fans as about Casey is apparent from his devoting a full half of the ballad to the at first "hopeful" and finally "stricken multitude" watching the game. Like William Carlos Williams in his much later poem, "At the Ball Game" (1949), which also focuses on the personality of the crowd in the bleachers, Thayer knew that a hero like Casey required a reverent and mindless audience. In the inflated language of the mock heroic, he describes their "deep despair" and "grim melancholy" at the lopsided "four to two" score. Buoying them out of those abysmal depths is the equally inflated hope that surges when they see Casey on deck with two men on base:

> Then from 5,000 throats and more there rose a lusty yell;
> It rumbled through the valley, it rattled in the dell;
> It knocked upon the mountain and recoiled upon the flat,
> For Casey, mighty Casey, was advancing to the bat.

The reference to "5,000 throats," combined with the later "ten thousand eyes" and "five thousand tongues," has paradoxical effect. By counting the anatomical parts separately, Thayer swells the numbers to suggest additional thousands of presumably thinking individuals in attendance; the paradox emerges when these tens of thousands behave as a single, nonhuman "storm-wave." They become one "crowd" that, as a collective unit, "the patrons of the game," leaves its individuality and rationality outside the gates. As William Carlos Williams describes the crowd in "At the Ball Game": "It is alive, venomous" (31). "Casey at the Bat" thus tellingly renders not only the toppled hero, but also, as it were, "all the king's horses and all the king's men" who comprise his society. Thayer's vivid depiction of the game and the crowd creates the poem's third effect mentioned above, the universal significance of the folk hero's fall.

Certainly by subtitling his work, "A Ballad of the Republic," Thayer indicated his interest in connecting Mudville with the larger universe of humans. In fact, the Casey/Mudville tale is sandwiched between the subtitle's allusion to the "Republic" and a similar allusion at the end: "Oh, somewhere in this favored land the sun is shining bright; / The band is playing somewhere, and somewhere hearts are light." Even though that "somewhere" is ignorant of, and thus indifferent to, the Mudvillean re-enactment of essential tragedy— that is, the fall of a hero and its effect on his society—nonetheless, its homely particularity, its very triviality, makes it emblematic of everyone's tragedy, anywhere, a conclusion that partially accounts for the century-long controversy concerning the poem's authorship (Murdock, 9–20). The work's universality also helps explain the numerous adaptations and analyses of "Casey at the Bat" that have circulated to the present time (see Gardner and Murdock, passim), including a short story in *Sports Illustrated* by Frank Deford commemorating the ballad's centennial in 1988.

Consequently, even though Thayer's mock heroic ballad will stay within the province of children's tales, like Mother Goose's Humpty Dumpty, for example, it is not because it is absolutely devoid of artistic, and therefore cultural, substance. Like all folklore and much popular literature, the poem makes us "acutely experience" its hero's defeat, and it gives the sense "that what happens on the diamond resonates beyond the ballpark" (Knisley, 24). That interactive resonance between ballplayer-hero, fans, diamond, and beyond vibrates powerfully in the baseball fiction that followed "Casey at the Bat."

RING LARDNER AND OTHERS

If Thayer, with help from William DeWolf Hopper, gave America its first major baseball fiction, it was Ring Lardner who nearly singlehandedly transformed the sport from a casual motif in juvenile stories to a formal nuanced metaphor serviceable to serious literature. In the process, Lardner continued the demythologizing of baseball and other popular heroes begun in "Casey at the Bat," a trend that later gained increasing momentum both in fiction and in the popular perception of the actual life of the sport. "Perhaps more than anyone else, it was the great sportswriter, Ring Lardner, who portrayed ballplayers as they were," writes baseball historian David Quentin Voigt, who thinks Lardner's "candid sketches" not only "humanized players by smashing romantic myths" but also "hurt" baseball's image "by exposing the narrow range of player interests and tastes" (*From Gentleman's Sport*, 70). In contrast, one of Lardner's early biographers, Donald Elder, believes that baseball provided Lardner's fiction with "an ordered world" having "definite rules of conduct" and a clearcut "code of honor" (205–6). Elder suggests that the ballplayers satirized in his fiction were those who failed to show the proper respect for baseball's inherent "order" and "ethical ideal" (205). Whatever the perceived effect, no one disputes Lardner's germinal influence on baseball fiction.

Readers of widely different tastes have recognized his pioneering effort as an "innovative chronicler of American games, comic players, and their foibles" (Messenger, 108). Virginia Woolf was one of the first serious readers to note that Lardner had "talents of a remarkable order," and she linked her recognition of his literary skill to his important "interest in games." She realized that writing about games solved "one of the most difficult problems" faced by writers in a young nation with only embryonic institutions and no civilizing traditions: "Games give him what society gives his English brother" (Woolf, 118). Poet John Berryman also paid tribute to Lardner's achievement of finding a stimulus to the fictive imagination on the baseball diamond. Like Woolf, he saw that the sport's "conflict" and "tension" produce just the "precariousness" that "makes a good subject" for fiction (Berryman, 420). These and other appreciators also praised Lardner's "mistrust" of "popular heroes with self-inflated egos" (Messenger, 109).

Disagreeing with these views was F. Scott Fitzgerald, an admirer, friend, and later neighbor of Lardner, whose final assessment is negative specifically because of Lardner's focus on baseball. Fitzgerald felt that his friend's work "fell short of the achievement he was capable of" because Lardner's most important formative experiences as a young man were found among "a few dozen illiterates playing a boy's game"; to Gatsby's creator, baseball lacked the "novelty or danger, change or adventure" needed for a writer's "schooling" (36–37). Mark Harris, through his character Henry Wiggin, is equally harsh. In *The Southpaw*, Wiggin dismisses Lardner's baseball fiction as "not about baseball at all" because of its satirical invective, which clashes with Wiggin's idealization of the sport (Harris, 32). Whether favorable or not, opinion about Lardner's baseball fiction has invariably been intense. It may be that the dichotomy of intense opinion stems from a split within Lardner himself, for he was both a "modern Puritan" who "admired order, restraint, and solid achievement" and also a "modern humorist" who found his raw material in the "disorder...spiritless profligacy, and wasted lives" of his Yankee compatriots at play (Messenger, 128; and see Oriard, 90–95).

One of the major reasons for the intensity of opinion about Lardner, and one which also explains the effectiveness of his work, stems from his skillful evocation of vernacular speech. This skill aligns him directly in the path of literary descent from Mark Twain and the local colorists of the late nineteenth century, especially those, like Bret Harte and George Washington Cable, who are remembered for their ability to evoke the sounds and cadences of several varieties of spoken American English. In this context, critic Richard Chase's observation about Twain's literary language applies:

Hemingway's well-known pronouncement that "all modern American literature comes from one book by Mark Twain called *Huckleberry Finn*" states a large truth, even though literally it is untrue. Wherever we find...a style that flows with the easy grace of colloquial speech, that gets its directness and simplicity by leaving out subordinate words and clauses, we [know]...this is the language of Mark Twain....[It] is literary because it is sustained beyond the span of spoken language to meet the requirements of a long story and because it is consciously adapted to the purpose of a novel. (Chase, 139–40)

Chase's remarks also describe Lardner's baseball fiction, which actually depends on the realistic rendering of American dialectal speech for its thematic substance and parodic realism. The linguistic realism serves as vehicle for the author's mordant irony about Yankee tastes and values. Maxwell Geismar has even gone so far as to suggest that Lardner's realistic treatment of idioms is so evocative in divulging his characters' subconscious motivations that it belongs within the same tradition of stream-of-consciousness fiction associated with Joyce, Woolf, and Stein (Geismar, 91).

Another key attribute of Lardner's prototypical style as a baseball fiction writer relates to the starkness of his narrative realism. Avoiding the embellishments of a figurative prose style, as well as the omniscient stance of a detached observer-narrator, Lardner achieved much of that realism by relying on first-person narration in most of his work. Moreover, because a great many of his stories are narrated by average, folksy characters employing what C. K. Messenger calls a "deadpan, laconic" vernacular (109), the realism hinges on a consistent point of view. In *You Know Me Al*, for instance, Jack Keefe describes what he sees and feels with the literalism of the innocent bumpkin he is. Similarly, in "Alibi Ike," the narrative style accounts for the consistency of viewpoint, which in turn contributes to the plausibility of its psychological study of Frank X. Farrell, a self-justifying baseball player who "never pulled a play, good or bad, on or off the field, without apologizin' for it" (Lardner, *Short Stories*, 35). "My Roomy" and "Horseshoes" also exemplify the way in which Lardner manipulated his chosen subject, his characters, and their distinctive idioms to disclose the symbolic and thematic thrust of his fiction.

Lardner's recognition of the centrality of recreation to U.S. American society provides his oeuvre with another of its definitive characteristics. Just as he saw beneath the functional aspect of colloquial speech to its esthetic mettle, he also perceived the complex significance of American leisure-time rituals. His finding Dionysus among the working bourgeoisie and frontier proletariat met slow acceptance among the literati of this infant nation. "There was a bias against Lardner because he was a sportswriter," wrote Jimmy Cannon, one of his peers, because "he was a guy who just wrote funny little stories about ballplayers" (in Wakefield, 84). Like Hemingway and Fitzgerald, however, who also wrote considerably about Americans trying to seize the day, Lardner wrote his most successful, well-focused satires about his idle, sporting compatriots earnestly spending their leisure "killing time," an activity described as "the essence of comedy" by philosopher Unamuno (who also thought the "essence of tragedy" to be the "killing [of] eternity" [Prologue to *San Manuel Bueno*, 1933]). Both *The Big Town* and "Gullible's Travels" concern couples vacationing among the snooty "high polloi" far from their accustomed niches in their everyday worlds. Similarly, "Carmen" and "Some Like It Hot" deal with persons "putting on airs" to try to obtain the social acceptance and prestige their boorishness denies them, an effort which, of course, seals them to their fate on the bottom rung of the status ladder. Accordingly, his sports stories (for example, "A Caddy's Diary" and "Champion") and, especially, the baseball fiction, demonstrate the writer's search for human truths among Americans captivated by and zealous about a divertissement that often ends up controlling them. "And the accuracy of this theme in Lardner's work," writes Geismar, an early appreciator, "strikes us more and more" with the passage of time (130).

Thus, Lardner's focus on baseball as subject and symbol arose from his awareness of the literary possibilities inherent in both popular idioms and popular forms of recreation. His unblinkingly realistic depiction of the sport's participants and its entire cultural ambience has the paradoxical effect of clarifying its metaphorical potentiality, because when he began to write seriously about baseball, it was "still in its romantic era" and even "on the threshold of glorification" (Seldes, 3–4). By presenting organized baseball starkly unadorned, Lardner helped reveal its manifold texture, which had been veiled by the curtain of romance woven by such baseball fictionists as

Gilbert Patten, "father" of boy hero Frank Merriwell, and Zane Grey, who experimented with a number of popular subjects before finding his forte in pulp westerns. "Into this parade stepped Ring Lardner with the pinprick report that baseball players were ignorant and childish and stupid. . . . He was doing a bit of debunking ten years before anyone had heard the word" (Seldes, 3–4). As a result of his precision in registering the vernacular, his narrative realism, and his concern with Americans at play, his work prefigured most later baseball fiction.

Although the truth of Jacques Barzun's observation that "[w]hoever wants to know the heart and mind of America had better learn baseball, the rules and realities of the game" may not be as encompassing today as it was in the first third of the century, its appropriateness for Lardner's era explains the writer's fascination with the sport and its social implications. He believed he knew the "heart and mind of America" down to its grimmest essence, and he sought to convey it authentically without any prettifying veneer.

In the epistolary novel *You Know Me Al*, first published serially in 1916 in the *Saturday Evening Post*, he zeroes in on the "rules and realities" of baseball as business, occupation, entertainment, and professional sport. Told entirely through the letters to Al from his "pal," protagonist Jack Keefe, a small town busher who gets called up to pitch major league ball for the Chicago White Sox, *You Know Me Al* follows the often distressing, sometimes joyful, usually funny vagaries of the young pitcher's career as he progresses form vigorous throwing ace to a prematurely rundown athlete with a woefully abused arm. The novel also presents the parallel disaster of Jack's personal life as country hayseed succumbing to the Dionysian temptations of the city before marrying Florrie, a gold-digger prototype of such later characters as Memo in *The Natural* and Angela Whittling Trust in *The Great American Novel*.

Lardner's chief strategy in *You Know Me Al* is to depict Jack Keefe's pathetic character and his narrow circle through reductive use of the first-person point of view. Remaining a small-time busher even after leaving the bush league, Jack seeks to portray himself in his letters as someone he is not, in Higgs's lexicon, the ideal sports hero and Apollonian paragon. The cliché of the title thus becomes the controlling irony of the book, as Jack's revelations about himself and the vicissitudes of big-league baseball give Al much more knowledge about himself than he intends, and it is negative.

I want you to be sure and come up here for the holidays and I will show you a good time. I would love to have Bertha [Al's wife] come to and she can come if she wants to only Florrie [Jack's wife] don't know if she would have a good time or not. . . . But be sure and have Bertha come if she wants to come but maybe she would not enjoy it. (Lardner, *You Know Me Al*, 84–85)

Such childish circumlocutions pervade Jack's letters, unwittingly proclaiming both his fundamental weakness of character and his wife's basic pettiness.

His boastfulness, miserliness, empty bravado, and stupidity slip through the self-aggrandizement and assertions of modesty, generosity, and maturity.

By the end of the book and the last letter from Canada, Jack's character is clearly known both to his pal Al and the reader. Self-knowledge and mature insight continue to elude him, however, making his fear of traveling to Japan for exhibition games at the end particularly poignant (216–17). Club owner Comiskey plans the trip in order to capitalize commercially on Japan's nascent interest in baseball, and Jack is one of the players who is deceived into going despite the probability of permanently ruining his disabled arm and also despite his genuine desire to stay with Florrie and their young son (190–210).

The tyranny of baseball as a commercial enterprise, suggested by the forced trip to Japan as well as by the many details showing management's predatory exploitation of its worker-players, glares through Jack's naïve descriptions of life in the majors. When he is sold to another club, for example, his utter lack of choice is all the more blatant because of the ingenuous way he describes his being bought, sold, and transported without any voice or recourse in the matter:

Well Al it's all over. The club went to Detroit last night and I didn't go along. Callahan [the manager] told me to report to Comiskey this morning and I went up to the office at ten o'clock. He give me my pay to date and broke the news. I am sold to Frisco. (49).

Here as elsewhere Lardner's delineation of the profit-making tyranny and pervasive social damage of American commerce gains its force from the victim's failure to perceive clearly either its causes or its agents. Later works examined in this study exhibit a similar cynicism toward baseball and business in America.

The depiction of baseball as sport in *You Know Me Al* ranges from references like those above, to Jack's description of a random assortment of memorable games and plays, to numerous allusions to actual players and teams. The latter allusions "enhanced the realism of Jack's letters," thereby underscoring the novel's theme of the busher's "overestimation of his own worth" (Messenger, 114). However, Lardner did not focus unduly on baseball data, for he knew that it was an absorbed part of his readers' experience, and that as long as he remained true to its outline he—in F. Scott Fitzgerald's phrase— "recorded the voice of a continent" (Fitzgerald, 36). Moreover, because his focus was on the personal and societal details and effects of professional sport as essentially a money-making industry, he was not compelled to expand on its athletic or statistical aspects.

In focusing on the personal, societal effects of baseball, Lardner let the narrative style shape the thematic message. Jack's first-person narration resembles a transcript of unedited colloquial speech, a technique that underscores his pathos as a character completely lacking in self-awareness:

Well I asked Callahan [the manager] would he let me pitch up to Detroit and he says Sure. He says Do you want to get revenge on them? I says, Yes I did. He says Well you have certainly got some comeing. He says I never seen no man get worse treatment than them Tigers give you last spring. (56)

Never aware of irony, whether directed at himself or others, Jack reveals himself transparently. The novel's distinctive features—its humor, its baseball realism, and the pathos of its bush league characters—are all riveted in the authenticity of the language. Lardner's contemporary, H. L. Mencken, recognized this technical feat when he commented that the writer "set down common American with the utmost precision, and yet with enough imagination to make his work a contribution of genuine and permanent value to the national literature" (Mencken, 424).

Typical of most baseball fiction, humor occupies a large part of Lardner's work, and it, too, derives much of its efficacy from his employment of a variety of American vernacular idioms. By masterfully individualizing the idiom to serve varying specific purposes, his humor embraces many different personality types from a broad expanse of U.S. society. For instance, Jack Keefe's choppy syntax, pervasive misspellings, and odd neologisms contrast with the more urbane, faddish locutions of the Hoosier businessman who narrates *The Big Town*. Likewise, the baseball player narrating "Alibi Ike" displays more worldliness than Keefe through his more complex, if rambling, sentences and his vividly detailed descriptions. All these idiosyncrasies of style contribute to the author's considerable comic achievement.

The humor in Lardner's work ramifies from his irrepressible use of satire and irony. These twin approaches undercut the materialism that had been prevalent in the United States at least since the Gilded Age, and that was loudly touted by the roar of the twenties in which Lardner lived. Blessed with the incisive eye of an artist, he was able to transform into fiction much of what he saw around him, but burdened by a pessimism rivaling Ahab's, Lardner frequently turned much of what he saw into unbridled invective. Thus, despite his not unimportant place in American literature, one of his prominent weaknesses as a writer stems directly from his savage, sometimes gratuitous use of parody, without assuming the "responsibility" of "suggest[ing] alternative and more humanizing worlds" (Messenger, 110). Another of his partisans acknowledges that his "one major fault" is his "habit of carrying a theme farther than it's worth" (Seldes, 6). Perhaps the problem is, as Allan Nevins suggested in an early review, that Lardner does not "identify himself with his characters . . . to show how even the moron has his relations with heaven and hell, to touch the deeper chords of life, love, and death" (in Patrick, 145).

Relatedly, his work also suffers from too staunch a reliance upon first-person narration, a stylistic technique that Henry James and others have pointed out as severely limiting a work's generic possibilities. Despite its idiomatic precision and sharply honed satire, *You Know Me Al* can, ultimately, expose only the pitiable limitations of the busher as a "blustering, abrasive lout" from beginning to end (29). It may be that the pettiness of Jack's immaturity, along with his evident intellectual limitations, preclude reader identification with him. P. A. Knisley, however, argues convincingly that the novel develops a readerly "sympathy" for the busher's oppressed professional situation and dismal personal condition (91–93). Still, the reader's sympathies remain limited, due to Jack Keefe's offensive social and psychological warts which assault one without pause throughout his narration.

Likewise, when Lardner's humor becomes tedious and repetitious—for example, in "Alibi Ike"—the problem again relates to the author's inability or unwillingness to create either omniscient or limited third-person narrators capable of some measure of mature detachment and judgment. Even in the omnisciently presented stories, however, the extensive, uninterrupted dialogue—which nearly replicates unvarnished first-person narration—impedes Lardner's evocation of the psychological subtleties and cultural nuances constituting his primary thematic interest. For example, "Travelogue," like so many of his short stories, brings the reader few deep insights because the author relies solely on the raw ore of the unenlightened, pathetic characters' words, without benefit of authorial comment, either directly through the characters or obliquely, through symbolism. In keeping with his oft-repeated statement, "I just listen hard," Lardner preferred to record what he heard and witnessed as literally as possible, and for him that preference resulted in unstylized first-person accounts (in Patrick, 38).

By parodying baseball and its constituent parts, Lardner also satirized certain American institutions and habits. He lampooned the myth that flourishing commerce in a democratic society produces even greater attention to democratic forms, as well as the myth that dedication to hard work leads to greater virtue and wealth. He showed that while club owners like Comiskey flourished at the box office, they tightened control over their players, whom they believed they owned like chattel. At the same time, the harder Lardner's players work, the poorer and more bitter most of them become. In this way, Lardner shed light on the complex multifariousness, especially the hitherto concealed unsavoriness, of the society that produced those institutions and customs. With few exceptions, subsequent writers of baseball fiction extended the radical implications of Lardner's vision to their logical and extreme ends.

LATER PULP FICTION

Except for Ring Lardner, the only other writer to undertake serious baseball fiction prior to World War II was Heywood Broun, another journalist. His baseball novel, *The Sun Field* (1923), presents a self-consciously "modern" love triangle involving a baseball hero modeled after the then nascent Babe

Ruth legend. In his characterization of Tiny Tyler, Broun anticipates the widespread appearance of the Ruth figure in later baseball fiction—for example, Roy Hobbs in *The Natural* and Gil Gamesh in *The Great American Novel*, as well as the controlling metaphor in *Babe Ruth Caught in a Snowstorm* and "My Life and Death in the Negro American Baseball League: A Slave Narrative." Broun's work also contains a bold female character, a hint of the bright and strong Holly Wiggen who appears much later in the Henry Wiggen tetralogy.

The plot of *The Sun Field* revolves around Tiny's unlikely marriage to the sophisticated Judith. Their love affair is closely chronicled by narrator George Wallace, the third member of the romantic triangle. Keyed to the vagaries of Tyler's marriage and to Wallace's status as a jilted observer, the novel is less interested in baseball than in the complex nature of modern, post-World War I adult relationships. The title conveys the author's interest in using baseball as a symbol for life outside the ballpark.

Judith invokes the "the sun field" when she tells George that "all of us alive should be set to looking straight at the sun and getting our light firsthand," rather than having it "reflected" through the "little cracked mirrors that we call novels and poems and paintings" (Broun, 65). But it is Judith, through her penchant for literary allusion and her cultivated taste for art, who proves to be the character most given to living in the reflected light of others' broken mirror shards. She even borrows the image of "the sun field," that is, that part of the outfield most exposed to the sun, from George, her impatient would-be lover. In her relationship with Tiny Tyler, however, Judith demonstrates a genuine and forthright passion for life directly experienced. Despite the promise, however, heavily weighted by Broun's treatment of amorous, religious, philosophical, and political themes, his slender novel's baseball metaphor collapses under the burden.

Although H. Allen Smith's *Rhubarb* (1946) does little to improve the overall quality of baseball fiction, its very existence—like that of *It Happens Every Spring*, discussed below—again gives evidence of the sport's literary possibilities and of the growing development of a fictional subgenre. Because *Rhubarb* is primarily a light farce about a cat who inherits a baseball team, with baseball holding an at best tertiary place in the tale, our main interest in it here is its overt allusion to the work of Ring Lardner.

H. Allen Smith's association of feline and sport recalls Lardner's "kitten and coat" saga in "A World's Serious" (1929), a short story which "convulsed" the country, according to F. Scott Fitzgerald (Fitzgerald, 24). His predecessor's astounding success with the ludicrous duo may have encouraged Smith to rely on it for his central comic premise. Also indicating the author's homage to his "old pal Ring" and to *You Know Me Al* are the letters he incorporates into the novel. Two long letters from Benny in chapter fourteen, and an even longer one in chapter twenty-one, present Smith's attempt to capture an authentic busher idiom. It fails, however, because the author's own voice, straining for a language he does not own or control, obtrudes through the misspellings, slang, and cartoon-like inanities to damage the vernacular tone. That Benny—unlike Jack Keefe—remains an uninteresting enigma even after three lengthy letters in his "own" hand indicates how short of Lardner's narrative mark Smith fell.

Only the baseball setting connects *Rhubarb*'s farcical cat-owned baseball team with the fantastic magic pitching fluid featured in Valentine Davies's *It Happens Every Spring* (1949). The storyline concerns the transformation of a nondescript university chemistry professor into pitching hero, Vernon "King Kelly" Simpson. Like the gimmick in *Rhubarb*, Professor Simpson's clever laboratory invention permits him magical success as a pitcher for the St. Louis Cardinals, including a no-hitter as well as the ultimate glory of pitching the winning game of a World Series. Davies based his novel on a movie comedy produced after World War II, when Hollywood sought to capitalize on the public's hunger for homespun Yankee entertainment as an escape from war news. *It Happens Every Spring*, movie and book, thus ignored serious moral or ideological postwar themes, and, like *Rhubarb*, it employed baseball merely as a facile gimmick.

For the plot and theme of *The Year the Yankees Lost the Pennant* (1954), Douglass Wallop chose a variation of the Faust story, whose moral complexities have long made it a favorite subject for artists and writers. In generic outline the story tells of an erudite doctor (or scientist) who sells his soul to the devil in exchange for perpetual youth, great knowledge, and power, often supernatural. Among the celebrated versions of the legend are Marlowe's drama *Dr. Faustus* (1593), Goethe's masterpiece *Faust* (1808, 1833), and Mann's modern novel *Doctor Faustus* (1947). Faustian characters and re-creations appear as well throughout American literature in figures like Melville's haunted Ahab, Hawthorne's obsessed Goodman Brown, and, in this century, in Faulkner's dynastic minded Colonel Sutpen, as well as, recently, in the Fausto Tejada character in Ron Arias' prizewinning novel, *The Road to Tamazunchale* (1975).

Despite the lofty literary history of its prototype, Wallop's novel, a comic work of popular entertainment, lies within the sub-literary tradition. Successfully reworked as *Damn Yankees*, a Broadway musical comedy, *The Year the Yankees Lost the Pennant* tells the story of inoffensive Joe Boyd, a protagonist notable for his rare integrity and honest conduct. The author closely links Joe's moral attributes to his uncommon love for baseball, which is depicted in idealized form as a model of scrupulous order, noble honor, athletic prowess, and simple beauty. Boyd's "honorable" moral purity makes him natural bait for the novel's Satan figure, Mr. Applegate, while his "worthy" desire to "deny the Yankees a tenth consecutive pennant" provides him with just the thirst for power needed to attract Applegate's ravishing interest (Wallop, 25–26). Steadfastly seeking to topple the Yankee dynasty by saving the Washington Senators' chances for winning the American League pennant thus constitutes the hidden infirmity in the protagonist's average-Joe decency.

The Faustian motif underlies the novel's thematic framework of appearance versus reality. Indeed, as summarized in chapter one above, the history of baseball has been fraught with confusions and deliberate discrepancies regarding the game's true origin, making Wallop's appearance/reality polarity an especially fitting theme. To convey the theme Wallop creates a series of contrasts and dualities which recur throughout the plot in mirroring patterns. They include Joe Boyd's double Joe Hardy, a celebrity ballplayer alter ego who fulfills his every boyhood fantasy, as well as Roscoe Ent, the foolish pitcher who serves as moral antithesis of Boyd and Hardy. There is also Boyd's wife Bess, who is doubled by Lola, the femme fatale who balances the plot by enchanting Hardy.

The link between the novel's thematic polarity and baseball lies in the problem of Joe Hardy's "birth," a plot twist which parallels the contrived myth of baseball's reputed American origins. It is neither entirely true nor entirely false that the magically created Hardy hails from Hannibal, Missouri— birthplace of Samuel Clemens, also known for his twin identity as Mark Twain. And, of course, Twain's oeuvre is famous for its many character doubles (for example, Tom Sawyer and Huck Finn, August and Emil in *The Mysterious Stranger*, Chambers and Tom in *Pudd'nhead Wilson*), thus adding texture to Wallop's tidy allusion to the traditional canon of American literature. At the end of the book Wallop compounds the problem of Hardy's strange birth with another confusion of appearance and reality: Applegate creates a replacement for Hardy, a player whose corporeal appearance identically matches Hardy/Boyd's even though his essential inner form is very different. Wallop thus concludes with a sly wink at the reader.

Equally sly—despite its serious implications—is the novel's treatment of good and evil. Applying this weighty moral polarity to baseball is a funny idea, and Wallop magnifies it by adding religious import in vast, mock-heroic proportions. The following glimpse into Boyd's thoughts reveals the comic tone:

It was his love of baseball. That was all. It was comforting, at any rate, to know that it wasn't because he had established a reputation for evil. And yet, the evil aspect... seemed the least consideration of all.... Could an aim so worthy as denying the Yankees a tenth consecutive pennant be evil?... It was a worthy case.... It was like a crusade. (25–26)

Applegate also knows that lurking inside Boyd's noble support of the Senators is the more selfish desire for public glory. As Joe tells himself, "a crusade, by its very nature, mean[s] glory as well as sacrifice" (26). True to his satanic nature, Applegate plans on breaking his pact with this ingenuous Faust figure by keeping the Yankees victorious even after he transforms Boyd into baseball star Joe Hardy, a deception that further complicates Joe's crusade.

Wallop underscores the deceptive "fine print" in Boyd's contract with the devil through the character of Roscoe Ent, the clown/pitcher who "had been signed on by the [Senators] as an attendance booster a number of years before" (70). The exact antithesis of the hero, Ent's behavior contrasts sharply with Boyd's, and thereby highlights the latter's supposedly greater virtue. Their polar roles effectively exemplify the novel's thematic concern with the conflict and gap between appearance and reality. Because of his antagonistic conduct, Ent appears to be the team's demon, while Hardy/Boyd appears to be its savior—its amazing folk hero, in Andreano's terms. Paradoxically, Boyd is, in fact, the devil Applegate's agent and Ent—until Hardy arrives—has actually been the team's box-office "salvation" (70).

A similar emphasis on the appearance/reality theme arises from Hardy's ephemeral romance with Lola, an embodiment of beauty as illusion. In one sense theirs is the perfect romance, epitomizing beauty and compatibility; but because they are wraithly products of Applegate's magic, they are actually "like two people without substance...they were two people who did not exist" (135). Hardy eventually recognizes the "evil" of knowingly yielding to the "temptation" of the enchanting illusion and of preferring it to the more painful, more mundane reality. Importantly, the clarity of his discovery that "the realities were just the opposite" shocks him into realizing the greater value of his "real" life as Boyd (203). By allowing the aptly named Boyd fulfillment of his puerile baseball fantasy, Applegate forces him into a world of play where the rules of conduct nonetheless have real consequences outside the diamond. Wallop thus neatly links America's religious heritage with baseball's commercialism through the biblical overtones of Applegate's name—he is a portal to knowledge like the apple in Eden.

Hardy/Boyd's mature but troubling insights into himself and life explain his extraordinarily deep sleep when he returns home to his wife Bess. Consistent with the book's reversals of appearance and reality, his first genuinely restful, healthy sleep in weeks is interpreted oppositely, as illness. In the same vein, Hardy's uncompromising "decency," which enabled him to resist Applegate, appears to be quite the opposite when after successfully wooing Lola, he "cruelly" rejects her. "And the thing about all this," thinks Boyd, "is that nothing ever gets better. Always and always worse" (221). Unlike Rip Van Winkle, his American literary prototype, Boyd cannot totally avoid responsibility in his escape from the daily routine of wife, family, and work.

Wallop fittingly concludes his story about the effect of life's paradoxes on a latter-day American Faust with a faint echo of Hawthorne's ambiguous query at the end of "Young Goodman Brown." At the end of his story, Hawthorne asks: "Had Goodman Brown fallen asleep in the forest and only dreamed a wild dream of a witch meeting?" Similarly, at the end of *The Year the Yankees Lost the Pennant*, as Boyd returns home after his mysterious absence while playing Hardy, we are told:

The house was dark. There was no bridge game tonight.
    He approached slowly, with an eerie feeling. Except for the change in the weather,

except that there was no sound of Old Man Everett watering his azaleas, this might be the same night in July. (250)

But ambiguity does not prevail. It is a different night, for both Joe and Bess are different, as Goodman and Faith were changed after their night in the forest. Unlike their Puritan counterparts, however, who were made tragically bitter by their encounter with evil, as comic figures the Boyds' transformation by "evil" brings wisdom, harmony, and a reawakening of their love. Thus, Wallop's protagonist progresses from the boyish immaturity of the Apollonian type to the self-indulgent naturalness of the Dionysian figure, and emerges finally as a true Adonic hero.

Although published within a year of each other, a vast thematic and stylistic gap separates *The Year the Yankees Lost the Pennant* and Eliot Asinof's *Man on Spikes* (1955). Lying within the boundaries of naturalism, Asinof's novel concerns the attempted rise to stardom of talented baseball player Mike Kutner, son of an immigrant coal miner. Employing a detached omniscient narrator, who relates Kutner's story through the perceptions of other characters important in his life, Asinof portrays Kutner as an old-fashioned player in the style of Ty Cobb, but one who cannot achieve the celebrity he so passionately desires because of a series of mishaps. For example, he is waylaid by three years of Army duty during World War II, and he is also bypassed by the owner in favor of a black athlete recruited begrudgingly as a token of integration.

Through these misfortunes Asinof exploits every convention of naturalism in its most extreme form. To begin with, the naturalistic motif of an arbitrary fate shaping human lives accounts for Kutner's career, which gets its start through a fortuitous coincidence that momentarily overcomes the handicaps of a squalid childhood. The determinism that gives naturalism its characteristic pessimism appears in the novel in the form of manifold oppressive external forces—primarily owners and fans that entangle Kutner in his hapless fate. Making decisions based solely on market economics, for instance, the distant owners exhibit little care for the welfare of their all too human workers. Similarly, the fickle fans celebrate players on the basis of a usually phony public image and not necessarily because of genuine talent or substance. Anchored to the tradition of fatalistic naturalism, then, *Man on Spikes* ends with the protagonist's failure in major-league ball, a failure made particularly poignant by Kutner's objective merit as a player and his hard-earned maturity as a person.

Returning to the escapist entertainment of *The Year the Yankees Lost the Pennant*, but with considerably less polish and depth, Bud Nye's fantasy, *Stay Loose* (1959), presents the same polarity of good and evil found in Wallop's Faustian tale. On the side of good are the "genuine appreciators of the national sport," including the narrator Ferriss Bracken and his loyal supporters (Nye, 15). On the other side are the avaricious entrepreneurs, like Bracken's boss, who view baseball strictly as commerce—"based on money chasing," in the words of a 1922 *Sporting News* editorial on baseball (in Crepeau, 105). Into this moral dichotomy Nye integrates the theme of the nobility of nature; baseball, synonymous in the novel with natural primal innocence, represents that which is most good. Accordingly, the title cliché, "stay loose," serves as a caveat against the negative tension that impedes effective game performance as well as a maxim against moral rigidity.

What keeps *Stay Loose* among the same cluster of frivolous farces as H. Allen Smith's *Rhubarb* is its computer gimmick which, to its credit given its 1959 publication date, was quite forward-looking. The gimmick features the computerized selection of an outstanding team of baseball players from a distant tropical island, a device that produces comedy of farfetched proportions. The amusing incongruity of baseball in the jungle also looks ahead to Roth's hilarious jungle scenes in *The Great American Novel*. Despite its too blatant handling of the theme of baseball as Edenic innocence, and the gratuitous frivolity of its humor, *Stay Loose* gives further evidence of the consistent interest in baseball among writers and readers throughout this century.

In the same style of comedy is *A Pennant for the Kremlin* (1964), by Paul Molloy. Borrowing part of his plot from *Rhubarb* in which a rich owner dies and leaves his baseball team to a cat and the remainder of his wealth to a woman, Molloy works from the well-worn notion that universal goodwill can be achieved more successfully when anchored to sports, especially baseball. Molloy thus mines the familiar belief that sports-based foreign exchange programs promote crosscultural understanding, harmony, and peace, a pleasant sentiment unfortunately undermined by the reality of geopolitics and centuries of unmitigated conflict.

A Pennant for the Kremlin develops the premise that so long as baseball is "popular with the people, the [dictator] Kaisers and the [communist] Trotskys" will "strike out" (Crepeau, 25). The important aim, suggests the novel in seriocomic tones, is "to arm" Russia with baseball, "the democratic game." To achieve this aim, the fictive owner of the Chicago White Sox bequeathes the team to the Soviet Union. He also leaves a sizable fortune to his faithful secretary Adelaide, who remains with the team throughout the trying, often funny, vicissitudes of Russian ownership. Consistent with baseball fiction's local-color origins and use of vernacular speech, Molloy makes a clumsy attempt to blend a contemporary busher idiom with Russian-accented English. Ultimately the story produces a few individual Russo-Yankee friendships, along with the knowledge that even baseball is powerless as an agent for peace.

Appearing a year after Molloy's book, *Today's Game* (1965) by Martin Quigley differs markedly in the realism of its style and the seriousness of its treatment of baseball. The plot of the novel is constructed entirely around one game that is crucial for the novel's protagonist, Barney Mann, manager of the professional Bluejays. Played against a team that bested Barney in an earlier player trade, the game caps a nine-game losing streak for the Bluejays.

Barney feels he must win the game in order to revitalize team morale and to enhance his coaching effectiveness. The novel's most compelling chapters deal with the actual playing of the game, and with Barney's clever strategy, which is counteracted by the equally astute opposing manager. By providing extensive, authentic details about baseball play and strategy, Quigley achieves a literary realism that compensates for some of his work's weaknesses of dialogue and characterization. That verisimilitude also offers a natural link between *Today's Game* and William Brashler's realistic *The Bingo Long Traveling All-Stars and Motor Kings*, published in 1973.

Brashler's novel tells a story about the Southern Negro leagues before World War II, a relatively unknown period of American history. The work's hero, Bingo Long, is a talented catcher for the Ebony Aces, a team owned by Sallie Potter, depicted as an unscrupulous owner, a stock figure in baseball fiction. Spurred by Potter's chicanery and greed, Bingo's qualities of leadership emerge early in the novel when he decides to create his own team from players disgruntled with Potter and other mercenary owners. Even though he is adept in managing the athletics of the team, the Traveling All-Stars and Motor Kings, Bingo lacks both skill in business management and emotional maturity. Before long, his team's success on the diamond spoils him and elicits his basic impulsiveness, indiscretion, and impracticality. The novel's end suggests that the team will disband and that the All-Stars will likely rejoin Potter or another of his commercial counterparts.

Bingo Long's failure as a team owner nonetheless proves his success as a player and individual. The meticulous descriptions of games, as well as the extended delineations of the team's trials outside the ballpark, repeatedly highlight both Bingo's athletic talents and his personal attributes. Oppositely, Bingo's maladroit attempts at managerial gains, usually at the expense of team members, undermine his development as a mature adult. For Brashler to have concluded the book on a happier note—for instance, with a victorious barnstorming tour—would likely have meant transforming the title hero into another Sallie Potter.

As an historical account of baseball's Negro leagues and barnstorming era, *The Bingo Long Traveling All-Stars and Motor Kings* fictionalizes much of the sociohistorical information chronicled by such observers as historian Robert Peterson in *Only the Ball Was White* (1970) and journalist Donn Rogosin in *Invisible Men: Life in Baseball's Negro Leagues* (1983). Both these books expose the insidious contamination of all aspects of baseball by the same fundamental racism pervading American society, and both focus on the achievements, remarkable and commonplace, of Blacks who were arbitrarily excluded from the sport. Accordingly, one point in Tony Tanner's excellent discussion of Ralph Ellison's now classic *Invisible Man* (1947) is especially salient here. "It is an aspect of recent American fiction," writes Tanner, "that work coming from [or about] members of so-called minority groups" applies directly to the "situation of people not sharing their immediate racial experience." *Invisible Man*, Tanner continues, is not "limited to an expression of an anguish and injustice experienced peculiarly by Negroes," but is "quite simply the most profound novel about American identity" written since World War II (Tanner, 50–51). Although Brashler's novel may not be as profound a work as Ellison's, still it is luminous in the authenticity of its portrayal of the world of organized Black baseball, a world of athletic experience and social complexity deliberately excluded from the "official" sport. *Bingo Long Traveling All-Stars and Motor Kings* thus illuminates other baseball fictions because, as Tanner notes about *Invisible Man*, to understand visibility we must first comprehend its opposite (54).

Like writers from Lardner to the present, Brashler models his characters after actual baseball players, but his models are not as well-known, or as rooted in the entire culture's folklore, as are such whites as Ty Cobb and Christy Mathewson. This ignorance about the black players exists even though their play in the Negro leagues matched and often surpassed that of the dominant culture's major league baseball (Voigt, *From Gentleman's Sport*, 278–79). Models for Brashler's characters include such superb players as Josh Gibson, Rube Foster, and Satchel Paige. He also incorporates into the novel aspects of the organized Southern Negro Leagues (for example, the Chicago American Giants and the Cuban Stars) as well as of the Black minstrel entertainment world (for example, the Tennessee Rats and Miami Clowns [Rogosin, 10 and 150]). Without reader recognition, however, these associations and allusions are meaningless—a fact that introduces an issue peculiar to this novel.

Brashler's novel differs significantly from other baseball fictions in that the world of Black baseball and its wider social ambiance is immediately distinguishable from that in other baseball novels, a distinction which automatically, and also quite dramatically, affects its form as baseball fiction. To begin with, no matter how good they were, Black players historically could not rise above their status within the Southern leagues, for the ostensibly "real" baseball world—organized, official, and rabidly white—was closed to them (Peterson, 120–21; Voigt, 279). Consequently, the dream of playing in the majors does not appear in *Bingo Long Traveling All-Stars and Motor Kings* even though that dream constitutes a grail-like motif in most other baseball fictions. Similarly, in contrast to the characters in other works examined here, Brashler's actors are more concerned with the game's financial rewards than with its less practical, romantic ones. His characters' preoccupation with game promotion and gate receipts is grounded in the history of black players struggling to eke out a living from the game.

Romance eluded the Black players even in the matter of statistics and record-keeping, that essential written aspect of the game, for official, definitive records for Black teams and players did not exist until after 1940. Because they were forced to earn their living by playing against both amateur and semiprofessional as well as professional teams, often several times a day, professional Black players were unable to record meticulously their games' statistics. Moreover, since one purpose of recordkeeping is comparison, and since the Negro leagues were not recognized by the Baseball Commissioner's office, there was little to be gained, either in image or profit, by keeping statistics on the Negro teams. In his fiction, then, Brashler remains faithful to his historical subject by ignoring the literary convention in baseball fiction of reporting statistics and presenting box scores.

Finally, *The Bingo Long Traveling All-Stars and Motor Kings* differs from other novels in this study in its narrative language and episodic structure. Told by an omniscient narrator, dialogue dominates the novel, a feature that reinforces its realism. The dialogue in Southern Black dialects offers an idiom seldom seen in other baseball fiction. Furthermore, as Michael Oriard points out, "many of the encounters of white and black in the novel appear as 'Marster and John' stories of the old South; tricking the trickster became a way of survival" (224). Brashler's is the only work considered in this study, other than Smith's *Rhubarb*, having an episodic structure and lacking a linear plot and overarching theme. Built entirely around Bingo's newly formed team and its barnstorming tour, it is essentially a fictionalization of historical accounts of traveling Black teams, a subject viable for use as a quest theme, but which Brashler employs discursively to present a slice of little-known American life.

Offering a fictional slice of another little-known part of American life and baseball history is *The Celebrant* (1983), a novel about Jewish participation in the sport. Written by Eric Rolfe Greenberg, it relates the story of the fictional New York Kapp family (originally Kapinski), immigrant jewelers who, primarily through the actions of two sons, become integrally involved in baseball's commercial and promotional activities at the turn of the century. Although the sons, Eli and Jackie (originally Jacob), share their work in the family business and a great love for baseball, they are temperamental opposites. Extroverted Eli, the older brother, handles public relations for the firm, and he projects the kind of gregarious bravado masking insecurity that is typified in American literature by George Babbitt and Willy Loman. His interest in baseball is both social and mercenary, in that he enjoys taking clients out to the ball game and engaging in "friendly betting," having "staked this out as his personal form of business entertainment" (Greenberg, 17). By contrast, introspective Jackie is the family jewelry designer, and the novel's artistic sensibility. He is the celebrant of the title whose devotion to the sport is idealistically pure, even sacerdotal. He recognizes what poet Rolfe Humphries describes in "Polo Grounds" as the "highly skilled and beautiful mystery" of baseball (Humphries, 84). Indeed, Jackie's attitude toward the game in the novel exactly captures the reverential tone evoked by Humphries's persona in the poem.

What makes Greenberg's an effective baseball fiction is his integration of the literal facts of baseball history into the story. In particular, he skillfully interpolates the biography of legendary New York Giants pitcher Christy Mathewson into his account of the Kapp brothers and the family business. One of the first five inductees into the Baseball Hall of Fame, Mathewson is regarded as one of the greatest pitchers of all time, and still holds the record for most shut-outs in a World Series (three in 1905). His extraordinary success as a pitcher furnishes Jackie Kapp with an appropriate symbol of perfection in a disappointing world he finds "mundane and tawdry" (178).

Mathewson towered over a circle of admirers. He was no less at ease in black tie than in his playing togs.... He did not so much fit into his surroundings as define them; he seemed innately, magnetically right in every circumstance. My father-in-law ... declar[ed] his delight at the scene Mathewson had engendered. Grover Cleveland had never created such a fuss! (83)

This excerpt from one of Jackie's many rhapsodies illustrates Matthewson's unique gifts in the eyes of the commercial artist and very frustrated ballplayer. Central to the pitcher's role as a noble Adonic hero in the novel are his remarkable nonathletic attributes as a college-educated man with cultivated, humane tastes. Greenberg displays this side of Mathewson by having him employ literary allusion—for example, from Shakespeare and Alexander Pope—as easily and precisely as a fastball.

The Celebrant succeeds because of Greenberg's plausible treatment of the way business and industry have been part of organized baseball from its very beginnings. By paralleling the rise of Matthewson's career with the growth and expansion of the Kapp family's business, by interweaving their reciprocal public relations activities and Eli's increasingly heavier gambling, and concluding the novel with Jackie's disgust at the Black Sox fix, the author gives his narrative greater literary texture and weight. But even as the novel exposes this aspect of the sport it also celebrates—and does so without contradiction—baseball's "unreality," the fact that it is "all clean lines and clear decisions," that each day "produces a clear winner and an equally clear loser" and the next day "the slate [is] wiped clean"; and that at the end, when all is gone and the flesh dies, "the legends abide" (86; 268). Greenberg's fusion of celebration with condemnation reflects the clear-eyed respect for the game expressed by most of the writers in this study. They know that sport, like art, offers the stylized paradigms of perfection needed to make human imperfection bearable.

BASEBALL MOVIES: A FOOTNOTE

This chapter's examination of baseball fiction, from the 1888 ballad "Casey at the Bat" to the 1983 novel The Celebrant, demonstrates the public's consistent enthusiasm for the game for nearly a century. Whether on the diamond of professional sport or the rough of sandlot play, whether in juvenile stories or adult popular fiction, baseball has had an undeniably rich life both as recreation and in re-creation. Its popularity can also be gauged by the fact that many of the baseball novels published in the last sixty years were later converted into screenplays. The following list, taken from *The New York Times Film Review*, Phyllis R. Klotman's *Frame by Frame—A Black Filmography*, and miscellaneous periodicals, suggests the cinematic variety of baseball's movie history.

Casey at the Bat. 1899.

Take Me Out to the Ball Game. 1910.

Baseball, an Analysis of Motion. 1919.

Giants vs. Yanks. 1923.

Babe Comes Home. 1927.

College. 1927.

Slide, Kelly, Slide. 1927.

Speedy. 1928.

Elmer the Great. 1933.

Alibi Ike. 1935.

The Pride of the Yankees. 1940.

The Babe Ruth Story. 1948.

The Stratton Story. 1949.

It Happens Every Spring. 1949.

The Kid from Cleveland. 1949.

Take Me Out to the Ball Game. 1949.

The Jackie Robinson Story. 1950.

Kill the Umpire. 1950.

Rhubarb. 1951.

Angels in the Outfield. 1951.

The Winning Team. 1952.

The Pride of St. Louis. 1952.

The Big Leaguer. 1953.

Fear Strikes Out. 1957.

Damn Yankees. 1958.

Biography of a Rookie [Willie Davis]. 1961.

Jackie Robinson. 1965.

Bang the Drum Slowly. 1973.

The Bingo Long Traveling All-Stars and Motor Kings. 1976.

The Natural. 1984.

Bull Durham. 1987.

Eight Men Out. 1988.

Major League. 1989.

Field of Dreams. 1989.

A great many of these movies, and certainly the most artistic of them, were adapted from novels and stories. But their fictiveness extends farther back to a mythic origin, for they were drawn from a sport originating in a folk game that itself derived from ancient ritual. Contemplating baseball in film gives an insight into the multileveled permutations and complexities of play and *Homo ludens*.

This chapter thus discloses the emergence of the baseball metaphor from the raw vigor and folk energies found in the actual playing of the game to the stylized reproductions of that play crafted from the fictive imagination. In bouncing from the fields of actual play to literary scenarios, the baseball figure examined in this chapter reveals some of the reasons a scholar like A. Bartlett Giamatti, who left the presidency of Yale University to head the National League, finds "it the most satisfying and nourishing of games outside of literature" (quoted in *Newsweek*, 6–23–86, 81). The sections on "Heroes" and "Fallen Casey—or, the Tragedy of Humpty Dumpty" explain some of the characteristics and peculiarities of the United States' breed of national heroes and the culture that defines them. In "Ring Lardner" the baseball metaphor appears in its first complex treatment in full-length adult fiction, a literary complexity apparent in all the later works discussed in *Seeking the Perfect Game*. Finally, this chapter's penultimate section, "Later Pulp Fiction," surveys baseball's appearance for over fifty years in sundry works of popular fiction, demonstrating the metaphor's timeless vitality and unique salience to American literature.

REFERENCES

Ralph Andreano, *No Joy in Mudville: The Dilemma of Major League Baseball* (Schenkman, 1965).

Eliot Asinof, *Man on Spikes* (McGraw-Hill, 1955).

Jacques Barzun, *God's Country and Mine* (Vintage, 1954).

John Berryman, "The Case of Ring Lardner: Art and Entertainment," *Commentary* 22 (July–December 1956): 416–423.

William Brashler, *The Bingo Long Traveling All-Stars and Motor Kings* (Harper and Row, 1973).

Heywood Broun, *The Sun Field* (Putnam, 1923).

Richard Chase, *The American Novel and Its Tradition* (Doubleday, 1957).

Richard C. Crepeau, *Baseball: America's Diamond Mind 1919–1941* (Orlando: University of Central Florida Press, 1980).

Valentine Davies, *It Happens Every Spring* (Farrar, Straus, 1949).

Frank Deford, "Casey at the Bat," *Sports Illustrated*, 69:3 (July 18, 1988), 52–75.

Donald Elder, *Ring Lardner* (Doubleday, 1956).

F. Scott Fitzgerald, "Ring" (1933), in *The Crack-Up*, Edmund Wilson, ed. (New Directions, 1956).

Martin Gardner, *The Annotated "Casey at the Bat"* (Clarkson N. Potter, 1967).

Maxwell Geismer, *Ring Lardner and the Portrait of Folly* (Crowell, 1972).

Eric Rolfe Greenberg, *The Celebrant* (1983; reprint, Penguin Books, 1986).

Mark Harris, *The Southpaw* (Knopf, 1953).

Robert J. Higgs, *Laurel & Thorn*, 1981.

Rolfe Humphries, "Polo Grounds," in *Collected Poems* (Bloomington: University of Indiana Press, 1965).

Phyllis Rauch Klotman, *Frame by Frame—A Black Filmography* (Bloomington: University of Indiana Press, 1979).

Patrick Allen Knisley, "The Interior Diamond," 1978.

Ring Lardner, *The Best Short Stories of Ring Lardner* (Scribner's, 1957).

The Portable Ring Lardner, Gilbert Seldes, ed. (See below under Seldes.)

Lardner, *You Know Me Al* (Curtis Publishers, 1914; reprint, Scribner's, 1960).

Fred Lieb, *Baseball As I Have Known It* (Grosset and Dunlap, 1977).

H. L. Mencken, *The American Language* (1919), 4th ed. (Knopf, 1955).

Christian K. Messenger, *Sport and the Spirit of Play in American Fiction*, 1981.

Paul Molloy, *A Pennant for the Kremlin* (Doubleday, 1964).

Eugene C. Murdock, *Mighty Casey: All-American* (Greenwood Press, 1984).

Bud Nye, *Stay Loose* (Doubleday, 1959).

New York Times Film Review 1913–1968 (6 vols.) (Times and Arno Press, 1970).

Michael Oriard, *Dreaming of Heroes: American Sports Fiction, 1868–1980* (Nelson-Hall, 1982).

Walton R. Patrick, *Ring Lardner* (Twayne, 1963).

Robert Peterson, *Only the Ball Was White* (Prentice-Hall, 1970).

Martin Quigley, *Today's Game* (Viking, 1965).

Donn Rogosin, *Invisible Men: Life in Baseball's Negro Leagues* (Atheneum, 1983).

Gilbert Seldes, ed., *The Portable Ring Lardner* (Viking, 1946).

Harold Seymour, *Baseball: The Golden Age*, 1971.

H. Allen Smith, *Rhubarb* (Doubleday, 1946).

Allen F. Stein, "This Unsporting Life: The Baseball Fiction of Ring Lardner," *Markham Review*, 3:1 (October, 1971):27–33.

Tony Tanner, *City of Words: American Fiction* (Harper & Row, 1971).

Ernest Lawrence Thayer, "Casey at the Bat," *San Francisco Examiner* (June 3, 1888): 4. Also see Martin Gardner and Eugene Murdock entries above.

Voigt, *From Gentleman's Sport*, 1966.

Dan Wakefield, "The Purple and the Gold," *Atlantic Monthly*, 234 (August 1974): 84.

Douglass Wallop, *The Year the Yankees Lost the Pennant* (Norton, 1954).

William Carlos Williams, "At the Ball Game," in *Selected Poems* (New Directions, 1949).

Thomas Wolfe, *The Letters of Thomas Wolfe*, Elizabeth Nowell, ed. (Scribner's, 1956).

Virginia Woolf, *Collected Essays*, vol. 2 (Harcourt Brace & World, 1967).

3

LITERARY FUNGOES

Baseball is really two sports—the Summer Game and the Autumn Game.
One is the leisurely pastime of our national mythology. The other is not
so gentle.

Thomas Boswell, *How Life Imitates the World Series*

ALLUSIONS TO BASEBALL IN MAJOR WORKS

In 1981, doomsayers from Shea Stadium to Chavez Ravine prophesied that
the baseball strike in that year spelled the sport's certain demise. To these
pessimists, the labor dispute, in revealing the game's basic entrepreneurial
nature, was to complete the destruction begun by the 1919 Black Sox scandal.
Yet, when the resolution was reached and the bases were swept from Montreal
to San Diego, major league baseball's vitality, and spectator enthusiasm,
showed little sign that the worst was at hand. The optimists suddenly appeared
in the bleachers, peanuts in hand, and patiently resumed their box-score
computations. To these eternal hopefuls, baseball had weathered another
bout of inclemency with bouncing high spirit, just as they had known it would.
To them, baseball deserves a star on Old Glory, for it is as resilient as the
New Eden that produced it (if we overlook baseball's actual origin in ritual)
and the grand old nation that nurtured it. It must be stressed that both sides
of this opposition—the doomsayers and the optimists—share one important
trait, a fierce devotion to the game and to their perceptions of its purity.

These polar opposites exist in literature as well. They appear in the works
of such classic American writers as Mark Twain and Ernest Hemingway, to
name but two of many, where they occupy an allusive infield within the
thematic diamond of the entire text. To be sure, many writers have "per-

formed," as critic Arthur Cooper has written, "in Allegorical Stadium, finding in our national obsession with baseball a metaphor for the American condition" (*Newsweek*, April 30, 1972). It is the development of that metaphor in discrete patterns of allusion, as opposed to controlling figures of theme, that will be surveyed here.

To a great extent, the authors considered in this chapter share in *la recherche du temps perdu*, which Proust immortalized with the famous bite into the madeleine dunked into noon tea. For these American writers, baseball serves to trigger the nostalgic *recherche*. With the exception of Philip Roth, they have not taken their interest in the game beyond allusion into book-length treatments of the subject; instead, they have stayed within the borders of literary reference, allusion, and, for some, apostrophe, using baseball in a narrowly figurative rather than in a broadly thematic way. Thus, it is in their allusive treatment of baseball that the works examined in this chapter are like fungoes. That is, they resemble an important miniature version of the baseball novels, short stories, and poems which comprise the larger body of serious baseball fiction, just as fungoing represents a tiny but crucial part of the game.

For the most part, significant references to baseball in the major works studied here fall within the opposing categories sketched earlier: those that present baseball nostalgically as the very métier of America's golden age, and those that present baseball ironically to reveal mundane, regrettable truths about American life and human ritual. Usually the nostalgic view of the sport is presented from a temporal distance by the narrator who, in stressing the ideal, finely structured game played in spring or summer, offers a pastoral or lyric celebration of the sport's uplifting possibilities. On the other hand, the ironists seem to find most of the culture's flaws and venalities reflected in the game, a perspective that frequently results in biting social satires, along with apocalyptic portrayals of doom.

THE IRONISTS

Among the ironists, Mark Twain is the first major writer to allude significantly to baseball in his work. In *A Connecticut Yankee in King Arthur's Court* (1889), he presents an extended passage in chapter forty that places baseball out of temporal and spatial context to underscore its utility for parody. The novel is set in the stereotypically dark and dreary Middle Ages; Twain's ingenious Yankee protagonist, The Boss, departs from medieval custom by replacing a jousting tournament with a baseball game. Recognizing the libidinal release within ritual forms of play, The Boss draws effective parallels between jousting and the nineteenth-century sport:

It was a project of mine to replace the tournament with something which might furnish an escape for the extra steam of the chivalry, keep those bucks entertained and out of mischief, and at the same time preserve the best thing in them, which was their hardy spirit of emulation. . . . This experiment was baseball. (405)

Twain's Boss also grasps the underlying significance of another essential of ritual, the human need for group membership and peer approval through conformity, a characteristic stingingly disparaged throughout Twain's work. If the "best thing" about these men is "their hardy spirit of emulation," one knows the author is being ironic, and that he finds them accomplices in "snivelization," his term for a rigidly conformist civilization.

Twain takes another shot at baseball—and thus humanity—when the Boss describes the game played between the "knights" and the "sceptered sovereigns," literal nomenclature having comic effect. "Their practice in the field was the most fantastic thing I ever saw. Being ballproof [because of their armor], they never skipped out of the way"; if a ball hit a player "it would bound a hundred and fifty yards" (406). Through such exaggerations in a different age and society, Twain could satirize the very essence of baseball, which he saw as "the outward and visible expression" of the "booming nineteenth century" with its "raging . . . drive and push and rush and struggle" ("What Mark Twain Said").

In the same zany spirit, the Boss describes the "umpire's first decision," which was "'usually his last,'" for the players and spectators "broke him in two with a bat, and his friends toted him home on a shutter" (406). Surfacing through the hilarity is the umpire's hapless demise, which illustrates the familiar fickleness of people who supposedly revere the social rules and systems of authority they devise, but who ignore or attack them with hypocritical regularity. Twain knew that for all the respect traditionally paid to baseball's elaborate system of rules and regulations, it is "KILL THE UMPIRE!" that usually brings the greatest spontaneous example of group unity, as the following poem graphically suggests.

> Mother, may I slug the umpire,
>   May I slug him right away,
> So he cannot be here, mother,
>   When the clubs begin to play?
> Let me clasp his throat, dear mother,
>   In a dear, delightful grip,
> With one hand, and with the other
>   Bat him several in the lip. (in Voigt, *From Gentleman's Sport*, 189)

Like the poem's anonymous creator, Twain makes ironic use of baseball's clichés by interpreting them literally, a technique later writers of baseball fiction adopt for both comic and tragic effect.

Twain's caustic commentary on the human herd instinct continues in Sinclair Lewis's celebrated *Babbitt* (1922), a novel about the limitations of Midwestern middle-class habits and values. In the three brief paragraphs Lewis devotes to George Babbitt's "determined" interest in the game, he shows the ritual of play transformed into a ritual of mindless conformity. As a "loyal" supporter of hometown Zenith's team, Babbitt "performed the rite scrupulously. He wore a cotton handkerchief about his collar; he became sweaty; he opened his mouth in a wide loose grin; and drank lemon soda out of a bottle" (154). Making explicit his anthropological parody of such well-practiced human rites, Lewis tells us "the game was a custom of his clan, and it gave outlet for the homicidal and sides-taking instincts which Babbitt called 'patriotism' and 'love of sport' " (154). As discussed elsewhere in this book, the connection between the national spirit and fervor for the game is a staple of baseball fiction. It alternately reflects a perverted sense of nationalism (as in *Babbitt*), or a wholesome feeling of cultural pride (as in James T. Farrell's work), or a compound expression of both (as in Greenberg's *The Celebrant*).

Fifty years after *Babbitt*, another fictional fan, Harry "Rabbit" Angstrom, sleepwalks through another arid baseball ritual in John Updike's novel, *Rabbit Redux* (1971). Updike underscores the sad irony of an undesired afternoon at the ballpark by presenting not just one or two, but three generations suffering each other through a baseball game. Since part of the ritual requires enjoyment and fun, ex-basketball player Rabbit and his son and father-in-law work hard at the pretense:

But something has gone wrong. The ball game is boring.... There was a beauty here bigger than the hurtling beauty of basketball, a beauty refined from country pastures. ... Rabbit waits for this beauty to rise to him ... but something is wrong ... the players themselves seem ... intent on a private dream of making ... big money ... they seem specialists like any other, not men playing a game because all men are boys [that] time is trying to outsmart. (83–84)

Such limited introspection occurs for the immature Rabbit only when motivated by external forces such as television shows or the baseball game; the ineffectual protagonist obtains personal meaning only from the social landscape, and not from anything within himself. Even his tender "yearn[ing] to protect the game from the crowd," a potentially noble sentiment, is actually only a yearning for the indulgence others used to grant him as a boy and later as a star athlete. He does not realize that his disappointment in the game "whose very taste ... was America" arises from the disease inside his "microcosmic self" (83, 406). As one critic writes, through Rabbit, Updike uses "the sports spectacle" imaginatively to "illuminate cultural dilemmas within the arena" such as, for example, "the decline of heroism, the submission to authority, the loss of idealism, and the lack of personal validity to experience"; this treatment also sheds light on the culture "outside the arena" (Messenger, 315).

In his *Dreaming of Heroes*, Michael Oriard perceives Rabbit's characterization as an incarnation of the "timeless American myth" of the ever-youthful hero. He writes that the "childlike athlete exists against the grain of American society" and all "necessary human maturity," and yet that boy-hero—whether Rip Van Winkle, Natty Bumppo, Huck Finn, Holden Caulfield, or any other male who is nonchalant about adult responsibilities—"embodies a wish-fulfillment deep in the psyche of the American public" (Oriard, 166–67). As a former basketball star, Rabbit is never forced to discover adult forms of gratification and achievement because, as a boy, he was excessively rewarded for this youthful ability to win and to typify the public's image of a hero. No wonder, then, that in *Rabbit Run* (1960) he decides middle age and maturity are "the same thing as being dead" (Updike, *Rabbit Run*, 106). Many of his friends and neighbors and, especially, most of his former fans, would agree.

Unlike Rabbit Angstrom, the protagonist of Canadian writer Mordecai Richler's *St. Urbain's Horseman* (1971) has a mature understanding of the fluid relationship between self and society and, parodic worrier-type that he is, he meditates on it endlessly. Film director Jake Hersh is a wholly contemporary figure who comprehends the sham that accompanies human rituals, and his performance in them is therefore knowing and clear-eyed. "Sunday morning softball on Hamstead Heath in summer was unquestionably the fun thing to do. It was a ritual" (222). As with other parts of his life, Jack takes his fun seriously, seeking to direct it with the control and artistry he exerts on his movies.

For the softball game in chapter nine, Richler paints on a broad, inclusive canvas not unlike Twain's baseball episode in *Connecticut Yankee*. From the chapter's one-word opening, "Summer," to his play-by-play descriptions, Richler includes every convention of baseball fiction: the starting lineup sheet; hissing fans (in this case, the "abandoned but alimonied first wives"); baseball jargon, especially the terms hinting at sexual double entendre; inning-by-inning box scores; and meticulously detailed accounts of key plays. In this way, Canadian Richler builds upon and expands the satiric function baseball as literary device holds for his U.S. counterparts. The obsessive urge to conform, parodied in Twain's medieval game and in both Babbitt and Rabbit's public displays of camaraderie, appear in *St. Urbain's Horseman* as Jake's unrelenting commitment to the Sunday softball ritual. Richler's descriptive monotony is deflating, while his baseball references—like those of Updike, Sinclair, and Twain—sharply satirize the culture whose anomie they disclose.

The closing stanza of Robert Fitzgerald's poem, "Cobb Would Have Caught It," captures some of the poignancy of the ironic handling of the baseball allusion seen in this section:

> Innings and afternoons. Fly lost in sunset.
> Throwing arm gone bad. There's your old ball game.
> Cool reek of the field. Reek of companions. (Fitzgerald, 80)

From its wistful title in tribute to the glorious, departed Ty Cobb to its faintly bitter tone, the poem's references to baseball couple a tender regard for the game with a weary cynicism toward its natural flaws. The "lost" fly, the "bad" arm, the "reek" of sweat, even the "old" ball game, all belie the lyrical notion that the sport is somehow safe, immortal, perfect. It "would" be perfect, the poem hints, *if only....*

THE PASTORALISTS

In structure as well as in tone, the pastoral treatment of baseball is quite unlike the ironic. Whereas the ironists usually present their baseball allusions in distinct episodes easily excised from their respective texts, the novelists who write nostagically about the sport are markedly diffuse about it. As the previous discussion discloses, baseball serves the ironists in a way comparable to the use Hawthorne made of the scarlet "A" he found in the Salem Custom House and which he used as seed for his bleak romance depicting the "persecuting spirit" of seventeenth-century New England. For the pastoralists, on the other hand, baseball embodies the finest ideals and possibilities of the prelapsarian, so-called the New World.

One such baseball pastoralist is James T. Farrell, who wrote so extensively about his passion for the sport that his baseball references, speeches, and tributes even produced a separate book, *My Baseball Diary* (1957). An expert on the game and its history, Farrell fills his fiction with allusions to real players and actual games; in fact, "at least half of [his] short stories contain some reference to sport, particularly baseball" (Oriard, 10). Knowledgeable appreciation of his work thus requires coming to terms with the meaning of baseball in it.

Farrell's use of baseball begins in his Studs Lonigan trilogy: *Young Lonigan* (1932), *The Young Manhood of Studs Lonigan* (1934), and *Judgment Day* (1935). One of his most credibly drawn characters, Studs grows up to become the tragic protagonist of Farrell's novels about the "pervasive spiritual poverty" he saw affecting the immigrant neighborhoods of America (Farrell, *Studs Lonigan Trilogy*, xii). Baseball offers Studs escape from his oppressive environment, but the rowdy youth "usually didn't give a damn about baseball," even though he does occasionally take advantage of its momentary relief from the squalor of his Chicago home (Farrell, *Trilogy*, 99). The same outlet exists for Studs's Irish-American compatriot, the more virtuous and imaginative Danny O'Neill. Danny avidly identifies with baseball heroes, and filters some of his youthful fantasies through his perception of the game.

Interestingly, even in *Young Lonigan*, the first part of the trilogy, Farrell links baseball to storytelling as a means of showing the divergence between the ordered way of art, which saves Danny, and that of unregenerate nature, which destroys Studs. For example, in the latter half of *Young Lonigan*, Studs and the younger Danny play a "goofy baseball game" on the sidewalk: "Danny beat Studs four times. Studs didn't like to get beat at anything, so he quit playing" (*Trilogy*, 104). Both reactions disclose inveterate responses to life. Studs's petulance foreshadows his immaturity as an adult, while Danny's growing zest for the game is transmuted into enthusiasm for other life experiences, and represents the latter's finer sensibility.

Each of the baseball references in the trilogy illustrate the antihero's emotional shallowness and moral laxity, traits that are underscored in the third novel. In *Judgment Day*, the failure of Lonigan's entire life is captured poignantly in the scene where he, now a grown man, walks out of a sandlot game in characteristic angry frustration:

He should have gone on playing. He would have gotten into his stride.... He imagined himself driving a home-run over the center-fielder's head and then making one-handed and shoe-string catches in the outfield. He shrugged his shoulders, laughed at his sudden interest in baseball. (*Trilogy*, 187)

The passage reveals the infantile Studs's typical self-deception, which allows him to mask his underlying, painfully felt insecurities.

In his indifference to baseball and his stunted development as a man, Studs is the antithesis of Danny O'Neill, the author's surrogate persona and hero of his next series of Chicago-based novels—*A World I Never Made* (1936), *No Star Is Lost* (1938), *Father and Son* (1940), and *My Days of Anger* (1943). Early in the first chapter of *A World I Never Made*, we see young Danny's first contact with a world outside the troubled O'Neill and O'Flaherty households that constitute his life. That contact comes from a Chicago White Sox game, which offers him a respite from domestic strife. Danny's total immersion in the game arises partly from its being the only ordered, cleanly structured experience of his life.

He watched the White Sox, wearing white uniforms and white stockings, as they trotted on the field. Hugh Duffy, the manager, slapped grounders to the infielders and they scooped them up and tossed the ball around. The outfielders shagged fly balls that were fungoed to them. (*A World I Never Made*, 33)

The boy takes in every minute detail, wide-eyed in awe of the neatness and clarity of the scene on the diamond. Even the rising suspense and excitement over Ed Walsh's possible no-hitter (which actually occurred) do not lessen Danny's absorption of even the game's minutiae: he "watched Ed Walsh, his foot on first base, putting on his sweater. With eyes of adoration, he followed Walsh's fingers while the big fellow buttoned his sweater" (39).

Through Danny's perceptions of baseball at different stages of his life, Farrell guides us into the autobiographical character's innermost self. In one scene, the writer meticulously describes Danny's preoccupation with Steel's Baseball Game, a board-game replica that, in its employment as metaphor, looks ahead to Coover's *The Universal Baseball Association, Inc., J. Henry Waugh, Prop.*

Most of the time when he was blue or worrying, he found he could forget everything by playing Steel's. He could imagine that instead of playing he was really watching a regular big-league game. He decided to pretend that the White Sox and the Cubs won the pennant and to play the world series between them. (*No Star Is Lost*, 233)

Here as elsewhere, Danny as spectator as well as omniscient manipulator of events foreshadows the adult Danny as writer, once again emphasizing the link between baseball and art.

In *Father and Son*, the adolescent Danny has to face the difficult realization that he has not grown up to be the success he had hoped; the novel deals with that and its attendant crises. At one point he thinks: "yes, after kidding himself about his destiny, and having the nerve to think that he would be a star like Ty Cobb or Eddie Collins, he was a miserable failure. . . . He didn't have what it takes" (*Father and Son*, 233). But the normal despair of youth resolves itself in his anticipation of the writing career, which is delineated in *My Days of Anger*.

In this novel, Farrell presents his surrogate, now a hard-working college student, in a stage of experiment, questioning earlier beliefs, testing and discarding them, as he forms the basis for his life as an artist. Relying on frequent allusions to literature and philosophy to indicate the breadth of O'Neill's intellectual growth, Farrell's references to baseball are few but carefully placed. Baseball still serves O'Neill as an escape from the nagging concerns of his family's finances, and again, as in the earlier novels, Farrell ties the hero's intense intellectual curiosity to the sport: "When he'd been a kid and had spent so much time playing baseball in Washington Park, he used to chew grass. . . . That was such a long time off now. . . . Distance was measured, also, in ideas, in intellectual growth. . . . How many years ago was it in ideas?" (*My Days of Anger*, 234). Unlike Studs, Danny matures into a deepened, rounded adult, a development that is symbolized throughout the tetralogy by his respectful response to the game.

Parallel to the intellectual development and sexual experimentation that characterize this last phase of O'Neill's story is a heightened awareness of the limitations of America's traditions, especially the discrepancy between the country's declared principles and its government's consistent debasement of those principles over time. Accordingly, Danny's skepticism and emergent socialism are reflected in baseball. In answer to a question about "what America is," for example, O'Neill "sarcastically" replies that it is a "nation of frustrated baseball players" (*My Days of Anger*, 309). The reply shows his growth from the childhood romanticization of sports heroes to an adult recognition of the dangers of overvaluing folk heroes who are merely mortal athletes.

For Farrell the baseball trope is, therefore, simple. A character's boyhood love of the sport is a sign of health; indifference or hostility to it is a sign of dissolution and decay. Even the prolific writer's twenty-second novel, *Invisible Swords* (1971), reaffirms the use of the figure. Here an uncompleted family baseball game mirrors the unhealthiness in the Martin family's interrelationships, the work's thematic crux. Thus, for over four decades Farrell's treatment of baseball manifests a singular consistency—one that typifies the nostalgic, pastoral appreciation of his youth's "consuming enthusiasm" (*My Baseball Diary*, 29).

Baseball also symbolizes a golden time for another prolific writer, Thomas Wolfe, who for the most part describes it and its context rhapsodically. The tribute he offers in one of his letters summarizes his affection for the game:

In the memory of almost every one of us, is there anything that can evoke spring— the first fine days of April—better than the sound of the ball smacking into the pocket of the big mitt, the sound of the bat as it hits the horse hide; for me, at any rate, and I am being literal and not rhetorical—almost everything I know about spring is in it—the first leaf, the jonquil, the maple tree, the smell of grass upon your hands and knees, the coming into flower of April. And is there anything that can tell more about an American summer than, say, the smell of wooden bleachers in a small town baseball park. (*Letters*, 722)

A baseball game contains for Wolfe all the elements needed to form "living art" (Wolfe, 1935, 202). He sees it as dynamic and unframed, but sculptured nonetheless: "The scene is instant, whole and wonderful," he writes evocatively in *Of Time and the River* (1935).

In its beauty and design that vision of the soaring stands, the pattern of forty thousand empetalled faces, the velvet and unalterable geometry of the playing field, and the small lean figures of the players set there, lonely, tense and waiting in their places, bright, desperate, solitary atoms encircled by that huge wall of nameless faces, is incredible. (202)

Wolfe's baseball idylls touch the edges of literary apostrophe and encapsulate perfectly the pastoral view of the game.

Moreover, the allusions to the game reinforce this effect throughout Wolfe's fiction. Through the character Pearl Hines, a high-risk "bush-league ball player" in *Look Homeward Angel*, and Nebraska Crane, the aging big-league slugger, in *You Can't Go Home Again*, Wolfe reawakens powerful feelings for a nobler past. Nebraska Crane is especially effective as a touchstone of "the very heart of America," with his first name suggesting the country's simple heartland and his last name conjuring up its native fauna. Taken together, their untypicality conveys the athlete's Cherokee ancestry, which is intended to account for Nebraska's appealing openness and his complete detachment "from the fever of the boom-mad town" as well as the "larger fever of the nation" (*You Can't Go Home Again*, 80).

As Robert J. Higgs observes, "Nebraska is a superb creation, at once an archetypal figure" and also a "real" individual (Higgs, 126). Part of his realness comes from Wolfe's unromantic delineation of the hardships ballplayers face behind the scenes of their public adulation. For instance, Nebraska explains how the monotony and exhaustion of constant train travel quickly destroy the glamour of major-league routine: "I can tell you what telephone post we're passin' without even lookin' out the window. . . . I used to have 'em numbered—now I got 'em *named!*" (50). He also tells of the physical and emotional pain that comes when faced with the hard-edged realization that age is replacing the opposing team as the more merciless opponent.

Despite these and similar acknowledgments of the quotidian realities of baseball, Wolfe's allusions to the game are, overall, eloquently pastoral. Toward the end of George Webber's conversation with Nebraska in *You Can't Go Home Again*, Wolfe carves a crystalline image of baseball that seems quintessential to his idyllic perspective: "for a moment George looked out at the flashing landscape with a feeling of sadness and wonder in his heart— 'it all comes back'" (69). "It" is the solid good in America's roots, origins, and formative energy that baseball embodies.

Lefty Bicek, the protagonist of Nelson Algren's *Never Come Morning* (1941), experiences a sense of wonder akin to that of Wolfe's George Webber when Lefty daydreams about a future as a major-leaguer. Unlike Wolfe's hero, however, Lefty is trapped in the sordidness of life in a Polish immigrant ghetto. Writing in the naturalistic tradition, Algren casts his main focus on the predacious urban environment, that tyrannizes countless "insignificant" people like Bicek, whom he describes as another "American outcast." Through the squalor of his environment and the meanness of his outlook, Bicek shares many personal characteristics with Asinof's Mike Kutner in *Man on Spikes*, and with Farrell's Studs Lonigan and Danny O'Neill. Algren also echoes Farrell's observations about his own work when he states that *Never Come Morning* "attempted to say . . . that if we did not understand what was happening to men and women who shared all the horrors but none of the privileges of our civilization, that we did not know what was happening to ourselves" (Algren, xiii).

In the early chapters of the story, baseball appears as the only affirmative, if fantasized, psychological release from Bicek's depressing surroundings. When he escorts a date to the amusement park, for example, he suffers such a revulsion to the engulfing sleaziness that escape to his private baseball idyll is his only alternative.

He stood in the spiked sand of the pitcher's box. The green stands and the cropped grass. And the striped sun on them. To be a man out there in the world of men. To be against men, and other men, with a hard-fought game behind him and the hardest hitter in front of him and the trickiest runner behind him; pouring sweat, the breaks going against him, aching in his great left arm every time he raised it; but forever in there trying, not caring that they were all against him. (Algren, 54)

Even though the ballgame offers, in Robert Frost's words, a momentary stay against confusion, Lefty's fantasy is colored with the same social paranoia that plagues him and his ghetto brethren. To Lefty, baseball represents a green and sunny world of possibility and choice. This explains why allusions to the sport disappear from subsequent chapters, after the now beaten protagonist has ceased dreaming and, instead, has succumbed to a seductive but ultimately futile life of crime.

Ernest Hemingway's contribution to the allusive treatment of baseball parallels Algren's, but without the same naturalistic doom. Hemingway wrote about baseball as he wrote about other physical endeavors such as hunting, fishing, boxing, and bullfighting. In his fiction, sport functions as a metaphor for life and for one's style of coping with it. To Hemingway, performance in the arena reveals the essence of the player's personality and integrity. Appropriately, in his last major completed fiction, *The Old Man and the Sea* (1952), the grizzled "Papa" of all macho artists singles out baseball as the "one thing," other than mind itself, that his hero Santiago has left when his body, now utterly fatigued, can barely move.

"Don't think, old man," he said aloud, "Sail on this course and take it when it comes."

But I must think, he thought. Because it is all I have left. That and baseball. I wonder how that great DiMaggio would have liked the way I hit [the shark] in the brain? (99)

Santiago/Hemingway's view of baseball crystallizes the pastoral view into a touchstone of perfection and hope against even debilitating odds.

A pastoral view of baseball also appears in Irwin Shaw's *Voices of a Summer Day* (1965). In this novel of Jewish identity, the sport functions as a unifying metaphor for the protagonist's recollections. While watching his son play baseball on an afternoon in 1964, Benjamin Federov recalls "the distinct, mortal innings of boyhood and youth" (12). By means of flashbacks to his youth interspersed with memories of more recent adult experiences, Federov's life emerges, mosaic-like, during the course of his son's game, which occupies three-fourths of the novel.

The American sounds of summer, the tap of bat against ball...the umpires calling "strike three and you're out"....The generations circle the bases, the dust rose for forty years as runners slid in from third, dead boys hit doubles, famous men made errors at shortstop, forgotten friends tapped the clay from their spikes with their bats as they stepped into the batter's box. (12)

Besides serving as a Proustian "madeleine" to catalyze a review of Federov's life, the game also works as a cultural motif anchoring him, a Jew burdened by the Diaspora, to America.

It was at the Polo Grounds that Benjamin, aged six, had watched *his* first baseball game at his father's side. An uprooted people, Federov had thought half-mockingly, we must make our family traditions with the material at hand.... So, bereft of other tribal paraphernalia, he took his son to the Polo Grounds. (152–53)

In view of baseball's specific role in the novel as an agent of cultural assimilation, the assertion by folklorist Tristram Coffin that "football or basketball or a piece of cake would have served" Shaw as well is totally groundless. Shaw depends on the sport's essence and history as *the* American game to provide the cohesive weave, in loco parentis as it were, for his tale of contemporary American life.

The baseball metaphor coheres the novel's nonsequential chronology and diverse historical references. Accordingly, Shaw ends his novel with a low-key allusion to the sport's crucial importance as an emblem of culture. After Federov's private drama at his son's game has elapsed, he returns home and is greeted by his wife's casual query, "What did you do all afternoon?" The novel ends with his simple reply, "I watched a ball game" (223). The pregnant suggestiveness of the understated conclusion emerges solely because the game he watched was baseball, the sport whose cultural density makes it a very metonym of his adoptive land.

Another writer interested in baseball is Jay Neugeboren, whose work gives "highly focused" accounts "of New York's Jewish boys and tradition-haunted old Jewish men, of its Black and Puerto Rican streetwise youth, of its bookies and its sports" (Candelaria, "Ethnic Fiction," 71). Author of an acclaimed first novel, *Big Man* (1966), about a college basketball player caught in a gambling fix, Neugeboren nimbly incorporates references to sports, especially baseball, into nearly all his five other novels, his short stories, and his autobiography. Because Neugeboren's characters and settings are typically etched in the poverty and gloom of overpopulated urban neighborhoods, his treatment of baseball resembles those of Farrell and Algren without, however, their pessimistic naturalism. That is, he employs the game as a pastoral emblem of order, stability, clarity, and hope in a chaotic world of poverty and deprivation.

Baseball figures prominently in *Corky's Brother* (1964), a collection of thirteen stories, mostly about slick urban adolescents, that "recreates an ethnically mixed Brooklyn neighborhood ... and pays loving attention to the forces that shape the young boys as they struggle to manhood" (Oriard, 201). A few stories in the book also deal with other subjects, from committed public school teachers seeking to spark a desire to learn among illiterate, drug-using New York students, to a Kentucky family's trip to a funeral, reminiscent of Faulkner's gently macabre humor in *As I Lay Dying*.

In "Luther," the first-person narrator, Mr. Carter, a white teacher at Booker T. Washington Junior High School, befriends Luther, a black student described by the other teachers as "a mental case." Their mutual enjoyment of baseball helps cement their relationship, even though mutual trust ultimately eludes them. Neugeboren exposes Luther's continuing distrust and suspicion in a scene at the Giants versus Cardinals game: the youth tries to bait the teacher with his proposal that instead "of an All-Star game every year between the leagues, what they ought to do one year is have the white guys against our guys" (15). When Carter refuses to be baited into an argument about race, Luther's real worry surfaces in his query, "would you have ever offered to help me if I wasn't colored?" (16). In the excitement of the game's ending, Carter does not specifically answer, forcing the reader to recognize that their bond is, in fact, based on both color and socioeconomic class.

Still, an afternoon of watching Willie Mays ("Ain't he the sweetest!" [16]) and the Giants win helps smoothen the social barriers. The baseball metaphor sharpens Neugeboren's theme about the dehumanizing effects of racism and poverty on everyone, especially its poorest victims, and it also helps focus the story's depiction of Luther's development from disruptive class clown and juvenile felon, to gang leader and convict, and eventually to a clean-living, activist Black Muslim.

Three other stories in the collection contain significant references to baseball, and because of their shared subject, setting, and characters, they comprise a narrative unit, despite their separation into discrete titles. The trio—"The Zodiacs," "Ebbets Field," and "Corky's Brother"—deals with adolescent boys growing up in Brooklyn. Baseball serves as an outlet for their natural restless energy and also as an entertainment in which players, especially the Dodgers of Ebbets Field, offer ideal heroes to worship. Narrated by Howie, an observant, athletic Jewish boy who functions as the author's surrogate, the stories describe an interesting array of teenage experiences, and Howie's sensitive responses to them. Like Neugeboren, who once described his first career ambition as becoming a professional baseball player, Howie is an avid, knowledgeable fan of the game who dreams of being a major-league star even as he proceeds studiously through high school and, eventually, a liberal arts college major (Neugeboren, *Corky's Brother*, 58).

In "The Zodiacs," for instance, Howie tells of the confrontation between his friend Louie Hirshfield, an unathletic whiz kid who organizes a boy's "professional" baseball team, and George Santini, "the best athlete in our school," who is also the "leader of . . . the most dangerous gang the world had ever known" (52). Although "tough" George publicly humiliates "sissy" Louie, the latter refuses to be intimidated, thereby becoming the unlikely leader who manages to elicit courage and team spirit from the other boys. Moreover, as Oriard points out, in this and other of his sports stories, Neugeboren "create[s] a muted tension" in his portrayal of the Louies and Howies—"the nonathletes and the athletes"—of America, one that "deals touchingly but unsentimentally" with the nonathlete in a sportsminded society (201–202).

Quite different in theme, "Corky's Brother," the longest story in the collection, presents Howie's sexual initiation in the context of the death of his

best friend's older brother. Although it focuses on Howie's and Corky's reactions to his death, the brother, a ballplayer in the Dodger farm system,
lives in the story as a hero worshipped because of his fame both as a paid
athlete and a sophisticated womanizer. Skillfully weaving sports and sex together to explore their manifold rite-of-passage significance, Neugeboren also
subtly threads the subject of death into the theme. At the funeral for Corky's
brother, for example, Howie suddenly becomes aware of his own mortality
when he "realize[s] that some day I would be standing in front of a box
looking in at one of . . . the guys I'd spent a lifetime knowing—and that some
day some of them . . . would be looking in at me" (253). Here again—as in
"Ebbets Field," which depicts a later period in Howie's life—the baseball
metaphor exists as a pastoral center of subjective gravity, at once attracting
the natural idealism of youth as well as exposing the baser commercial and
exploitative dimensions of the sport as part of society.

Isolated allusions to baseball, as well as to other sports, also punctuate
Neugeboren's novels. In *Listen Ruben Fontanez* (1968), the beleaguered
protagonist, Harry Meyers, an old and tired public school teacher, prevents
intimidation by his rowdier students by insulting them first: "Cowboys of the
schoolyard. Mickey Mantle, Willie Mays . . . Sandy Koufax would laugh at you"
(50). The baseball stars serve by common consensus as models of integrity
and strength. Receiving more than just passing reference in *Sam's Legacy*
(1973), baseball occupies a critical place in the novel, where it serves as the
subject of the long story-within-a-story titled "My Life and Death in the Negro
American Baseball League: A Slave Narrative," discussed at length below in
chapter five. By contrast, football garners greater space in *An Orphan's Tale*
(1976). Baseball nonetheless appears in it as another nostalgic image in
Charlie Sapistein's memory of the orphanage in which he grew up.

Illustrating especially well the pastoral use to which Neugeboren puts the
baseball metaphor in his novelistic fiction is *Before My Life Began* (1985),
about David Voloshin, a Jewish racketeer who manages to escape the violent
world of organized crime only to remain forever shaped and haunted by it.
Despite Davey's excellence as a high school basketball player, and several
pages devoted to descriptions of basketball games, the book's symbolic structure depends on baseball, especially on well-focused highlights from its quasi-
mythological past, for much of its texture. In particular, the Jackie Robinson
legend serves as a paradigm of perfection to the young David, whose only
immediate role models are a pathetically weak father and a pistol tough
uncle, both agents in kingpin Rothenberg's syndicate. Robinson's story of
pain, struggle, and, finally, success in white major-league ball exists as a
luminous source of nobility amid the scenes of unadulterated corruption,
bullying, killing, and fear that comprise the first two-thirds of the novel.

In the beginning only his friend, Beau Jack, the black custodian-watchman
of his apartment building, fully understands David's admiration for the
Dodger pioneer. The older man even tells him, "When I look in your face,

Davey, all I see is young Jackie . . . You got a look in your eyes like he must of had" (95). Later, when he gets married,

The last thing we talked about before Gail fell asleep was Jackie Robinson, and it seemed crazy to me, but in a wonderful way, that in the middle of our wedding night I'd sat on the edge of the bed in our hotel room . . . and told her the story of Jackie Robinson's life. And when I gave her the news I'd been saving—that Jackie's third child . . . had [been] named David . . . we toasted the new David and wished him long life and happiness. (189)

In its reference to other "Davids," the passage discloses Neugeboren's fusion of the baseball metaphor with Old Testament lore, an association that is developed more fully elsewhere. Accordingly, *Before My Life Began* ends with Robinson's death and funeral, played off against the protagonist's decision to return to face the past he had escaped to save his life.

In the process, David Voloshin, also known as Aaron Levin, confronts the fundamental question of his being and its meaning within the totality of existence:

What Aaron wants to do, he knows, is simply . . . to tell the story of Jackie's life to his son. He smiles. Would his father, or his father's brothers, be pleased with him? At Passover, when the children ask the reasons for certain rituals . . . the father answers by telling the story of the going forth from Egypt. (389).

By yoking baseball to Jewish religious tradition, and to the tradition of storytelling, Neugeboren skillfully reshapes the metaphor as an emblem of U.S. sociohistory into a fecund mythical symbol. In its mythopoesis, the baseball metaphor lyrically captures the sport's affirmative, pastoral aspect without ignoring the negative truths explicitly spelled out by the ironic treatment of baseball found in other works.

Like Neugeboren and Irwin Shaw, Philip Roth also links baseball to Jewish identity in his work. In *Portnoy's Complaint* (1969), for instance, baseball references appear intermittently throughout Alexander Portnoy's book-length conversation with his psychiatrist, Dr. Spielvogel. Alex perceives the game as his only nonsexual solace and emotional relief from the myriad pressures and anxieties caused him by his mother, his Jewishness, and his guilt. At one point he asks Spielvogel:

Do you know baseball at all? Because center field is like some observation post, a kind of control tower, where you are able to see everything and everyone, to understand what's happening the instant it happens. . . . For in center field, if you can get to it, it *is* yours. Oh, how unlike my home it is to be in center field, where no one will appropriate unto himself anything that I say is *mine!* (76–77)

Through such observations and recollections of boyhood baseball as a model of clarity and possibility, Alex describes to the analyst a large part of his alienation from his "suffering" father and "ubiquitous" mother. As critic Tony Tanner suggests, in the process of analysis "memory becomes an ambiguous phenomenon" for Alex, because he "disencumbers" himself by actually "cherishing" his many painful evocations of the past (299).

In many ways Portnoy can be seen as a postwar Danny O'Neill. Both urban intellectuals are haunted by their claustral upbringings, as well as by the inherited religions they cannot escape even through fierce denial. Furthermore, their overriding guilt over what they see as their failure to attain the level of achievement fantasized in youth expresses itself in the baseball metaphor. Portnoy, for example, exclaims at one point:

Oh, the unruffled nonchalance of the game! There's not a movement I don't know still down in the tissue of my muscles and the joints between my bones ... every little detail so thoroughly studied and mastered, that it is simply beyond the realm of possibility for any situation to arise in which I do not know how to move.... And it's true, is it not? ... there are people who feel in life the ease ... that I used to feel as the center fielder for the Seabees? Because it wasn't, you see, that one was the best center fielder imaginable, only that one knew exactly, and down to the smallest particular, how a center fielder should conduct himself.... I ask you, why can't I be one! (79–80)

Portnoy's solipsism, however, with its outrageous humor, sadomasochistic sexuality, and modernist sensibility, clearly distinguishes him from the simpler, more straightforward, and more accessible Farrell character.

Despite the golden hue of Portnoy's memory of baseball, it is not precisely the same for author Roth, who handles his character's remembered idyll with a shade of irony manifested in the self-reductive confessions his emotionally tortured, wildly funny protagonist gives. "I am a child of the forties," muses Portnoy, "of network radio and eight teams to a league and forty-eight [states] to a country" (265). In other words, contemporaneity and all its libertine experiments are too much for someone steeped in archaic values like Portnoy, who ignores or forgets his upbringing during Dionysian immersion in the delights of contemporary freedoms but who then must suffer the agony of his punishing guilt. Roth's interest in baseball as a literary motif in what Tanner describes as "this crucial novel" of modern America (295) goes far beyond Alex Portnoy's neurotic memories, for Roth has also written one of baseball fiction's major works, *The Great American Novel*, focus of the last chapter in this study.

In one sense the nostalgic record of baseball kept by the pastoralist baseball writers serves as antithesis to that written by the ironists. In another sense, however, the two views taken together form a more complete, more inclusive literary record of the game in America. "Tao in the Yankee Stadium Bleachers," a poem by John Updike, unites these two views in striking imagery:

The thought of death is peppermint to you
when games begin with patriotic song
and a democratic sun beats broadly down.
The Inner Journey seems unjudgeably long
when small boys purchase cups of ice
and, distant as a paradise,
experts, passionate and deft,
wait while Berra flies to left. (Updike, "Tao," 79)

By placing the sport and its multitudinous societal matrix in the Eastern mystical context of Taoism, the quintessentially Western Updike asserts the efficacy of seeking the One, the All, the Way, in the homely but sacred and profane contradictions and unities of baseball. In this he is like all the other writers in this study who have measured America's strengths and weaknesses, its beauty and its evil, its possibilities and its hopelessness, on the solid quaternity of a diamond and the cosmic sphere of a baseball.

REFERENCES

The original version of this chapter, now substantially revised, appeared first in *Midwest Quarterly*, 23 (Summer 1982): 411–425.

Nelson Algren, *Never Come Morning* (1941; reprint, Harper & Row, 1963).
Thomas Boswell, *How Life Imitates the World Series* (Penguin, 1983).
Cordelia (Chávez) Candelaria, "A Decade of Ethnic Fiction by Jay Neugeboren," *MELUS* 5.4 (Winter 1978): 71–82.
———. "Jay Neugeboren," in *Dictionary of Literary Biography: Twentieth-Century American-Jewish Fiction Writers* (Detroit: Bruccoli Clark, 1984): 181–188.
Tristam Potter Coffin, *The Old Ball Game*, 1971.
Arthur Cooper, "Murderers' Row," *Newsweek* (April 30, 1973): 83.
James T. Farrell, *Father and Son* (Cleveland: World, 1940).
———, *Invisible Swords* (Doubleday, 1971).
———, *My Baseball Diary* (A. S. Barnes, 1957).
———, *My Days of Anger* (Cleveland: World, 1943).
———, *No Star Is Lost* (Vanguard, 1938).
———, *Studs Lonigan Trilogy* (Random House, 1938).
———, *A World I Never Made* (Vanguard, 1936).
Robert Fitzgerald, "Cobb Would Have Caught It," in *In the Rose of Time* (New Directions, 1943).
Ernest Hemingway, *The Old Man and the Sea* (Scribner's, 1952).
Robert J. Higgs, *Laurel & Thorn*, 1981.
Sinclair Lewis, *Babbitt* (Harcourt, Brace & World, 1922).
Christian K. Messenger, *Sport and the Spirit of Play in American Fiction*, 1981.
Jay Neugeboren, *Before My Life Began* (Simon & Schuster, 1985).
———, *Corky's Brother* (Farrar, Straus & Giroux, 1964).
———, *Listen Ruben Fontanez* (Boston: Houghton Mifflin, 1968).
———, *An Orphan's Tale* (Holt, Rinehart & Winston, 1976).
———, *Parentheses: An Autobiographical Journey* (Holt, Rinehart & Winston, 1972).
———, *Sam's Legacy* (Holt, Rinehart & Winston, 1973).

Michael Oriard, *Dreaming of Heroes*, 1982.

Mordecai Richler, *St. Urbain's Horseman* (Knopf, 1971).

Philip Roth, *Portnoy's Complaint* (Random House, 1969).

Irwin Shaw, *Voices of a Summer Day* (Delacorte Press, 1965).

Tanner, *City of Words*, 1971.

Mark Twain, *A Connecticut Yankee in King Arthur's Court* (1889; reprint, Hill and Wang, 1960).

———, "What Mark Twain Said," *New York Sun* (April 9, 1884): 2.

John Updike, *Rabbit Redux* (Knopf, 1971).

———, *Rabbit Run* (Knopf, 1960).

———, "Tao in the Yankee Stadium Bleachers," in *The Carpentered Hen and Other Tame Creatures* (Harper & Row, 1956).

Voigt, *From Gentleman's Sport*, 1966.

Thomas Wolfe, *Letters*, (Scribner's, 1956).

———, *Look Homeward Angel* (Scribner's, 1929).

———, *Of Time and the River* (Scribner's, 1935).

———, *You Can't Go Home Again* (Harper & Brothers, 1940).

4

EDUCING THE MYTH

Thanks to myths, one discovers that metaphor rests on the intuition of logical relationships between one domain and other domains, into whose ensemble metaphor reintegrates then the first domain.... Far from being added to language in the manner of an embellishment, every metaphor purifies language and restores it to its primary nature.

Claude Levi-Strauss, *The Raw and the Cooked*

If popular baseball fiction from "Casey at the Bat" to *The Year the Yankees Lost the Pennant* helped to transfer the game from the actual diamonds of, for example, Three Rivers Stadium and Chavez Ravine to the "sun field" of literature, then the allusive use of baseball—the literary fungoes just discussed—further indicates the efficacy of the sport as fictive emblem. Whether in the naturalism of *Men on Spikes*, the romanticism of Nebraska Crane's portrayal, or the psychological realism of *Before My Life Began*, baseball's utility as metaphor is well established. Clearly, serious writers have long used the sport as a concise metonym of American culture, usually through ironic or pastoral treatment and sometimes through both, as in the works of Jay Neugeboren. The possibilities suggested by the literature treated in chapter three are developed and extended especially well by two contemporary writers, Bernard Malamud and Lamar Herrin.

Both Malamud and Herrin integrate a mythic dimension into their handling of the baseball metaphor that adds cultural texture and emotional resonance to their novels. What is more, their explicit foregrounding of archetypal myth contains both anthropological and mythopoeic perspectives. In anthropology myth is understood as the anonymous, primitive representations of a given culture's traditional beliefs and values as they have evolved over time. My-thopoeia, on the other hand, views myth as conscious, stylized, aesthetic practice usually arising in art and intellectual thought as a reaction against the perceived excesses of modern empiricism, technology, and commerce. Malamud's and Herrin's novels encompass both perspectives, filtered through the membranes of their respective imaginations and individual apprehensions of baseball history and lore.

Chapter one has already demonstrated the natural correspondence be-tween myth and baseball through its summary of the game's origins in prim-itive ritual, grassroots forms of recreation, and American mercantilism. Chapters two and three presented the correspondence as it appeared un-evenly in some popular works and more effectively in others, notably *The Celebrant, Voices of a Summer's Day*, and *Before My Life Began*. Continuing the logical progression, this chapter examines the permutations of myth as two contemporary writers have applied them in strikingly different treatments. Malamud synthesizes baseball legend with classical Greek myth in *The Nat-ural* to depict Roy Hobbs's blind anti-heroic quest. In contrast, Herrin portrays his protagonist, pitcher Dick Dixon, in *The Rio Loja Ringmaster* as a sensitive thinker (like Henry Wiggen) who associates himself with the sacred ring-master of the classical Aztec ritual-game *tlachtli*.

Regardless of approach, however, both writers refine baseball fiction by educing the immanent mythic aspects of the evolved folk pastime. Huizinga's observations about the centrality of play to the formation of myth are partic-ularly germane here.

In myth, primitive man seeks to account for the world of phenomena by grounding it in the Divine. In all the wild imaginings of mythology a fanciful spirit is playing on the border-line between jest and earnest . . . . Now in myth and ritual the great instinctive forces of civilized life have their origin: law and order, commerce and profit, craft and art, poetry, wisdom and science. All are rooted in the primaeval soil of play. (Huizinga, 4–5).

The following pages investigate precisely how Malamud and Herrin link "the forces of civilized life" to baseball's "soil of play" to produce their works. In Michael Oriard's words defining "sports fiction," these novels find their es-sential "vision" of the individual's "condition in the basic meaning of the sport he plays, formerly played, or watches" (Oriard 6).

THE NATURAL (BERNARD MALAMUD)

"Though *The Natural* [1952] is not Malamud's most successful, nor his most significant book," writes Philip Roth, "it is at any rate our introduction to his world" which has somewhat of an "historical relationship to our own but is by no means a replica of it" (Roth, "Writing American Fiction," 151). The historical relationship of Malamud's world to our own arises largely from the author's use of classical myth and baseball folklore as a thematic infrastructure for the strange story of Roy Hobbs, who exists simultaneously as a baseball flash-in-the-pan and as an eternal folk hero. Resembling Joyce's patterning of *Ulysses* upon the themes and plot of Homer's *Odyssey*, *The Natural* is built on a framework of allusion that deepens the story of a simple American sports hero, transforming it from the frivolous celebrity journalism of popular culture into an allegory of modern American life. Given its mature handling of America's apotheosis of professional athletes, *The Natural*'s portrait of Roy Hobbs is as classic a rendering of an American type as Twain's Huck Finn, Ellison's nameless Invisible Man, and Arthur Miller's Willy Loman. As Leslie Fiedler recognized in an early review of the novel, Malamud "found, not imposed, an archetype in the life we presently live; in [*The Natural*] the modern instance and the remembered myth are equally felt, equally realized and equally appropriate to our predicament" (quoted in Andreano, 20).

Malamud's novel presents Roy's story through a traditional linear plot with a clearly delineated beginning, middle and end. The short "Pre-Game" section opens the novel and introduces Roy Hobbs as an innocent country boy in a man's athletic body. He sets out, like Robin in Hawthorne's "My Kinsman, Major Molineux," to conquer the city, that microcosmic symbol of a vast and distant world—in Roy's case, the world of major league baseball. Quickly hobbled by a series of trials typically associated with an ingenue's initiation into the adult world of error, corruption, knowledge, and suffering, at the conclusion of "Pre-Game" Roy is unexpectedly shot in the stomach by femme fatale Harriet Bird's "silver bullet" (33). The shooting forces him to take a fifteen-year hiatus from ballplaying, a lapse that Malamud does not show in sequence in the narrative but which appears as intermittent recollections folded into Roy's later thoughts and dreams.

The second section, "Batter Up!," comprises the rest of the novel, and it mirrors in expanded form the events of the first section. It takes the now thirty-four-year-old Roy fully into the world of adult experience through his career as a New York Knight, his love affairs with Memo and Iris, and his contamination by the pervasive greed and selfishness around him. In keeping with the rite of passage pattern, he descends even further into error and suffering before eventually attaining, at the very end of the book, a slight degree of the self-awareness and knowledge that mark the mature person. But, like his baseball precursor, the Apollonian hero Jack Keefe in *You Know Me Al*, the insights come too late to translate into happiness and joy.

While *The Natural* may be, as this summary suggests, fairly simple in its unfolding of plot, its incorporation of multiple threads of myth and symbol is experimental and complex. As Oriard correctly points out, Malamud was "our first writer" to demonstrate that the "character of the hero" along with the "relationship of country and city, youth and age, masculinity and femininity in American sport are explicitly mythic concerns" and that these "major concerns of sport" actually "define the essential myths of Western civilization" (Oriard, 211). Even though initial reaction to the novel was largely favorable, some readers were troubled by the treatment of so mundane a subject as baseball within the lofty and classical framework of myth (see for example, Baumbach, Hassan, Richman). Over three decades of interpretation and criticism, along with a motion picture based on the novel, have led to exhaustive analysis of the book's mythic structure and motifs and have had the consequent salutary effect of producing greater understanding of—and respect for—the author's vanguard achievement in the Fifties.

For excellent mappings and explanations of *The Natural*'s use of myth, the reader is encouraged to refer to the articles by Earl Wasserman (1965) and Frederick W. Turner (1968), as well as to the sections on the novel in Oriard's *Dreaming of Heroes* (1982) and Knisley's "The Interior Diamond" (1978). The latter two discuss the topic within the context of sports fiction. In light of these fine analyses, a summary of the novel's mythic patterns will suffice here to introduce the baseball metaphor and its relation to myth. What should also be borne in mind are the related concepts already developed in this study—for instance, the game's primeval origin in ritual, its evolution within the nation's folk habits and sociohistory, and its steady manufacture of folk heroes for an appreciative democratic and secular society.

Five major archetypal patterns provide the primary foundation of *The Natural*. They are not presented in distinct units; Malamud conflates them in constantly shifting combinations of circumstance and experience to convey the universality of Roy's story. The five mythic patterns are: fertility and vegetation; death and rebirth; underdog versus giant; the questioning hero; and the archetypal ideal of feminine/masculine balance.

To begin with, Malamud develops his hero's baseball career according to natural vegetation cycles, drawing a parallel to ancient fertility myths and rituals. Starting in the springtime of his youth, the narrative then moves through the summer of his active career, the autumn of the World Series when, like the season, he "'falls," and it ends in the winter of his demise as a player and Iris Lemon's injury. Accordingly, the young Roy's disappearance after the trauma of the "Pre-Game" shooting mirrors certain ancient legends in which fertility gods (for example, Osiris and Demeter) descend to the underworld until the next season of regeneration. When Roy returns to begin his professional career, he brings fresh vigor and promise to the Knights team during an especially "dry" and luckless season. The team name gives explicit emphasis to the idea of baseball players as questers.

Related to his role as a source of mythic vegetation and fertility is Roy's status as a secular embodiment of a reborn savior. When he returns, now healed of his wound and eager to save the Knights' chances for a pennant, he resembles such resurrected deities as the Teutonic Odin, the Mesoamerican Quetzalcoatl, and Jesus Christ, among others. As manager Pop Fisher laments, the thirty-four-year-old rookie is "starting with one foot in the grave" (38), a casual phrase that underscores the motif of burial and resurrection.

Although this motif usually implies a spiritual renewal and redemption through suffering, Roy's limitations as a self-centered mortal and celebrated adolescent baseball star effectively blind him to any insight or spiritual transcendence. For most of his life he learns nothing from his pain. For example, even though Memo Paris resembles the destructive Harriet Bird, Roy's lust for her, fueled by her manipulative teasing, causes constant misery: "in his dreams he still sped over endless miles of monotonous rail toward something he desperately wanted. Memo, he sighed" (72). Ironically depicted, then, Roy's rebirth is more an unthinking repetition of painful habits than a resurrection that redeems and brings joy.

A third archetype central to Roy's story is that of the questioning knight in search of a Holy Grail. Like any number of ancient and medieval questers, from Gilgamesh and Odysseus to King Arthur and Tristram, Roy travels a perilous journey. There are natural hurdles that forever plague him, like the impenetrable night on the train (7–8); tests of strength and skill, like the impromptu duel with Whammer Wambold en route to Chicago (21–25); and great suffering, such as his car wreck (99) and his endless troubles with women. Also like the prototypical questers on whom he is modeled, Roy possesses a magical weapon—his bat Wonderboy. It endows him with a suitably phallic symbol of worth and manhood (compare King Arthur's sword Excalibur), one from which he must wean himself if he is to discover his own inviolable human strengths.

In keeping with Malamud's symbolic reversals, however, irony motivates the rendering of this myth as well, for Roy's grail ultimately turns out to be not the fame and public success he had singlemindedly pursued, but his infamy and memorialization as a Black Sox-style fixer (190). His quest is exposed as what it has persistently been: a chronic self-indulgence for immediate gratification, however corrupt, painful, or destructive the means or the ends. This is the point of the novel: Roy does not learn from his mistakes, so he repeats them over and over again. The mythic structure informs the reader of the universality of Roy's story.

The biblical story of David and Goliath, with its quintessential conflict between underdog and giant, provides the basis for a fourth mythic pattern in *The Natural*. That disproportion of power holds special relevance to sports in general and baseball in particular because the very nature of athletic competition frequently places the responsibility for victory upon the shoulders of one athlete who must, when game point is at stake, face the entire opposition alone. The problem is that when the David and Goliath myth is applied unexaminedly to sports celebrities, it produces grand expectations of heroic, even godlike behavior that most sports figures are not remotely capable of approximating. The "tragedy of Humpty Dumpty" rings through here (see chapter two).

First seen in his early encounter with Goliath-like Whammer in the "Pregame" section, Roy briefly becomes "King"—that is, *le roi*—by striking out the bigger, more experienced, and famous slugger. Later, when he returns to the game with the handicap of age and only bush-league experience, Roy is again an underdog pitted against the powerful star, Bump Baily, whom he vanquishes on the field and in the bedroom, like the natural he is. Yet, like another natural, his folk forebear, Casey of Mudville, Roy cannot conquer himself; as the coach says to Pop Fisher, Roy is "less than perfect" because he likes to "hit at bad ones" (67). Nor can this little "roi" conquer the Goliaths of the baseball business and related crime world, represented in the characters of owner Judge Goodwill Banner and "Supreme Bookie" Gus Sands, who maintain their hegemony over the game and are quite willing to debase it to satisfy their greed. Therefore, like the biblical David who murdered to obtain Bathsheba, Roy manages to slay the outer giants, but he simply cannot overcome the dragons inside him.

One reason Roy cannot overcome them relates to the novel's fifth important archetype, the ideal of feminine/masculine balance and unity. Known from antiquity in Eastern philosophy as the doctrine of yin and yang, the female/male complementarity was described in the West by Jung as a synthesis of animus and anima. In brief, the concept holds that reality is in a constant state of flux caused by the interaction of two primal, fundamental forces, the feminine and the masculine, working in conjunction with the world's other first principles (time, space, the elements, and so on). The feminine force, yin, is described as passive, negative, weak, yielding, while the masculine yang, is described as active, positive, strong, and assertive (Reese, 636–37). Jung believed that both female and male egos possess degrees of the opposite principles and which, in love, are matched through the double projection and possession of each other's animus and anima (Reese, 274). Importantly, each force needs the other to realize its true form, thus making complementarity, interaction, and synthesis basic essentials of life.

The Natural explores this primal opposition and complementarity in a number of ways, but most noticeably in its female characters and their relation to the hero. The novel's three most negative women are Harriet Bird, Memo Paris, and Roy's mother, whom we meet through his brief recollections of her. These characters appear as grossly distorted creatures arising from Roy's infantile perceptions of them. He sees them solely in terms of his own selfish physical gratification, and their distorted natures, which he helps produce, ultimately diminish his potentially heroic stature. Malamud renders his antihero's shrunken heroism and infantilism through erotic fantasies of birdlike women and gargantuan food binges:

[The] naked . . . lovely slid out of a momentary flash of light, and the room was dark again . . . but the funny part of it was when she got into bed with him he almost cried out in pain as her icy hands and feet . . . slashed his hot body, but there among the apples, grapes, and melons, he found what he wanted . . . .

The hamburgers looked like six dead birds . . . . Eating them . . . they all tasted like dead birds. They were not satisfying but the milk was. He made a mental note to drink more milk. (51, 152)

Unable to release his true feelings, to actually cry out his pain, Roy as sucking baby dehumanizes his women into food he can feast on without restraint. These scenes disclose the dis-ease of profligacy common to the Dionysian hero as defined by Higgs. This image of disease finds sexual analogue in Harriet's and Memo's "sick" breasts.

Were he mature enough to free his anima, Roy could possibly help them restore their animus in mutual complementarity, but overgrown boy that he is, he contents himself, as in the nightclub scene of magic tricks, merely to continue "plucking" magical "duck egg[s]" from Memo's décolletage (90). Despite Harriet's attempt to elicit nobler things from him (26–28), and despite Memo's mnemonic mirroring of Harriet's basic destructiveness, he remains orally fixated in his identity as a "thirsty" and "hungry" orphan son of a "whore" (148). Consequently, so long as he remains immature and selfish, his narcissism controls him, and he remains obsessed with Memo and Harriet.

Iris Lemon, on the other hand, represents the sunny, open goodness of a flower (iris) combined with the occasional bitterness (lemon) that comes from facing honest truth. Associated with the purgative purity of water, Iris begins Roy's baptism into true manhood in the swimming scene at the lake, were he impregnates her healthy, yielding body (128). Nevertheless, he foolishly resists his own instincts and rejects her—despite her relative youth, natural beauty and his attraction to her—because she is a grandmother, an image of old age that repulses him as it repels any immature person who seeks to deny the inexorable passage of time. Furthermore, still obsessed with Memo, he is incapable of comprehending either Iris's solidity or the significance of her yang passivity—the natural and perfect complement to his yin:

"We have two lives, Roy, the life we learn with and the life we live with after that. Suffering is what brings us toward happiness."

"I had it up to here." He ran a finger across his windpipe... "and I don't want any more." (126)

Iris's final scene at the rigged ballgame succinctly encapsulates the novel's female/male archetype. Hobbled by his guilt and self-loathing for agreeing to the fix, Roy angrily aims to hit the surly fan, Otto Zipp, but the ball, deflected upwards, strikes Iris instead (179). He is suddenly forced into epiphanic awareness of the pattern of all his childish mistakes and, worse, his failure to "learn anything out of my past life" (190). He realizes at last that by compulsively indulging his physical appetites in seeking to become the stereotypical hero of popular myth, he has actually hurt himself and others. In other words, he has denied his animus—his feelings, fears, and vulnerabilities. The act of burying his broken bat Wonderboy at the end of the book illustrates the death of the tragic hero's boyhood and the birth of his newly acquired self-knowledge (188). Consequently, one critic's claim that all Malamud's novels are "fables or parables of the painful process from immaturity to maturity" certainly applies to this, his first novel (Tanner, 323).

The overt interlacing of Roy Hobbs's tale with these five mythic patterns emphasizes the appropriateness of placing baseball on the eternal wall of universal archetypal tradition. In writing this book Malamud recognized that baseball, like all organized sport, is mythic both in its immanent meaning as a sport and social ritual and also in its peculiar re-enactment of universal myths. *The Natural* exploits those mythic dimensions and, additionally, pursues another important function of sport as myth—its recreation of a particular culture's myths, legends, and history—for the novel is about baseball in America.

Besides serving as subject and setting of the novel, baseball itself functions archetypally in it. Malamud relies on the sport's central place in American life in developing Roy as a folk hero—in Andreano's conception of the term— determined to fulfill first his Apollonian dreams and then to gratify his Dionysian desires, in Higgs's definition. Accordingly, baseball as archetype in the novel ramifies from two broad sources: first, baseball history and lore, and second, baseball fiction.

The first category is dominated by two basic facts of baseball history and lore that have been so seminal they have evolved their own elaborate myths: the Black Sox scandal and the Babe Ruth story. As explained above in chapter one, the throwing of the 1919 World Series between Chicago and Cincinnati by eight Chicago players marks a crucial milestone in U.S. sociohistory. Before the scandal, baseball's contamination by business and gambling were regarded as minor but necessary evils in a supposedly free-market economy. Afterwards, the painfully disillusioned public could no longer keep up the pretense of the sport's arcadian purity, for, as the Chicago *Herald and Examiner* reported when the players were legally acquitted, "the national character" had been "profoundly" harmed (Asinof, *Eight Men Out*, 274).

Alert to its mythic importance to contemporary U.S. culture, Malamud integrates into his novel exact details of the scandal, along with the general premise that corruption is endemic to American business and industry—and, because it is both, also to sport. For example, he models the pennant race between the clearly superior Knights and the "flat failure" of the Pirates (167) on the vastly disproportionate abilities of the "lucky" Reds and the highly talented White Sox in the 1919 Series. Similarly, Roy's $35,000 agreement with the Judge to help lose the playoff game replicates some of the terms of the Black Sox fix (162–68). The protagonist's similarity to illiterate Southerner "Shoeless" Joe Jackson, one of the indicted Black Sox eight, is suggested by Roy's hayseed rusticity and farmboy candor, Jackson's most prominent personal traits. Malamud echoes the famous line "say it ain't so, Joe," reputedly asked of Jackson when news of the scandal first broke, on the last page of the novel. Roy, accosted by a boy with a newspaper account of "Hobbs's Sellout," wants to say "it ain't true," but he can't look "into the boy's eyes" and lie. Instead, he lifts "his hands to his face and" weeps "many bitter tears" (190). That it is the Judge, owner of the Knights, who arranges the fix against his own team suggests the extent of the degradation of business ethics Malamud seeks to expose.

As Knisley astutely points out, the mere presence of either Jackson or the Black Sox scandal in a work of fiction "precludes the possibility of a workable pastoral." The scandal marks the sport's "fall from grace," and permanently erases the idea of an innocence "separate from and more complete," and implicitly more perfect, than the wider society around it (Knisley, 179). *The Natural* can only be read as tragic, then, and its occasional comedy, frequent excursions into fantasy, and strange cast of melodramatic rustics and crooks reflect Malamud's search for authentic material from which to construct American tragedy.

The second element of baseball history and lore that Malamud employs archetypally in the novel is the story of Babe Ruth (discussed fully in chapter five). Baseball historians generally credit the advent of George Herman "Babe" Ruth, along with the appointment of Judge Kenesaw Mountain Landis as commissioner, "with rescuing professional baseball and restoring it to good repute after the Black Sox" scandal (Seymour, 367). Ruth's extraordinary career, especially his home-run hitting and insouciant style, excited the public and attracted the fans back to the stadiums. His high-living, carefree decadence off the diamond also made him one of the first modern media celebrities of popular culture (Creamer, passim). Malamud fittingly chose to model his fictional "natural" on the legendary Sultan of Swat: indeed, reporter Max Mercy dubs Roy "El Swatto" (Malamud, 91).

Even though Ruth was not an orphan, as the familiar legend holds (his orphanhood was cooked up to cover up his juvenile delinquency), he did spend time in a boy's home (Creamer, 55). Thus Hobbs, like the Ruth legend, is a psychological orphan, estranged from his parents and eternally obsessed with compensating for that primal loss through gratification of primitive physical desires. Like Ruth, Hobbs is "adopted" by surrogate fathers, Sam Simpson and Pop Fisher, who incidentally are the first in a long list of "symbolic father figures" in Malamud's canon (Tanner, 333). Accordingly, both historical figure and fictional character compensate for their psychic deprivation through gluttony and sexual promiscuity. The incapacitating "bellyache" that sends Roy to the hospital after an especially gluttonous binge (Malamud 153) recalls the "stomach-ache heard round the world" that afflicted the Babe during the 1925 season. Their unharnessed randiness with women also links the two.

As ballplayers, both men also accomplish the singular feat of converting their talents from outstanding pitching to outstanding hitting (like Tiny Tyler in Broun's *The Sun Field*), just as both are deemed Atlas-like in keeping baseball afloat at a time of declining fan interest. Hobbs's love-hate relationship with the fans (90–92), in particular with Otto Zipp (178–79), resembles

Ruth's similar reputation, which is best illustrated by the heated atmosphere of his famous "called shot" in the 1932 World Series (Rosenburg, 94–97). Malamud incorporates into his story the faith-healing scene, in which Roy "hits one" for the dying boy ("everybody knew it was Roy alone who had saved the boy's life"; 118) to suggest the legends, largely unfounded, about the Babe's fondness for and mystical cures of children (Seymour, 429; Creamer, 55).

By using these and other details from the actual George Herman Ruth biography later embroidered into the Babe Ruth legend, Malamud reveals the inadequacy of America's popular heroes as either paradigms of conduct or as spirits of regeneration. No matter how extraordinary a player Ruth was, and no knowledgeable observer can argue against his magnificent record, he, like most other celebrities of popular culture, lacked the noble qualities of superior intelligence, virtue, self-sacrifice, and leadership embodied by such epic heroes as Achilles, Gawain, and King Arthur. Interestingly, Ruth's adulation by the public was recognized as "monstrous" and "senseless" by a few perceptive writers who frequently called attention to the pathetic social values reflected in such blind hero worship. Frank Lane, for example, wrote in *Baseball Magazine* in 1925:

To the public that surveys its heroes through a golden haze of misplaced admiration [Ruth's suspension in 1925] came as a distinct shock.... This haze of inaccurate, almost meaningless bunk endowed Ruth with virtues utterly alien to his nature, and ... obliterated ... those crude defects, those fatal weaknesses which made him an object rather of apology and pity than of admiration to those who knew him best. Ruth's popularity has become a thing monstrous, formless, senseless. It has been exaggerated beyond all the bounds of sanity. (Lane, "A Fallen Idol")

Through use of the Black Sox motif and the Babe Ruth legend, *The Natural* exposes both the insanity and the pathos of deifying and hero-worshipping mere mortals, especially those whose deification rests solely on the laudable, but hardly noble, ability to hit a ball or steal a base.

The Natural's employment of baseball history and lore is not limited to the Black Sox and Babe Ruth parallels. Malamud braids other motifs from the sport and its culture into the narrative. For example, besides resembling the Arthurian Excalibur, Roy's fantastic bat also has an analogue in Big Betsy, the well-publicized bat of hall of famer Ed Delehanty of the Philadelphia Phillies, a player whose career also ended tragically when he disappeared after a drinking binge in 1903. Harriet Bird's shooting of Roy on the train compares to the 1949 shooting of another Philly player, Eddie Waitkus, by a woman in a hotel; unlike Roy, the injury effectively ended Waitkus's career. Malamud alludes obliquely to pitcher Walter "Big Train" Johnson through the scene at the railroad tracks when Roy's hard pitching strikes out the veteran (23). These motifs, then, enhance the novel's use of the Black Sox affair and Babe Ruth's story and link the fictional Roy to the tradition of "true-life" American heroes.

As mentioned earlier, the baseball archetype in *The Natural* also stems from a second broad source, baseball fiction. Whether the juvenile romances of the Frank Merriwell genus, the hard-hitting satires written by Ring Lardner, or the literary fungoes of allusion in nonbaseball fiction, the novel exhibits close kinship to previous literature about the sport. For instance, both the Whammer and Bump resemble Lardner's boastful busher Jack Keefe, and even Roy shares professional and personal qualities with Lardner's character. Both are puerile country greenhorns who lack the maturity to gain the self-knowledge needed to successfully cope with the fame, money, and women their remarkable athletic talent awards them. Additionally, the gold-digging, self-possessed Florrie Keefe may be seen as Memo's precursor; both Keefe and Hobbs show incredible blindness to their parasitic natures. *The Natural* also shares outrageous magical gimmicks with such earlier farces as *Rhubarb* and *It Happens Every Spring*—for instance, Wonderboy, the Knights' impossible statistics and records, Roy's fantastic magic act, and so on. It may be stretching the comparison to suggest that Roy's gruesome memory of his mother's torture of the cat (176) might serve as an indirect reference to the cat in *Rhubarb*, but the facts are there. Thus, while not as powerful in its impact on *The Natural* as baseball history and lore, baseball fiction before 1950 nonetheless influenced Malamud's work if only by its presence on the American literary landscape.

The novel's expressions of baseball as archetype having been catalogued, and the universal mythic patterns on which Malamud embedded that archetype having been inventoried, the question arises, what does myth achieve in *The Natural*? What is its function and cumulative effect? One answer is that by placing baseball alongside world myth Malamud exposes the thinness of U.S. America's traditions, the banality of its customs, the crassness of its values, and the utter doom of its future. As the "national pastime," baseball expresses the nation's diseased and debased heart and soul, illustrated in the novel through the author's inversion of familiar myths to show their undersides. Roy is like the savior David slaying Goliath when he strikes out Bump, but in giving his surrogate father Sam Simpson a fatal injury in the process, Roy recalls the tragic Oedipus, who could become king only by killing his father. He is like the peerless Lancelot and other Round Table knights in almost singlehandedly advancing the team's standing, but like Lancelot's obsessive love for King Arthur's wife, Roy's obsession with Memo destroys him. One event in Roy's life counters another; one achievement undermines another, and these antitheses are stressed by Malamud's use of myth.

The novel accordingly addresses the familiar discrepancy between the myths about baseball and the reality of the professional sport. Nearly every one of Malamud's baseball references has a severely degraded or tragic dimension. As discussed above, Babe Ruth's pathetic side as an uncouth glutton and libertine are emphatically foregrounded in Roy's portrayal, just as the depictions of Judge Banner and Gus Sands have their historical analogues in a number of active participants in the sport like Judge Kenesaw Mountain Landis, Ban Johnson, and Arnold Rothstein. The shooting of Eddie Waitkus and the disappearance of Ed Delehanty, both historical incidents of a sordid nature, hold a similarly allusive spot in the novel. It is thus not surprising for Malamud to conclude the narrative by standing Roy Hobbs not alongside noble questers, heroes, or kings, but among the fallen athletes of the Black Sox scandal. In its use of myth, *The Natural* is ironic through and through in exposing the myth of Yankee innocence.

In his first as in his later novels, Malamud stresses the importance of locating our identity in genuine American myths arising from the folk. He also works the theme that suffering is essential to the development of an individual's maturity and dignity. Roy's "rise and fall" becomes a "parable concerning the fate of those youthful energies and abilities" needed to "revitalize and maintain" society, writes Tanner (324). But since Roy is ultimately an ironic hero and is eventually displaced by the rookie Youngberry (as he himself replaced the Whammer and Bump), his revitalization of his community (like Ruth's and "Shoeless" Joe's and, indeed, like that of any narrow sports figure) is shortlived and fraudulent until the folk and their leaders learn hard truths from their experience. Through Roy's example, played against the dense texture of world myth, the novel suggests that the struggle to attain heroic stature by an individual or a society is doomed to tragic failure if nothing is learned from what came before: a cumulatively shared, collectively suffered experience.

THE RÍO LOJA RINGMASTER (LAMAR HERRIN)

Unlike Roy Hobbs in *The Natural*, Dick Dixon, the protagonist of Lamar Herrin's *The Rio Loja Ringmaster* (1977), learns from experience. Indeed, he learns and relearns but his outlook is so obscured by an Apollonian ideal, rooted in earliest childhood, that for much of his early adulthood he fails to gain self-insight from the repeated lessons. Partly because his experiences are exceedingly manifold, and partly because he is single-mindedly keyed to winning—that is, to victory narrowly comprehended—he fails to see the relevance of any experience beyond the particular situation at hand.

Like Roy, however, indeed like all the protagonists of the major baseball fictions examined in *Seeking the Perfect Game*, Herrin's hero Dixon is a pitcher, a position of strong appeal to writers primarily because of the pitcher's structural role in the game. Moreover, the pitcher's lone centeredness on the mound in the middle of the diamond may be seen as a "sacred space" (Gardella, "The Tao of Baseball," 28), making pitchers ideal author surrogates. Another integral aspect of the pitcher's lot is pain. According to Jim Lonborg, Philadelphia pitching ace, "Pitching is a totally unnatural motion.... It's a hazard just to be throwing"; it is fundamentally a contradictory act between body and mind (in Boswell, 192). Lonborg believes that the essence of pitch-

ing "is to be aggressive" while also simultaneously being "relaxed and comfortable," which means that "out on the mound, every pressure, every internal
tension" produces relentless "stress" (Boswell, 195). The pitcher's role as
suffering hero therefore makes him a perfect focus of reader sympathies, and
it explains their predominance as central figures in baseball fiction. Joining
pitchers Jack Keefe, Roy Hobbs, and Dick Dixon later in this study are Mark
Harris's Henry "Author" Wiggen, Jay Neugeboren's Tidewater, Robert Coover's Damon Rutherford, and Philip Roth's Gil Gamesh—all of whom share
with Herrin's protagonist an uncommonly strong tendency towards self-ex-
amination and, in some instances, even solipsism.

Dick Dixon's self-absorption stems both from the unresolved childhood
ghosts haunting his memory and from his related tendency to fashion reality
into prettier patterns of illusion. The son of a couple whose dissonant marriage negatively shapes his future relationships with women, Dick is haunted
by the boyhood pressures of a do-good Christian mother, Susan, leading him,
a perfect boy soprano, "from city to city . . . to entertain the wounded [World
War II] veterans" (40). Summed up in her "one golden rule: Never offend
and please when you can," Susan's bourgeois conventionality propels him
decidedly away from Yankee orthodoxy to a bohemian way of life (41). The
other childhood ghost in his life is his father, Sarge, whose former life as a
professional baseball player offers little to offset his being a flimsy parent
(Herrin, 42–44).

Dick's chief memory of Sarge's rare assertion of his parental role occurred
when he was ten years old, and his father insisted on interrupting the normal
Sunday church routine to take the boy, a zealous fan, to his first major-league
ballgame (44). Depicted as a rite of passage for pubescent Dick, the trip is
pegged in his memory alongside the fear that his beautiful soprano voice
seemed "about to change before every song he sang" (38). The fear gave
every performance at this crucial time in his life an "added punch because
it might be his last" (38). The trip to St. Louis to watch a doubleheader
between the Cardinals and Dodgers consequently resonates in his memory
as a milestone, his first major league experience. Unfortunately, the trip is
also a major debacle for the boy.

St. Louis and the game expose him to the sordid side of adult male life as
his father, unfettered from spouse and far from home, eagerly teams up with
a sleazy business associate. Distressed by what he perceives as the two men's
flickleness and disloyalty when they quite realistically assess the home-team
Cardinals and conclude this "just ain't their year," his enjoyment is further
tarnished by the sweltering rough and tumble of the crowds at the stadium.
He finally decides it is definitely "no place for a Stan Musial home run" (50).

Despite the wonderful surprise of the underdog Cardinals' victory, his joy
is totally ruined when his father and sleazy companion, drunk to a stupor,
lose Dick in the crowds. In classic rite of passage symbolism, Herrin paints
the scene of the boy's search for his father in the poignant tableau of Dick's

anxious descent from the bleachers onto the field, where he finds himself frighteningly alone. Not only does this scene remind us of his earlier panic at the thought of his voice changing in the middle of a high note, it also parallels the novel's scenes of the lonely pitcher trying to pull a victory ballgame against great odds. "In the deepening shadows of Sportsman's Park he stood there, powerless to make a sign" (57). Dick recalls the serenity of only yesterday resting in the familiar security of his bedroom, surrounded by baseball cards. The sudden shift of circumstances caused by his father's drunkenness angers him to extreme rage "a rage, a rage that jarred him like the approach of a freight train" (57). The powerful force of the approaching freight train, like the similar railroad metaphor in *The Natural*, represents the forceful pressures and problems of the unavoidable adulthood Dick is entering. Lodged within the scene's image of his puerile but vehement rage at the horror of this first taste of adult reality is a hint of the inner confusion and personal anguish that will stalk him as a man. At this point we realize that this chapter's flashback to Dixon's youth is designed to remind the reader of the apparently universal feelings, fears, and confusions associated with early adulthood.

Also depicted in his recollection of the St. Louis episode is another important Dixon characteristic: his wise practicality. Even in the midst of harsh adversity, his uncommon maturity and common sense surface to set everything—if not his psyche—right. For example, in the St. Louis game scene, when he finally finds Sarge and friend "sprawled out" on the bleacher stairs, he immediately begins brushing off both men's clothes, removing peanut shells and popcorn from their hair. He even gives his father a few experienced gentle slaps to sober him (58). Here is our first glimpse of Dick's natural tendency to take charge in grim situations and to serve as a solid brace for weaker companions. Later in the novel, we see him similarly supportive and sensible regardless of the nature and extent of his own pain.

This profile of the hero's character is placed within a storyline modelled as a bildungsroman, that genre which recounts the development of a sensitive individual seeking to learn the nature of the world and to find his way in it. Presented nonsequentially, the book opens on a scrubby Mexican "beisbol" diamond with the adult Dixon, a major-leaguer "on sabbatical" from his club, "sweating out a no-hitter" as pitcher for the local San Lorenzo team. The game is being played in the cactus-trimmed village of Río Loja against the home team, called *el Ultimo Esfuerzo* (ultimate effort), and Dixon faces the last three outs in the ninth inning of a possible perfect game. Herrin stretches out the hair-raising inning's remaining pitches to fill the entire chapter by describing Dixon's anxiety and excitement. Struggling not to jinx himself with thoughts of the perfect game, Dixon reminds himself of past pitching performances when he had coasted through enough log-jammed ninth innings in the major leagues to make the one in Río Loja laughable. But reason does not quell his overcharged emotions, and he quickly finds himself praying to the "God of strikeouts" (7). He aches for the big one, the perfect "000000000" which has eluded him for two decades of "watching the ball pass the batter in a tracer-gleam of light" (3).

However, just as he is about to accomplish it, to his horror a limousine full of *federales* intent on a Yankee-bust appears on the perimeter of the field, intervening after eight and two-thirds innings and a mere two strikeouts away from his highest goal. To save his skin, he must lose the game and escape the ballpark with the other *norteamericano* on the San Lorenzo team. Only a breath separates him from the tawdry Rlio Loja present and the yearned-for perfection of eternity, but after nearly nine full innings pitched faultlessly, suddenly there is no time for the consummating perfection (18).

Here, in the agony of a just-missed perfect game, Dixon must settle for a reality more tawdry and much messier than the glorious ideal lodged in his imagination. This of course resonates in the reader's mind alongside the debacle of his first Cardinals game with his father and, later, the shambles of his first marriage. That is precisely Herrin's point: despite the florid rhapsodies to baseball as art by its devotees—Lamar Herrin included—baseball, like life, is a difficult and experienced process, not a pristine and unlived work of art.

The early sections of the novel situate Dixon in Mexico and explain how he ended up there. We learn what brought him south as we follow his own sorting out process, which consists of piecing together "my past by threes," an "exercise" he uses to try to overturn a "will" made listless by the nearly-reached perfect game. As we accompany him in his flashback in three-year blocks, we discover that not even the immense satisfaction of his former years as a U.S. major-leaguer with the Cincinnati Brewmasters, nor a World Series victory over the Yankees the previous year, can ease Dixon's frustration with his life and failing marriage with alcoholic Lorraine. Despite being one of the Series heroes, he feels empty.

Into this personal void arrives a letter from Lester King, Dick's college mentor and once intimate friend (until he learned that Lester was among Lorraine's long roster of extramarital lovers), congratulating him on his Series performance and inviting the sports star to visit him at "the top of the world—Eagle Creek, Wyoming" (268). Instead, Dixon writes to Lorraine to confirm her divorce terms and promptly heads the opposite direction, south to Mexico. He is not altogether sure why he makes this decision, but he is sure it is not simply "schoolboy" contrariness (271). All he knows for certain is that Mexico is "available." By "available" Dixon seems to mean that rugged and graceful Mexico in its unremarkable simplicity, allows him the freedom of his true identity (212). The country offers a freedom previously unknown to the man still fettered by his mother's demands for small-town Missouri decorum, his father's career failures, his mentor Lester's intense intellectuality (partly responsible for his seeking his "guiding metaphor" in baseball), and by his wife's alcoholic need to control him. In Mexico he is Ricardo Dixón, a composite of all his selves, happily free to be himself.

The talented gringo pitcher finds several important parallels between himself and his Mexican teammates, whom he sees as "short on God-given talent" but fearless and charming nonetheless (7). He shares their basic toughness—the kind that tames the desert's hostile mesquite and the "horrorscape" of cactus with "adobe homesteads and ragged gardens," with "pulque and tequila," as well as with such practiced human rituals as plaza courtships and bullfighting. He thinks the *mejicanos* themselves recognize that at the core of all such rituals is their plaintive cry against the limits of mortality. Their toughness appeals to Dixon's earnest practicality and common sense. For the same reason, he is naturally attracted to Juan Antonio, the local San Lorenzo *beisbolista* team manager. Juan's love for *beisbol* perfectly matches his resilience, and both enable him to assemble a team from local material so bush it would make backwoods Missouri slowpitch look major-league. The two men find each other and instantly fill each other's mutual needs.

Besides giving him the chance to play serious, if bush-league, baseball, Juan Antonio also gives his blessing to Dick's marriage to his daughter, Consuelo. The wary pitcher describes Consuelo as "so good. So giving . . . [and] delicately removed and so passionately here, so incredulous of greed and so won by generosity"; he sees her as the perfection for whom his entire life has been an apprenticeship (38–9). Unlike Lorraine and Terry and Mercedes, the "three women he'd given his love to" in the past (145), Consuelo elicits from him an unselfishness that fulfills him because it fulfills her too. Accordingly, near the end of the novel when he can finally think about his first wife, his parents, and erstwhile friends without bitterness, his ruminations disclose mature insight into and genuine respect for the simple fact of family life: "It doesn't come to much, but it's all here" (291–92). Ultimately what Ricardo and Consuelo achieve together—captured aptly by the obvious but never trite symbol of themselves as a family—is completeness.

Herrin foreshadows the affirmation of the novel's ending effectively in chapter two with the ritual of the *romero* bath that marries the couple. The bath, a nuptial baptism, substitutes for the church wedding forbidden them by Dick's divorce. Carlota, the steely family matriarch, insists that her daughter and the gringo perform the Spanish custom of *el baño de romero* to wash away the stains of the past. The bath traditionally prepares a couple for marriage. Carlota tells them that the God of mountains, winds, and stars blesses the *romero*, and that before it occurs not even kissing on the lips is permitted (32). Like *The Natural*'s pivotal nighttime beach scene, with its cleansing and baptismal imagery, which serves to spiritually bind Roy Hobbs and Iris Lemon, the bath of *romero* ritually unites Dick and Consuelo within a primal tradition that helps them discover one another. The epiphanic moment secures his love for her "as if she were the only yea in a world of nay" (35).

The *romero* scene reveals another important resemblance between *The Natural* and *The Río Loja Ringmaster*, allusion to myth. Introduced by the title and developed in the first chapter which, as mentioned before, is out of sequence with the story's chronology, Herrin's allusion is to Aztec myth and history. The author sought an indigenous Mexican ball game, not soccer imported from Spain or baseball from the United States. He found his aboriginal sport in the ruins of oblong Aztec courts built for the game of *tlachtli*, whose rarely attained object was to strike a small rubber ball with the hip or knee to make it pass through a designated ring (9). In his near-perfect game against *el Ultimo Esfuerzo*, Dixon identifies himself with the "Aztec-tlachtli ringmaster, a player who earned that title by striking the small rubber ball through the ring." For this grand feat, the ringmaster would even be entitled to the clothes and other portable property of the spectators in attendance (11). As Dixon nears his loftiest lifetime goal of pitching, the perfect game, he thinks of himself as "The Río Loja Ringmaster," a title quite worthy of Cooperstown in his mind (11).

Although not referred to again significantly in the novel, the ancient *tlachtli* and ringmaster legend, placed strategically as they are in the title and opening chapter, shape the novel's theme. Written as a bildungsroman, the book portrays Dixon's rite of passage from innocent idealist (the hymn singing choirboy) to wounded pessimist (the cuckold) and then to wry cynic (his self-preserving machismo among the *beisbolistas*). Herrin employs the ringmaster myth to emphasize Dixon's self-absorption. At this point, his highest aim in life is to pitch a perfect game; it is what he thinks he was sent to Rio Loja and, even, into this world to do (16). The arrogance of this single-minded hubris leads him to draw parallels between himself and such conquering heroes as Hernando Cortés and Abner Doubleday. But he does not limit himself to the historical legends—he also compares himself to such deities as Quetzalcoatl and Icarus.

At the start of his story, the "rings and zeros" of baseball's perfect game are what motivate him—just as they motivate another Apollonian character, Wallop's Joe Hardy in *The Year the Yankees Lost the Pennant*. Subsequently, however, after finding his tortuous way in the world as lover and cuckold, celebrity major-leaguer, expatriate American, Mexican bush-leaguer, and eventually as consolation's (Consuelo's) husband, Dixon discovers another—a higher—form of "ringmaster" tale with which to compare himself. In the epilogue he reviews his life, its debits and assets, and, in the playfield of his imagination (Candelaria, 1987: 1), pitches a final game. In this "crazy" game, the players consist of all the intimates of his life past and present, including his wife and son, who heckle him with endearments into altering his pitching style to allow hitters to hit the ball, not miss it. In other words, he redefines winning. To his amazement, he "forget[s]" who he is and begins to measure himself by "how hard and far they hit" the ball. A line drive sends a "mellow glow" through his bones. He begins to throw "affectionate lobs" to the plate totally unconcerned with winning (304). His success in this game "as it should be played" (305) soon leads to chants calling for the "Ringmaster! the Ringmaster!" As a consequence of his playing to allow others to win, to make them look good, even at his expense, "the mysteries unfold" to show him "the other half of my world" (305). What he finds there is a fun-loving, uncompetitive, and love-filled part of himself that had previously been shielded from open expression by his self-centered hubris (304–305).

While Dick's instinctive nurturing tendency and his capacity to care deeply about others might hint at a feminist or androgynous perspective on the author's part, it does not appear to be the case. In general, Herrin's characterizations and plot exhibit only modest awareness of feminist insights and values. For the most part, *The Río Loja Ringmaster*, like most of the titles examined in this book, reflects an unexamined, unquestioned portrayal of conventional gender norms. This is not to say that enlightened writers must avoid characters reflecting conventional ideas of masculinity and femininity. Rather, character portrayals that are products of a genuinely enlightened feminist consciousness disclose awareness of the conventions and, importantly, of their stultifying limitations.

As with Malamud's modernization of Greek myth in *The Natural*, Herrin's use of Aztec legend serves a similar purpose in *The Río Loja Ringmaster*. Both contemporary writers draw, in Levi-Strauss's words, "logical relationships between one domain and other[s]" (*Raw and the Cooked*, 191)—that is, between the experience of modern American life and the values and morals of ancient myths. Both find useful to their mythopoeic fictions the primal mythmaking of antiquity, for it allows them to place their twentieth-century concerns within ageless, universal dramas of shared humanity. By choosing baseball as their framework of allusion, Malamud and Herrin confirm the centrality of the national pastime within American identity. By embedding their baseball fictions within the larger texts and textures of world myth, they affirm the ethnopoetics of literature—that is, the direct applicability of the ancient human past to our unfolding present.

REFERENCES

Ralph Andreano, *No Joy in Mudville*, 1965.

Eliot Asinof, *Eight Men Out*, 1963.

The Baseball Encyclopedia, 6th ed, 1985.

Jonathan Baumbach, *The Landscape of Nightmare: Studies in the Contemporary American Novel* (New York University Press, 1965).

Thomas Boswell, *How Life Imitates the World Series* (New York: Penguin, 1983).

Cordelia (Chávez) Candelaria, "Seeking the Perfect Game in the Playfields of Fiction," *American Book Review* 9.2 (March-April 1987): 1–9.

Robert W. Creamer, *Babe: The Legend Comes to Life* (Simon & Schuster, 1974).

eter Gardella, "The Tao of Baseball," *Harper's* (Oct. 1986): 25–26.

Ihab Hassan, *Radical Innocence: Studies in the Contemporary American Novel* (Princeton: Princeton University Press, 1961).

Lamar Herrin, *The Río Loja Ringmaster* (Viking, 1977).

Johan Huizinga, *Homo Ludens*, 1955.

Patrick Allen Knisley, "The Interior Diamond," 1978.

F. C. Lane, "A Fallen Idol," *Baseball Magazine* (November 1925): 558, 565.

Claude Levi-Strauss, *The Raw and the Cooked: Introduction to a Science of Mythology*. John and Doreen Weightman, trans. (Harper Torchbooks: 1969).

Bernard Malamud, *The Natural* (Farrar, Straus and Giroux, 1952).

Michael Oriard, *Dreaming of Heroes*, 1982.

W. L. Reese, *Dictionary of Philosophy and Religion: Eastern and Western Thought* (Sussex, England: Harvester Press Ltd., 1980).

Sidney Richman, *Bernard Malamud* (Twayne, 1966).

John M. Rosenburg, *The Story of Baseball* (Random House, 1962).

Phillip Roth, "Writing American Fiction" (1961), in *Reading Myself and Others* (Farrar, Straus and Giroux, 1975).

Harold Seymour, *Baseball: The Golden Age*, 1971.

Tony Tanner, *City of Words*, 1971.

Frederick W. Turner, "Myth Inside and Out: Malamud's *The Natural*," *Novel* 1.2 (Winter 1968): 133–39.

Earl R. Wasserman, "*The Natural*: Malamud's World Ceres," *Centennial Review* 9.4 (Fall 1965): 438–60.

Paul Weiss, *Sport: A Philosophic Inquiry* (Carbondale: Southern Illinois University Press, 1969).

5

ESTABLISHING THE METONYM

Baseball is a *pastoral* sport, and I think the game can be best understood as this kind of art … it creates an atmosphere in which everything exists in harmony.

Murray Ross, "Football Red and Baseball Green"

If before 1950 Ring Lardner was the first name that surfaced when talking of baseball fiction, then after 1956 Mark Harris became synonymous with the genre. Harris's Henry "Author" Wiggen books appeared in the 1950s, and their popularity and warm critical reception eventually led to a complete tetralogy of novels narrated by the pitcher protagonist. The originality and fine quality, especially of the first two books, *The Southpaw* (1953) and *Bang the Drum Slowly* (1956), won their author considerable regard. Ralph Andreano, for example, likes Harris's "uniting" of "larger thematic values to the folk legend and mythology of baseball" to produce in Henry Wiggen a character of "towering strength" (Andreano, 21). Likewise, Norman Lavers finds Harris an effective chronicler of Americana. Later, the extremely favorable notice given the 1973 movie version of *Bang the Drum Slowly* further brightened Harris's place in the field of baseball fiction. Directed by John Hancock and starring Robert De Niro as catcher Bruce Pearson and Michael Moriarty as Wiggen, a role played on television by Paul Newman in 1956, the movie has been described as "the best sports movie anybody has ever made" (Cook, *National Observer*, np). More recently, baseball specialist Peter Bjarkman called it, with *The Natural*, one of "the two finest baseball films ever produced'" (Bjarkman, 4). Mark Harris, himself, deserves much of the credit for this achievement, for he wrote the screenplay.

Unlike Bernard Malamud and Lamar Herrin, who educe the mythic mean-ings of baseball in America by linking it to ancient Greek and Aztec myth and legend, Harris conveys the centrality of the sport metonymically by lo-cating his fictions squarely in the postwar U.S. present. His stories focus on the game and its players, on their personalities and relationships as adult athletes earning their living playing, in F. Scott Fitzgerald's phrase, "a boy's game" (Fitzgerald, "Ring," 36–37). Harris's "more 'personalized' versions of our human tragedy" (Bjarkman, 4) help establish baseball as a precise me-tonym for America in all its historical, social, and psychological aspects. Me-tonymy—wherein a part of something substitutes for the whole—assumes an unbroken consistency between part and whole ("hands" for "workers," for instance)—unlike metaphor, a figure of speech that relies on discontinuity for the freshness of its comparison. All the works examined in this chapter capture especially well baseball's role as a cultural metonym, that is, as a slice of American life that uniquely reflects and represents the whole.

In addition to the Harris tetralogy, this chapter covers the baseball fictions of two lesser known writers, John Alexander Graham and Jay Neugeboren. Despite their relatively low profiles, both are accomplished writers who have contributed significantly to the body of baseball fiction. Graham's *Babe Ruth Caught in a Snowstorm* and Neugeboren's *Sam's Legacy* joined Roth's *Great American Novel* and Brashler's *Bingo Long Traveling All-Stars and Motor Kings* to make 1973, the year of their publication, a benchmark date in the development of the genre. After 1973 there could be no question of baseball's peculiar precision as an important American literary symbol. Neugeboren's *Sam's Legacy*, with its subtle fictionalization of the Babe Ruth legend in the novel's stunning story-within-a-story, achieves a sophistication seen only in the finest works of baseball fiction. As discussed in chapter two, Neugeboren has a profound understanding of the singular role of sport in American life, and his appreciation of baseball's unique cultural niche is particularly deep. Like Harris, Graham and Neugeboren offer solid metonymic interpretations of the game.

THE HENRY "AUTHOR" WIGGEN BOOKS (MARK HARRIS)

Set in the vivid present of the 1950s, *The Southpaw* and *Bang the Drum Slowly* depict a preassassination, preVietnam, contemporary United States resolutely intent on shedding all traces of Hitler, the Nazi Holocaust, and Hiroshima, even as it is ignoring the killing in Korea. A collective yearning for a pastoral America of prelapsarian Edenic bliss, this intention actually reveals innocent ignorance because, in fact, neither past nor present pain can be shed simply because they hurt; the healing of wounds requires clear identification of an injury and its attendant problems. Both *The Southpaw* and *Bang the Drum Slowly* reflect the society's pastoral yearning, but they are not mindless, escapist idylls. Instead, their pastoral qualities are keyed to the baseball theme, described in the Murray Ross epigraph heading this chapter: it is "a *pastoral* sport," an art "in which everything exists in harmony."

The exact nature of the pastoral harmony differs slightly in each Harris novel, but it derives from the same two fundamental sources. The first and most important is Harris's undisguised love of and devotion to baseball, an affection that beams through even the darkest, most cynical passages in his fiction. In an essay for *The Nation*, he once expressed this admiring sentiment by saying that baseball ultimately is much "more than commerce," it "is a mystique" (126). The second source accounting for the pastoral quality of his novels is his choice of a first-person narrator, Wiggen, who is both a self-conscious author surrogate and a major-league ballplayer. The commitment and concentration demanded of major-league athletes is undeniable. The two traits are depicted as faith in the game in *The Southpaw*, and as respect for its essential purity in the later novels. These fundamental sources ramify into mature expressions of the pastoral, in which the neat harmony of baseball becomes the medium through which Harris telescopes his perspective on the complex, dynamic sprawl of American life.

The pastoral mode complements the author's intention to write a "good book about baseball," a refrain that echoes throughout his first baseball novel (Harris, *Southpaw*, 181). Through his narrator, young Henry Wiggen, who is just beginning his transition from high school to professional ballplaying, Harris indicates early that his "good book" will differ markedly from Ring Lardner's reductive satire.

Lardner did not seem to me to amount to much, half his stories containing women in them and the other half less about baseball then what was going on in the hotels and trains. He never seemed to care how the games come out. He wouldn't tell you much about the stats but only about bums and punks and second-raters that never had the stuff to begin with. (34–35).

He also intends to avoid the "corny crap" of juvenile baseball literature (35). Nevertheless, the vernacular first-person narration in Harris's novel harks back to Jack Keefe in Lardner's busher stories and also to their literary prototype, Huckleberry Finn.

These similarities suggest Harris's debt to his predecessor, but he also radically departs from the Lardner mold. Despite the infelicities of grammar ("punctuation freely inserted and spelling greatly improved by Mark Harris," says the book's title page) common to both Keefe and Wiggen, the latter possesses an intelligence and insight foreign to Keefe. This immediately distinguishes the two, for Lardner's caricature of Keefe, evinced through the busher's unwitting self-parody, precluded even the most basic self-knowledge. Wiggen develops self-knowledge to heroic proportion during the course of *The Southpaw* and later installments. In addition, Wiggen's romance with bright, sensitive Holly Webster contrasts sharply with Keefe's troubles with petty, dishonest Florrie. Harris avoids caricature by creating realistically rounded characters endowed with more talents and interests than those contained in a dugout and by placing them in vividly drawn baseball settings.

Because character development unfolds as Henry tells his autobiography, the only development we are sure of is his. Nonetheless, Harris makes it clear that his narrator's protagonist is a rich composite of the three intimate relationships comprising his personal life. They are Holly Webster Wiggin, her uncle Aaron Webster, and Henry's father, Pop.

At the end of chapter one he introduces "1 other person I had better mention fairly quick is Holly Webster, Aaron's niece, now Mrs. Henry W. Wiggen by marriage" (18). Aaron Webster, an astronomer who directs a scientific observatory, is the Wiggins' neighbor in fictive Perkinsville, New York. Henry's relationship with Holly, like his ties to Pop and Aaron, forms an important anchor in the novel's plot. She is the only character from his boyhood who accompanies Wiggen through the four novels of the series, and she represents a constant force, both as character and symbol, in his life. But he doesn't know that at first: "As a kid I couldn't stand her . . . because of some idea I must of had that your real ballplayer steered clear of girls" (25). The transformation of their familiar childhood friendship into romance puts an end to that idea very early in the book.

Then all the beating stopped, and the various hearts in their various places all drifted back to the 1 single slot where they belonged, and the cold got warm and the shivering stopped and my sweating body dried, and I was peaceful. I begun to hear the sounds of the night . . . and I seen stars in the window. (46)

The tranquility described in this passage comes after their first "fornication," as he calls it in an attempt at euphemism. Their peaceful union sets the tone for their relationship and removes it from the expected sturm und drang normally associated with youthful romance. In plotting their story in this way, Harris sacrificed a tenable strand of plot and risked losing reader interest. On the other hand, the well-calculated risk widened his canvas to deeper, subtler characterizations and portrayals of the game, inside and out, and also allowed a richer depiction of Holly.

Holly's influence upon Henry helps shape the three most important areas of his life—ballplaying, personal values, and writing—and his attitudes toward them form the essence of his character. Although he receives most of his basic baseball training from his father, he claims that Holly "give me the best advice anyone ever give me concerning baseball and how to play it. She said, 'Henry, you must play ball like it does not matter, for it really does not matter. Nothing really matters'" (135). The tenor of her simple assertion serves as the basis for his existential values and the unflappable temperament that come to define him especially on the diamond, where his coolness as a competitor makes him a major force.

Similarly, her devotion to literature, like Aaron's, which he hardly appreciates at first but which has affected him by osmosis, later becomes a direct influence. She reads Shakespeare, Donne, and other "great poets" to him, and she encourages his writing, saying that "any lunkhead can play baseball but he has got to be something special to write a poem" (63). When he begins to take his writing seriously enough to share it with her, she gives helpful editorial advice. Like his mother, who died when he was two and who loved literature enough to choose Whittier, "after the poet," for his middle name, Holly provides the title for his first book:

"You are a lefthander, Henry. You always was. And the world needs all the lefthanders it can get, for it is a righthanded world. You are a southpaw in a starboarded atmosphere. Do you understand?" (307)

Through Holly, then, Harris accomplishes a major first in baseball fiction, the creation of a fully three-dimensional female character whose role in the sports hero's life builds not from cultural stereotypes of femininity but from human complexity and artistry. This makes "the southpaw" even more untypical in his taste.

A large measure of Henry's "southpaw" unconventionality stems from his pacifist and socialist values, which, in the forgetful fifties' yearning to make the present into an idyllic ideal, alienate him from society's norms. Holly's uncle Aaron Webster, a lifetime neighbor and friend, is responsible for the logic and sophistication of Henry's outlook. Aaron, an astronomer who has heeded his own drummer steadfastly over the years, reinforces the boy's "natural tendency" toward individualistic behavior that cuts against the grain (37). The older man's example of principled unorthodoxy prepares the future star to be confident in his views and tough in his differentness. On the second page of the novel, for instance, Henry describes Aaron's defiance of "the Government" in its attempt to use the Observatory for military purposes: "there was a good deal of discussion in the papers . . . [about] Aaron holding fast and finally winning out, and you have got to admire him for that" (15). Later, when the glare of Henry's celebrity status complicates his life, Aaron's strength guides him: "It is partly on account of Aaron that I am lefthanded in the first place [for] Aaron said there wasn't nothing wrong with being lefthanded" (14–15). Enhancing his excellence as a pitcher, Henry's left-handedness also symbolizes the political and personal qualities that set him apart (for example, 39, 103, 237) and, in *Bang the Drum Slowly*, that makes him a major-league player representative in labor negotiations.

Octogenarian, astronomer, and homespun intellectual, Aaron's characterization is cut from the cloth of Thoreauvian radicalism. Like his biblical namesake, elder brother and spokesman for the stutterer Moses, and his American forebears, Daniel and Noah Webster, Aaron is blessed with the gifts of reason, cultivated mind, and originality. The townspeople of Perkinsville, however, find him an unpatriotic eccentric, deservedly isolated from their mainstream. Even Henry fails to fully appreciate his mentor's mettle until, as a prodigal son on the wing, he is far from home.

It was not until more than 2 years afterwards, riding the lobby in Chicago 1 evening, that it all flashed in my mind, and I said upstairs to Perry Simpson later that night, "Does it not strike you as queer that at half past 5 in the cold morning it was not Bill Duffy nor Mayor Real nor Mugs O'Brien nor Jack Hand nor Mr. Gregory N. Oswald [all ostentatiously public supporters of hometown success Wiggen] that give me my send-off, but Aaron Webster in his raggedy green jacket?" (74).

As the national pastime, baseball is traditionally linked with frontier individualism and pioneer determination, even though it is often quite opposite— just another corrupt business practice in America. Through the character of Aaron Webster, Harris emphasizes the existence of these hardy virtues outside the national chauvinism.

If Holly and her uncle help sculpt Henry's moral and ideological difference from other baseball players, it is his father who encourages his natural athletic abilities. The schoolbus driver in Perkinsville and part-time caretaker of the Observatory, Pop had a highly successful start as a pitcher in semiprofessional baseball. But after two winning seasons, "Pop simply up and quit . . . and I do not know why, and over all the years I have pumped him . . . but never got a good [answer]" (17–18). Harris plants this seed of mystery early in the story and apparently forgets about it because we never do learn, here or in the three remaining books, why Pop gave up his promising career.

At any rate, Henry grows up under Pop's close tutelage, with baseball at the center of their relationship. Pop's customary greeting to his son ("How's the old flipper?") whether in person, on the telephone, or in letters, captures the indispensable nature of baseball to their filial bond. Importantly, Pop's advice to Henry about the special relationship between pitchers and catchers foreshadows the central plot conflicts of the second and third novels in the series, when Henry is an established major league player. "Pop always told me you are libel to find yourself pitching to a lunkheaded catcher, and if that happens you have got to set him straight" (29). Bruce Pearson in *Bang the Drum Slowly* and Piney Woods in *A Ticket for a Seamstitch*, catchers for the New York Mammoths team, prove Pop's talent for preparing his son for his future as a pitcher, even to the extent of uncanny prophecy—handled by Harris, it should be said, with subtle realism.

Together, Holly, Aaron, Pop, and Henry comprise a well-integrated family whose individual lives and interactions with one another constitute important narrative strands of interest. These strands deepen the texture of the Wiggen tetralogy by securing the baseball story to hearts, minds, and souls not directly involved with the Mammoths' organization or the sport. They—a housewife, a scientist, and a schoolbus driver—are the reader's metonymic touchstones to the rest of America.

The novel's Chapter 11-A conveys their familial cohesion dramatically, and it also raises another important theme, the act and craft of fiction. The entire chapter presents the actual technical construction of Henry's bildungsroman, "The Southpaw," the text in our hands. "I begun this book last October, and it is now January, and I doubt that I am halfway through," writes Henry in the early morning hours right after reading his handwritten manuscript to Holly, Pop, and Aaron (99). Disgusted and tired, Henry has had his confidence shaken by their penetrating criticism and advises "any sap with the itch to write a book—do not begin it in the first place" (100). Despite his wounded ego and grumbling defensiveness, the reader realizes that he has taken their editorial advice and cut pages of irrelevant digression and added careful descriptions of the "inside" game of baseball. He has also conformed to their suggestion to stick to the main point: "I went up to the Mammoths in September of the second summer, and I pitched 1 inning of relief against Boston" (109). Humorously developed, Chapter 11-A, which replaces his original Chapter 12, broadens the characterizations of the protagonist and his three childhood foils and also draws attention to the craft of fiction itself.

The chapter calls to mind Huck Finn's pronouncements about writing in Mark Twain's classic: "if I'd 'a' knowed what a trouble it was to make a book I wouldn't 'a' tackled it, and ain't a-going to no more." Like Huck's, Henry's complaints about his struggling apprenticeship as a writer both anticipate and deflect any disfavor the real novel, *The Southpaw*, receives. Unlike Twain's work, whose closing chapters suggest a protagonist essentially unchanged by his adventures, Harris' novel presents a narrator gaining insight into himself and others through the difficult act of writing his autobiographical novel. Moreover, Harris' pervasive concern with the craft of fiction in this novel anticipates its appearance as a central theme in *Bang the Drum Slowly*.

If fictionmaking itself contributes to the narrator's personal development, it is by focusing on baseball in its quintessence that Harris achieves his intention of writing "a good book about baseball." As critic Tony Tanner observes, "once the writer has defined a subject of sufficient bed-rock firmness, he can build on it theatres of possibility, houses of fiction, and indeed any other sort of metaphorical domicile in which his constructive energies and delights can find full play" (Tanner, 440). Baseball has been Harris's bedrock subject. Although he has written nine other books of fiction and contemporary satire, his place in the national literary hall of fame stems from his pacesetting originality as a baseball fictionist. His understanding of the sport, its meaning in American life, both as play and as commerce, and his ability to distill this into literature place Harris alongside Lardner, Malamud, Coover, and Roth as masters of the genre.

From the beginning of *The Southpaw*, Harris creates a feel for the inside game of baseball, that played by managers and players and controlled by corporate owners. For example, Pop's explanations early in the narrative of how to account for external factors that affect hitting and pitching, like the weather and stadium differences, prepares Henry, and the reader, for the esoteric counsel he later receives from Mammoth manager Dutch Snell and player-coach Red Traphagen. From them, as well as from his own careful observations, he refines his baseball intellect and intuition—the traits that attracted him to the game in the first place.

I also played basketball . . . [and] consider it pretty nearly as dreary as football. . . . Where do your brains come in, and your speed and your lightning strategy, and your careful decisions? Compared to baseball what they call these contact games are about on a level with a subway jam. (67)

Henry's lengthy descriptions of spring training, game strategy, the machinations of Mammoth owners Lester T. Moors, Jr., and his daughter, Patricia, and other behind the scenes activities describe a varied and complex milieu— quite unlike, incidentally, the narrow-minded ignorance of Lardner's pawnlike players—and provide a striking local-color texture.

Other examples of this local color include detailed descriptions of rookie training camps, contract negotiations, and the full official roster of the Mammoths ball club. Interesting in themselves, these features also show Henry, the shrewd but brashly callow youth, discovering the professional world of baseball he had idolized but known only from a distance. When he first meets his boyhood idol, Sad Sam Yale, on the practice field, for instance, he is so overcome with enthusiasm that he gushes, "Sam, you are fatter than in your pictures" (82). Ignorance and overconfidence compete for mastery over his innocence in a variety of early encounters in his new world, traits reminiscent of other green conquerors in fiction like Robin in "My Kinsman, Major Molineux," Holden Caulfield in *Catcher in the Rye*, and Roy Hobbs in *The Natural*.

Part of Harris's skill in portraying Henry lies in the precarious balance of contradictions he ascribes to his hero. Henry is at once ungrammatical but literary, athletically tough but pacifistic even in self-defense, youthfully virile but loyal to Holly, one of the team's boys but comfortably secure in his "lefthand" opinions and convictions. These contradictions play off each other in his autobiographical meditations, and they also play off the other characters, especially Holly and the experienced catcher and worldly-wise cynic, Red Traphagen, assuring that the portrayal's delicate balance remains realistically drawn. What keeps Henry from seeming impossibly romanticized as an Adonic hero is the busher, grass-roots authenticity of his first-person voice.

The Southpaw also effectively depicts the camaraderie of the players and their dugout and locker-room interactions. On the first night of his third Mammoth spring training camp, for example, Henry, always the early bird, has the "blues, lonesome as the moon and not a soul to talk to." He is saved from melancholy by the arrival of Traphagen, who is becoming an important mentor for him.

I remember to this day Red standing there and asking me how my flipper was. I do not know why I remember yet I do. You will have pictures in your mind of certain ballplayers doing certain things. . . . And Red I remember . . . all loaded down in his gear, his mask and cap throwed off. . . . Then he asked me up to his room, and we went . . . .

He was in good spirits that night, *and I forgot how lonesome I was* . . . . He told me some stories about [individual players] . . . and some about the club in general. (130–31, italics added)

Here as elsewhere, Harris captures the men's relations with a naturalness that arises from mature insight and painstaking technique. Similarly, he develops another thread of plot and characterization through the Mammoth quartet formed, casually at first, by Henry and three teammates: "Coker and Canada and Perry [and me] . . . had not sung in a bunch since Q.C. [Queen City, the Mammoth farm team] the summer before. If we was rusty it never showed" (156). Henry's fondness for music and singing becomes a central metaphor in *Bang the Drum Slowly*, but even in this first novel, he thinks that when you add up all the things that made the club [successful] . . . you have got to take the singing into consideration, for it done something, just like Dutch's lectures . . . just like the hard work . . . done something to make us what we was" (157).

Well-rounded and complete in itself, *The Southpaw* also looks ahead to the next two novels in the series. Henry Wiggen introduced in the first novel as pitcher, lover, and nascent writer, becomes "Author" Wiggen, father and insurance agent, in *Bang the Drum Slowly*. The appearance of Bruce Pearson as a minor character in the first book prepares us for his major role in the second novel, just as the music imagery in the first builds to a major structure of allusion in the second. *The Southpaw*'s relatively enlightened depiction of women, particularly in its 1950s context, sets up the theme of *A Ticket for a Seamstitch*, the third novel. *It Looked Like For Ever*, however, lies apart, separated by two decades' time and lacking the vigor and intensity of played baseball as evoked in the first two books of the series. Nonetheless, it continues the Wiggen saga through the year of his forced retirement from the Mammoths. In sum, *The Southpaw* delineates all the essential characteristics of the Henry Wiggen character, which, added together, produce an appealing Adonic hero.

Bang the Drum Slowly (1956) is about dying and death, which is one way of saying it is about living and life seriously understood. Harris insists on this point throughout the second novel in the series, underscoring how easy and convenient it is to overlook death's constant presence within the rush and routine of life. Forgetting that, the novel says, diminishes the quality of living and whatever meaning life holds. In the words of Wiggen's teammate Bruce Pearson, the "doomeded" catcher, "Probably everybody be nice to you if he knew you were dying" (140). Such is precisely the team's response to the news of his terminal illness, and for once it "was a club, like it should of been all year but never was but all of a sudden become" (241).

If *The Southpaw* is Harris's "good book about baseball," then this novel is his contribution to the timeless tradition of memento mori—that is, art that reminds us we must die. Recognizing the fact of our mortality should improve our conduct and values, suggests the novel, by focusing our cares upon the truly essential and important. Throughout *Bang the Drum Slowly*, Harris plays variations on this central theme. Still, at the end, when Pearson is dead and buried, narrator Wiggen confesses to not having sent "the scorecard from Detroit" the sick man had requested before going home to Georgia to die; the reader realizes that human mortality itself produces the imperfections that make flawless conduct impossible.

With the same earthy warmth and naturalness seen in *The Southpaw*, the novel conveys a realistic feel for baseball's atmosphere from dugout to diamond. Wiggen's narration as a baseball insider, with many of the same interests and limitations of his teammates, continues here with the easy humor and occasional raucous comedy of the previous novel. Nevertheless, his distinction as a southpaw, both philosophically and athletically, deepens the texture of *Bang the Drum Slowly*. His greater maturity now that his rookie years have past and he has enjoyed twenty-plus game-winning seasons further enrich the work's texture.

Besides writing a memento mori within the local-color setting of baseball, Harris has other things in mind as well, and clues to them appear in the novel's opening epigraph. A passage from Wright Morris's *The Huge Season*, the citation alludes to Hemingway's *The Sun Also Rises*.

"Take this book here, old man—" and held up one of the books he had swiped from some library. Along with the numbers I could see Hemingway's name on the spine. "There's a prizefighter in it, old man, but it's not about a prizefighter."
"Is it about the sun rising?" . . . .
"Goddam if I know what it's about," he said.

Harris makes two important points with the Morris epigraph: that *Bang the Drum Slowly* is about more than just baseball, and that multiple allusions provide a framework for the novel.

It might seem odd that the same writer who sought to write "a good book about baseball" in *The Southpaw*, now wants to be perceived differently. Yet, in his concern that his second book not be read narrowly, Harris is answering the critics who dismissed his first novel solely because it was about baseball. Wiggen's mentor Red Traphagen speaks directly to this point after reading the first two chapters of his manuscript, the fictional "Bang the Drum Slowly" within the Harris book:

"But it must be more about Pearson being doomeded, which is what we all are . . . the people that read it will think it is about baseball or some such stupidity as that, for baseball is stupid, Author, and I hope you put it in your book, a game rigged by rich idiots to keep poor idiots from wising up to how poor they are. . . . Stick to death and Pearson." (207)

Traphagen's hardboiled advice reminds the reader again that the seasoned catcher is Mark Harris's alter ego. Wiggen tells us in each of the first three books that for a time Traphagen quit baseball "went to San Francisco and taught in the college there" (342), a direct parallel to Harris, who taught at San Francisco State University for a number of years. Like Aaron Webster, Traphagen represents the rare independent thinker, whose views and actions do not gain their validity by conforming to the dominant consensus (compare Mark Harris's viewpoint in his mordant satire, *Killing Everybody* [1973]). His biting comments throughout the novel underscore Harris's concern that *Bang the Drum Slowly* be understood in all its nuance and texture.

The second point communicated by the Wright Morris epigraph concerns the structure of multiple allusions through which Harris builds *Bang the Drum Slowly*. In addition to the Hemingway reference within the epigraph, Harris employs numerous musical images. The book's title, for instance, originates in the folk ballad, "The Streets of Laredo":

> O Bang the drum slowly and play the fife lowly,
> Play the dead march as they carry me on,
> Put bunches of roses all over my coffin
> Roses to deaden the clods as they fall. (212)

Rookie Piney Woods sings the ballad in the clubhouse after the word is out about Pearson's illness, and, later, the catcher sings bars of it to himself in the hospital. In the ballad, the dying cowboy's belated recognition of the fact of death leads to his solemn preparation for burial, just as the fact of Pearson's Hodgkin's disease forces him to concentrate on his playing for the first time in his life. As a result, his last season is his best. Similarly, his dying motivates his teammates to treat him, a busher they had been used to dismissing as "plum dumb," better than ever. Their changed attitude leads them to discover that he actually deserves legitimate respect. The allusion to "Streets of Laredo" thus stresses the novel's theme: mortality allows hardly enough time to live a decent life, because the fact of death seems to require dying before it can genuinely be understood.

Like the title's allusion to song, other musical imagery in the novel serves as refrains on death and dying to counterpoint the daily details which fill the Mammoths' life. Pearson's favorite tune is "Yes! We Have No Bananas," usually a spirited, frothy number, but in the context of his condition the negatives in the song produce a surprising moroseness. In the same way, Piney Woods adopts as his personal theme song for his motorcycle "Come, Josephine, in My Flying Machine," which Harris quotes in full. Because of Pearson, the song's references to flying and soaring take on spiritual meaning, a point related to Queen City farm club manager Mike Mulrooney's repeated comment that Bruce is "just passing on." Similarly, when Pearson's mother dies in Georgia midway in the novel, the song "M-O-T-H-E-R," sung by Wiggen, reasonates with unexpected melancholy. Even the normally innocuous "Happy Birthday to You" acquires a baleful note as teammates Goose Williams and Horse Byrd, the only ones besides Wiggen who at first know about the disease, cope with the secret (221). Overall, the allusions to popular song show conventional feelings being transfigured by the team's expanded consciousness about death.

An analogous allusion to popular art is the interpolation of material from *The Southpaw*, as Wiggen emphasizes his other role as "Author," the team's new nickname for him. He frequently alludes to scenes and comments in *The Southpaw*, citing page numbers for both first edition and "quarter book." To introduce Pearson in *Bang the Drum Slowly*, for example, he simply excerpts two pages of description from his first novel. He also refers to the actual process of writing and editing more often and matter-of-factly than he did in *The Southpaw*. "I wrote Chapter 3 and then again 4 . . . like Red [Traphagen] said to, writing right there in the room with Bruce not 8 feet away. He . . . will never read the book itself when it is done, which might be any day now with luck and quiet" (486).

On a figurative level, Wiggen's repeated allusions to his first book correspond to his "keeping a book"—this is, a scouting report—on the opposition. Harris draws an exact analogy between a pitcher's meticulous observation of other teams and an artist's keen sensitivity to the world. For example, through his close observation of Pearson when he accompanies the catcher to his home in rural Georgia, Wiggen learns that, although he lacks worldly sophistication and even baseball intelligence, Pearson is wise to nature in ways difficult to communicate in the concrete and asphalt vocabulary of cities and urbanites.

He traveled according to rivers . . . . he knew which way they went by the way they flowed, and he knew how they flowed even if they weren't flowing . . . even if they were froze. . . . Moving south he noticed cows. . . . He knew what kind they were, milk or meat, and what was probably planted in the fields . . . and if birds were winter birds or the first birds of spring coming home. (346)

This awareness kindles new respect for Pearson on Wiggen's part, and it leads him to work more diligently in helping the mediocre catcher learn to keep a book on the other teams. After persistent prodding from "Arthur," his rustic mispronunciation of "Author," Pearson resolves to "keep a book . . . [to] have more confidence and brains" (30), and confronted with death, his concentration helps him accomplish his resolve. His playing improves, and suddenly the other teams "were keeping a book on him, which they never done before" (81). Accordingly, through the novel's framework of allusion, coupled with the accentuated baseball custom of keeping a book on players, Harris fore-

grounds the act and art of fiction. In the process he demonstrates that fictionmaking is not limited to writers or other artists.

Henry "Author" Wiggen's development as a character from the Perkinsville
rookie to the New York Mammouth MVP, from the occasionally petulant
southpaw to the coolly patient pitcher, stems only in part from his fraternal
commitment to Bruce Pearson, or his professional commitment to writing.
Other things in his life also change, and he grows with them. Now contentedly
married to Holly, their first child, Michele, named after Queen City manager
Mulrooney, is born by the end of *Bang the Drum Slowly*. Fatherhood helps
season him. In addition he has added another sideline besides writing to
augment their income and to prepare for the future, when his arm "goes."
He sells Arcturus Life Insurance, mostly to ballplayers, an occupation he finds
worthwhile because it allows writing off most of his daily expenses as tax
deductions and because it provides the often spendthrift players longterm
investment income. "The reason they call it 'Arcturus' is because [it] is the
nearest star, or else the brightest. I forget" (339). Actually, derived from the
Greek word for "guard," Arcturus is the name used in astronomy to refer to
the brightest star in its particular constellation. Its aural link to "Arthur,"
Pearson's mispronunciation of "Author," makes the symbolism blatant: Wiggen is the dying catcher's guardian, just as he has grown to be, as player and
person, the brightest star in the Mammoth constellation.

As in the first book, the references to astronomy are reminders of Holly's
uncle, Aaron Webster, and his observatory. Accordingly, Harris uses space,
stars, and astronomy in a way comparable to Herrin's and Malamud's use of
myth in their baseball novels, and to Coover's use of remythified ritual (see
chapter six below). They demonstrate the efficacy of baseball in all its multiplicity as an expansive symbol of American culture, of its myths and rituals.
Harris, likewise, deepens the baseball metaphor by suggesting a metonymic
parallel between the mathematical order and balance of the game and the
mathematical harmony of the cosmos. Moreover, just as baseball (and life)
are frequently punctuated by spontaneity, surprise, and aberration, so too is
the cosmic order of things. Constituted of stars (for example, Babe Ruth),
satellites (for example, farm teams), and orbits (for example, roster rotations),
baseball—suggested Harris in the early fifties—is undeniably fitting as a literary subject for even the most solemn universal themes, and it ought not
be relegated to the children's room of literature. *Bang the Drum Slowly* and
The Southpaw prove the metonym.

A Ticket for a Seamstitch (1957) lacks the rich fullness of the first two
"Author" Wiggen books. In its first edition, the novella suffered printing
gimmicks to appear longer than its size. The plot of the work, which would
have made a distinguished short story, is simple, as Harris himself indicates
when he summarizes it in a 1962 essay, "Easy Does It Not."

A young lady writes to Henry Wiggen from "somewhere out West" to tell him that
she will be in New York to watch the Mammoths play on July 4th. Henry is her hero.

But Henry has a wife, and for this reason (he says) he attempts to transfer the young lady's affection from himself to Thurston Woods, inevitably called "Piney," a twenty-year-old catcher with a passion for women of the Hollywood type, fast motorcycles, and low-slung automobiles.

We follow the young lady's cross-country journey. . . .

Piney Woods discovers, when the girl arrives, that she is no beauty. His dream has overshot reality. He begins to discover, however, that love and charm may reside even within a form less divine than Hollywood specifies. He learns, too, that mechanical progress may ring hollow. (113)

Not only does the author summarize the plot of *A Ticket for a Seamstitch* in three brief paragraphs, he also manages to convey its principal themes in the last three sentences quoted.

The book presents the Mammoths during their championship season the year following Bruce Pearson's death. Author's roommate is another catcher from Georgia, Piney Woods, an immature, outwardly unfeeling busher. His childlike anticipation of the arrival of the "seamstitch," a factory seamstress, constitutes the work's focus. In a double pun on Piney's nickname, Harris depicts him both "pining" for the girl he's never met and as hardheaded as the pines in the backwoods of his Georgia home. Piney's one-tracked preoccupation with his seamstitch fantasy suggests his immaturity, especially the stunting of his emotional development. Harris conveys this quality of his personality through Piney's macho attraction to fast motorcycles and conventional-looking starlet-type women: Piney carries a picture of himself in a helmet on a motorcycle with a girl behind him and hangs it up on hotel walls as the team travels from city to city.

To further capture the character's emotional adolescence, Wiggen emphasizes Piney's ability to hang paper on walls by friction, "rubbing a hunk of paper down the wall . . . and slapping it on so it hangs by friction. I tried 750 times to work up the friction like Piney does, but I simply can not" (537). Symbolically, the "friction" represents Piney's inability to get along with his teammates and to exhibit proper respect for either manager Dutch Schnell or most of the experienced players. In the same vein, we are told that Piney "never smiles, but he might sometimes laugh, and then when he does he bites it off quick for fear you might catch him at it" (538).

When the seamstitch arrives, Piney is paralyzed with disappointment at her "average" looks, "a tiny bit on the heavy side," and his basic immaturity nearly consumes him. He takes refuge in the womblike comfort of the club's bathtub where he falls into a deep sleep, "snoring, his jacket for a pillow and his hat over his eye" (587). But the reality that his youthfulness has successfully helped him evade finally intrudes upon him when manager Schnell runs the water over him and wakes him up. A later intrusion comes from Wiggen who tells him, "You got no friends. . . . You got only dreams," and then threatens to destroy Piney's prized guitar unless the cocky catcher gives the seamstitch the chance she deserves (140). Thus pressured, he reconsiders, and the book's remaining two pages leave the impression that Piney has begun to grow up.

Although considerably thinner as a novel than its predecessors, *A Ticket for a Seamstitch* contains outstanding descriptions of baseball action. In meticulous detail, Wiggen describes himself trying to warm up his arm on a cold April day and concludes, "my fast-ball will not sink until I am warm" (530). Later, he describes Piney's batting in the same game:

It was Piney drove home our first run. Canada [Smith] drew a walk to open the second, and Vincent Carucci moved him along, and Piney slammed one back through the box into center. He was hitting in the 7 spot in front of Coker [Roguski]. Piney is a sweep hitter . . . . He aims at the pitcher's hat. I hate that kind . . . . [H]e aims his drives down through the middle. . . a sweep hitter, that a pitcher throws outside to and sometimes loses. (533)

He also offers vivid description of Schnell's seasoned management style, which consists of techniques ranging from harangue to gentle persuasion, from the levying of fines to locker room humiliation, and from personal reminiscences to lavish compliments. Throughout, Wiggen evokes the action of the game and its context with camera precision and informed realism.

Written over two decades after the original trilogy, in 1979 Harris published *It Looked Like For Ever*, the concluding book in the Author Wiggen series. Bothered by an unpredictable fastball and a pesky prostate, Wiggen, now thirty-nine, faces the end of his era as ace hurler for the Mammoths: after nineteen seasons, they drop him from the roster. He refuses to accept the inevitable end of his major-league career with the grace and ease that Holly, in her usual wisdom, urges; he postpones retirement by foolishly making himself available to teams from New York and Washington to California and even Japan. His release by the Mammoths, a surprise to no one but himself, resurrects his adolescent insecurities about his talent and virility, which is to say that his unexpected joblessness elicits the unadulterated egotism that we recall from his *Southpaw* youth. With the single-minded fervor of one denying the reality of his midlife crisis, Wiggen insists on proving himself all over again as fit to play major-league ball.

In some ways, *It Looked Like For Ever* answers baseball enthusiast Roger Kahn's call for "a major novel" about the forced shift in perspective veteran players experience. Wrote Kahn in *The American Scholar*: "It is a harsh, jarring thing to have to shift dreams at thirty, and if there is ever to be a major novel written about baseball, I think it will have to come to grips with this theme" (347). The fourth of the Author Wiggen books comes to terms with the veteran athlete's midlife dilemma by showing the protagonist struggle through denial, depression, anger, and rebellion. After pursuing the Japanese offer, trying network announcing, and suffering through his first oldtimers game, Wiggen finally accepts his place at the end of his professional career and embarks on another stage in his life.

Following his style in the earlier novels, Harris employs here a montage of allusions to other works, songs, newspaper accounts, rosters, and so on. He opens chapter three, for example, with a magazine account of his hero's retirement:

RETIRED. Henry Wiggen, 40, released by the New York Mammoths after 19 years of singular toil; 27th winningest pitcher in baseball history . . . whose southpaw career spanned Korea and Vietnam and whose popularity fluctuated with the popularity of the wars he opposed; he refused to tour for the entertainment of American troops abroad. . . . Sometime author of autobiographical volumes . . . one of his rules of writing: "If you do not know how to spell a word make your very best guess at it." Shrewd, venturesome, instinctual, he was once described by baseball's intellectual Red Traphagen as having "the instincts of a crayfish"; Author Wiggen boasted that he never threw to the wrong base or invested a bad dollar; to his home in Perkinsville, N.Y. (*It Looked Like For Ever*, 34)

The journalistic notice distills the highlights of his life into an obituary-like thimble of information, which depresses him both because of its finality and because several months are added to his age. Other symbols of mortality disturb him as much. For example, the slogans on the T-shirts of the younger players, and his daughter's compulsive jingles, underscore for the "younger older player"—as he insists on describing himself—his apparent decrepitude in a youthful world.

The sharpest salt for Wiggen's menopausal wounds occurs, however, on the playing field, the very diamond he hates to leave, for the old-timers game. The game opens with 55,000 people singing "When Your Hair Has Turned to Silver," which Harris quotes at length. Wiggen's rage at being there compels him to pitch meaner and harder to the *real* oldsters than is decent. The rival who beat him out of the manger's job rebukes him: "This is just a *game* . . . . not an actual game. Use your sense" (243). But at this point Henry has lost his senses, and is too immature merely to play; he is busy proving to himself all over again that he is a pitcher. Harris thus shows his Adonic hero experiencing the same challenges and paradoxes familiar to others graying in the resolutely young environment of contemporary America.

Henry's foil in *It Looked Like For Ever*, the counterpart to catchers Bruce Pearson and Piney Woods in the preceding novels, is his youngest daughter, Hilary, a prepubescent child resembling a precocious American Alice in Wonderland. Harris constructs the novel's conflict by intertwining Henry's midlife trauma with Hilary's developmental troubles. Her major problem is a sudden onset of screaming tantrums, which disturb the family's tranquility to such an extent that she is under the care of a "physchiatrist" (Wiggen's spelling). As we follow the poignancy of the fading southpaw's stubborn refusal to face the reality of nature, we also observe Hilary's stubbornness in making the transition from child to teenager. His struggle is poignant, with occasional touches of humor, whereas hers is comic, with frequent scenes of hilarity.

Like Malamud's *The Natural*, this novel treats youthfulness as a time of recurrent trial and error. As Roy Hobbs finally realizes at the very end, if you do not learn from your mistakes you will repeat them. The original trilogy demonstrates that Wiggen learns this, especially during the *Bang the Drum Slowly* period, and grows healthily from busher to professional to mature adult. The fourth and presumably concluding novel shows how life—nature and one's own psychology—insists on tripping even the Adonic hero on his way to immortality. "When I first come [to the majors] it looked like for ever," Wiggen remembers, "I was confident, I was strong, I had enthusiasm, I had motivation . . . I could throw a baseball harder than almost any body. The rivers flow and the bridges stand while the people come and go" (71). Eventually, of course, he is jolted to his senses, and he gives up burning all his bridges and trying to swim against the rivers' flow.

His epiphany occurs when he is knocked out by a baseball smash hit to his head by a young hitter named Muddy Rivers. This is a perfect epiphany because it emphasizes his dense refusal to accept the reality of life flowing on and also because the catalyst is the unavoidable reality of a baseball in the brain. Before this discovery, however, his self-assurance had increased to megalomaniacal proportion: "What happened was that word got round that as soon as I entered the ball game every thing was over, no body in the world could hit me in short relief, I was the master" (270). But as he is forced to acknowledge, on his hospital bed with a fractured skull and loose retina, he foolishly ignored nature's warnings that age has deteriorated his reflexes and eyesight: "I should of knew the night in Montreal when the ball come back so fast . . . . [or] when I made the error in Boston" (272). Baseball itself taught him the hard lesson that "[b]aseball was over. For the first time in my life I was no longer a baseball player. . . . The rest of my life all ready begun. It was here" (275).

The fourth of the Author Wiggen books thus comes to terms with the veteran athlete's midlife dilemma by showing Author, a character endeared to the very heart of baseball fiction, struggling through depression, denial, anger, and rebellion to finally accept both his middle age and his place at the end of professional baseball's road. He needed the struggle to enable him finally to recognize that the end of that road could be the start of another path of vigorous life. *It Looked Like For Ever*, the title, accordingly sharpens into focus as a double entendre. At first Henry thinks his role in baseball looks like forever, but after begging to get on the roster of other teams, after Japan, and after his spot in the television announcing booth, he realizes that what is really "for ever" is change. The "It" of the title is the change inherent in finite mortality.

In the character of Henry "Author" Wiggen, Harris adds to American literature a voice and presence as distinct in outline as Huckleberry Finn, Quentin Compson, Henry Waugh, and Alexander Portnoy. Like those characters, Author possesses an uncommonly keen insight into the people and habits that comprise his environment. Although each character is actively part of his environment, at the same time each is a critical observer of it. Despite the occasional harshness of their critiques, the fact that they are even offered indicates that their true source is affection. In Harris's case, for all his satirical invective against the documented corruptions of baseball and America, government and society, he suggests, through Author, that public success, benevolence, and indulgence in the magic of pure play can indeed unite in harmony if human proportion is kept in sight.

THE RESURRECTION OF BABE RUTH (JOHN ALEXANDER GRAHAM AND JAY NEUGEBOREN)

To try to be cogent on the subject of Babe Ruth in baseball is to confront the complexity of literal history interwoven with both authentic folk legend and contrived commercialism. There is ample evidence documenting his extraordinary performance on the diamond and amazing antics off it, just as there are careful accounts describing both the development of the Babe Ruth legend and the exploitation of his name solely for entrepreneurial profit (see Creamer, Smelser, Seymour, and Voigt). Yet, to understand the totality of his impact on the game (which, because of its embeddedness, is to say on American society) depends on a willingness to consider and comprehend a person in mythic proportion. Another way, therefore, of establishing the metonym of baseball as America in *Seeking the Perfect Game* is by surveying the fascinating treatments of the Babe Ruth story in baseball fiction.

George Herman Ruth (1895–1948), known as Babe Ruth, the Sultan of Swat, and the Bambino, was born to parents whose vague negligence helped send him to an orphanage at an early age. His mother died when he was seventeen, and his father, a Baltimore saloonkeeper, was killed in a brawl after Ruth had achieved major-league success. He was thus, contrary to popular belief, not an orphan, even though at seven he was sent to a Catholic institution for wayward and homeless boys because of his incorrigible delinquency. After twelve years in and out of the orphanage, where he found helpful father surrogates, he was discovered by the Baltimore Orioles, at that time part of the International League, and at nineteen began his career in the majors.

When he first came up, Ruth was a big, strong youth over six feet tall and weighed 195 pounds. He had not yet acquired the big belly that gluttony gradually grew for him. But he had the large head, blue-black hair, prominent flat nose, massive shoulders, powerful biceps and wrists, incongruously slender, ballerina-like legs and ankles, and pigeon-toed walk so easily distinguishable and familiar to fans. He needed no number on his uniform for identification. (Seymour, *Golden Age*, 429–30)

His starting salary as a rookie left-handed pitcher in 1913 was 600 dollars for the season. His potential power was quickly recognized, however, and the

Boston Red Sox bought him from the Orioles in 1914. So began the American League career which was to span over two decades and numerous baseball records, several of which stand to this day.

In the six years Ruth pitched for Boston, as the *Baseball Encyclopedia* informs us, he had three eighteen-or-better winning seasons, including a phenomenal record twenty-nine consecutive scoreless innings in two World Series for an incredible 0.87 ERA. At the same time, he was honing his batting talent and, by 1918, was so effective that Boston used him as an outfielder on his pitching off-days to help the lineup with his hitting, ultimately producing a batting average well over .300 in each of four seasons. His status in the majors established, in 1920 the New York Yankees bought Ruth for the unheard of price of $425,000 to help make them pennant contenders. The Red Sox needed the trade to pay off its huge indebtedness. The megadeal paid off, and for the first time ever the Yankees won the American League Pennant in 1921.

Ruth's record fifty-nine homeruns in a season hit that same year stood for over three decades, and he still holds the number two spot (after Roger Maris). His overall statistics as a major league hurler include an overall 2.28 ERA with a .671 winning percentage, and he ranks in the top eight in two categories of lifetime pitching in the World Series. As a slugger he compiled a .342 lifetime batting average with a .326 average in ten Series games, and he is still number one lifetime batting leader in both homerun percentage and slugging average. His 714 total homeruns held first place for four decades and is still ranked number two (after Hank Aaron). This remarkable performance only touches the highlights of Ruth's extraordinary career. Only when the numbers are studied in detail and explicitly within the context of his multiple successes as pitcher, batter, and outfielder, only then is it possible to appreciate the *sui generis* nature of his achievement.

Apart from political leaders, Ruth was probably the first major national celebrity in modern America. He was "unmatched as a baseball gate attraction . . . . [and] idolized wherever he went . . . . Thousands swarmed to the parks, many of them not so much to see the game, a reporter [once] said, as to watch Ruth 'in a supreme moment' " (Seymour, *Golden Age*, 427). Opposing teams even began to require by contract that he play when the Yankees came to town; communities all over the nation began declaring "Babe Ruth Days," and, in the manner of London's royalty-watching press, several newspapers began running regular boxed columns covering "What Babe Ruth Did Today" (Seymour *Golden Age*, 428).

While his athletic accomplishments might have justified the sports fans' adulation, wrote many a journalist, his unrestrained hedonism, which included debauchery, gluttony, and alcoholic intemperance, hardly merited their adoration. Harold Seymour observes that even though Ruth's biography outwardly resembled the Horatio Alger story, the parallel ended with the rags to riches saga. "Unlike the Alger heroes, Ruth did not struggle . . . [through] hard work and patient penny-pinching, and his [libertine] lifestyle was foreign" to their rigid morality (*Golden Age*, 433).

Ruth's matchless record-setting performance and his enviable freedom to do as he pleased made him an irresistible magnet for baseball. More than anyone or anything, he is credited for saving the sport after the Black Sox scandal (see Creamer, Seymour, Voigt, et al.) Likewise, owner Jake Ruppert built the landmark Yankee Stadium, "the house that Ruth built," because of his titan's drawing power. That power also explains the candy bar named after him which further established his celebrity and continued the exploitation of his name by business.

For all these reasons, and the many others that grew out of his particular appeal to millions of fans, the Babe Ruth myth embodies three fundamental American beliefs. First, it exemplifies the democratic notion that even the lowliest picaro can, through native talent and/or abundant industry, become the equivalent of king. Second, it personifies the critical frontier idea that wealth and fame in massive doses confer a regal mantle of privilege allowing the rich and famous to enjoy without restraint every liberty and self-indulgence desired, regardless of consequence. The logical result of these two beliefs produces the third: exceptional individual achievement resulting from individual effort is valorized over attainments derived from either community or tradition. While U.S. America's societal rhetoric might sanctify traditional ideals of community, family, law and order, in the end the people venerate so-called self-made "kings" who flagrantly violate those ideals—like Babe Ruth, Ollie North, John Wayne, and Muhammed Ali.

This book has shown that baseball fiction reflects, illumines, and sometimes even anticipates the cycles of change in American society and its values. Especially true of its treatments of the Babe Ruth legend, serious baseball literature metonymically captures, with dramatic precision, the reality of American life, history, and lore. Writers of every style and philosophical persuasion have resurrected Babe Ruth in their fictions as hero and villain, as busher and knight—but mostly as quintessential embodiment of the American Dream, however the Dream is perceived. The resurrection of Babe Ruth, from Casey in Mudville, to Damon Rutherford in *The Universal Baseball Association*, to Gil Gamesh in *The Great American Novel*, and countless others repeatedly vivifies the myths of the eternal return and the Second Coming.

Two writers who mine this vein of baseball fiction as a metonym of the United States with singular originality and power are John Alexander Graham and Jay Neugeboren. In his *Babe Ruth Caught in a Snowstorm*, Graham parodies the Bambino's busher image to a reductive extreme in a modernistic novel of black humor. Yet, despite the parody, his respect for Ruth's greatness ultimately explains the novel's title. More straightforward a storyteller, Neugeboren in his *My Life and Death in the Negro American Baseball League: A Slave Narrative*, the embedded fiction within his novel, *Sam's Legacy*, transforms the Babe Ruth legend by recasting it outside the context of White America's major leagues and inside the Negro leagues. In the process, Ruth's deified image is cut to human proportion, and the skill of the Black players is humanized in striking ways.

Compressed within the title of Graham's novel, *Babe Ruth Caught in a Snowstorm* (1973), are its definitive thematic elements. The title refers to a "most treasured possession" belonging to Mr. Slezak, one of the book's narrators. The treasured memento is a gimcrack Babe Ruth paperweight.

It was nothing more than a glass ball on a plastic stand, the kind of thing that often has the Statue of Liberty or the Empire State Building inside, and when you shake it up you get a snowstorm. Only this one, instead of the Statue of Liberty, had a figurine of the former Yankee great finishing up one of those breeze-rippling home-run cuts he was justly famous for. (36)

That the legendary baseball hero, whom some compare to such heroes as Dionysus, the biblical Samson, and the Old English Beowulf (Smelser, 300), is trapped in a blizzard inside a plastic bubble holds no irony for Slezak. He finds the paperweight "neat!" Graham intends irony, however, and he uses the memento's incongruities to define his major theme concerning the discrepancy between actual and ideal in modern American life. Baseball's suitability as subject for this theme has already been noted; it primarily stems from the sport's genesis fable about a Cooperstown origin, which is discrepant with the game's evolution from earlier precursors.

Graham's plot is simple: innocent bliss becomes decadent alienation when fame and public success are confused for the attainment of a principle. The innocents in *Babe Ruth Caught in a Snowstorm* are the Wichita Wraiths, a baseball team created by Slezak, a peripatetic businessman and baseball fan. Slezak recruits his team in novel fashion, through a newspaper ad announcing "baseball players wanted for professional team. Must love the game. Ability not necessary—only desires counts" (12). The ad attracts a motley assortment of misfits and average Joes, who at first participate with an abundance of feeling and generosity. They enjoy the sheer pleasure of playing and are indifferent to fame or money.

During this period of the Wraiths' Edenic innocence, Slezak manages the team democratically, allowing each player to make his own coaching decisions. No one notices the absurdity of this misguided generosity at the beginning, because the team lucks off to a good start. The democratic approach yields astonishing team cohesiveness and outstanding athletic performance— both more remarkable when considered against the players' assorted neuroses and ludicrous athletic weaknesses. In its infancy, the team achieves a momentary utopia in which community and individuality unite in happy harmony. Graham captures this period in a delightful scene showing the Wraiths selecting their non-uniforms.

Chinchin got his all-white one, and George Lloyd got one in a very light purple ('lavender,' he said) . . . . Loomis's was nothing special, but it was the cheapest one they had . . . . Folger's uniform had no sleeves, so as to show off his arm muscles, and his pal Harrington got one with an ascot tie attached that he thought was cool. (63)

Owner-manager Slezak is understandably distressed by their crazy-quilt appearance, but his anxiety is eventually mitigated by the enthusiastic reception of the fans.

In the novel's opening section, "Our First Year," Graham creates a mock pastoral in which the Wraiths, like rustic clowns of yore, frolic in simple bliss. This section introduces the diverse cast of characters and their peculiarities before the action quickens with the onset of their unexpected and wholly implausible success as a professional baseball team. When the plot thickens with victory after victory, the novel's ironic comedy bursts through the introductory pastoral mask. As one of the players, Petashne, exults,

Nothing could stop us. We were hot again, hotter than before! More and more fans flocked to join the Yak Pack. Newspapers gave us all kinds of play. Kids stood in line for our autograph. There was even a big rash of Wichita Wraith bumper stickers. (95)

The incongruity of the inept team's fantastic luck buttresses the humor and builds suspense for the inevitable downfall. Without knowing it, they are caught within a bubble of illusion that will soon overcome them like a storm.

As a result of the team's success, the action shifts from arcadian harmony to a period of baffling chaos, personal isolation for the players, and lack of faith in the Wraith organization. The club is now replete with managers, coaches, accountants, lawyers, and public relations experts—all worldly specialists who separate the devoted Slezak from his team. Success also demands increasingly ridiculous methods of "staying in the black," a punning reference repeated throughout the story to profitmaking within their now "dark state of being." For example, on the advice of his expensive advisors and without telling the players, Slezak sells their new stadium's underground "earthspace" for use as a garbage dump. He also agrees to the development of an apartment complex on top of the stadium roof, thus sandwiching the team literally and figuratively between commercial enterprises far removed from the original motive of play which brought them together.

Slezak, too, changes—although, keeping with the theme regarding the disparity between what is and what seems to be, Graham maintains ambiguity about Slezak's original nature. Did success corrupt him, or was he always thus? Graham does not provide a clear answer. Throughout the team's turmoil, Slezak insists upon accounting for his destructive actions by distorting them with reports of continued harmony. He twists the calamities into bizarre accounts of happiness that invariably conclude that all's well with the team and players. When he is asked to explain what's wrong, Slezak stonewalls with the finesse of an accomplished latter-day politician. Graham thus exposes the lying duplicity commonplace in executive boardrooms.

The novel's structure also reinforces its twin themes, that success corrupts and that reality is not as it appears. The narration is the most visible example, as Graham combines the point of view of several characters in binary opposition. The book opens with a frame chapter, "The Scuffle," narrated omnisciently; the third-person narrator reappears only in the closing chapter, "The Upshot." In modernistic fashion, "The Upshot" serves as epilogue to the events that preceded "The Scuffle." The forty chapters in between comprise the body of the plot, and they are alternately narrated in chronological order by Slezak, the owner, and Oscar Petashne, a player. Consequently, chapter one is chronologically the novel's ending, a discrepancy resembling Slezak's and Petashne's contrasting versions of events. Both characters narrate their segments in the first person intentionally to confuse the reader about the source of a given perception or opinion. Their viewpoints converge at the beginning, during the Wraiths' period of harmony, but they gradually diverge when things get bad and when Slezak starts to distort reality to give his avarice a proper appearance. Obviously, because both narrators are fallible and inhabit the same deranged world, neither is reliable. The only parts of the fiction we know to be "true" are the violence of the apocalyptic ending, omnisciently narrated at the beginning, and the cover-up in the epilogue. The narrative form clearly offers a structural analogue to the theme that appearance and reality are constantly discrepant.

Graham employs other distinctive ways of foregrounding his theme. The team name, for instance, would indicate that it is based in America's heartland—Wichita, Kansas—when, in fact, the Wraiths play out of Braintree, Massachusetts, a town famous as "the birthplace of John Hancock and John Adams, those symbols of Revolutionary spirit." It is also "the town where the hold up and murders occurred that resulted in the executions of Sacco and Vanzetti," symbols of innocent victims of persecution (Cooper, "Murderer's Row," 84). The team's name, coupled with the historical events associated with Braintree, signifies American equivocation. That is, the spirit of the American Revolution was undermined by the executions of Sacco and Vanzetti, leaving the nation's heartland with only wraithlike reminders of the nation's founding spirit. The existence of Wichita Wraiths in Massachusetts also calls to mind the modern transfer of teams like the New York Dodgers to California or the Milwaukee Braves to Georgia.

The prime agents of duplicity in the novel belong to a uniquely American métier, advertising and public relations. If there is a single sinister element in the novel it is the PR agents who surround Slezak, and who emerge as the only ones profiting from the problems they create. Indeed, Graham depicts them as dramatic stereotypes of villainy. Their influence pervades everything in the story; their role in the conclusion's violence is certain, yet, appropriately masked. They are responsible for the Tranquility Department, the euphemistically identified military arsenal Slezak is "forced" to establish when the players go on strike. With characteristic efficacy, the stock characters cover their tracks smoothly, leaving Slezak to carry the blame alone (5).

Paralleling Graham's use of dramatic stereotypes in the stock villains is his use of theatrical masks to intensify the ending's "bedlam and violence." The two devices actually merge, for the disguises used by the professional team hired to beat the Wraiths are part of the ubiquitous promoters' scheme, even though Slezak again appears to be solely responsible because of his enthusiasm for the idea of staging a game between players and management (252–53). The masks also underscore the Wraiths' deterioration into mere shells of their original selves. Relevant here is the controversy over what type of turf to use in the new stadium. Trying to persuade team captain Petashne of the superiority of artificial turf when the players prefer natural grass, Slezak artfully insists that "this is *actual simulated* grass, not just some plastic turf. Looks and feels like the real thing" (149). But the players demand the *real* "real thing," and the owner brings in a smooth-talking salesman from the turf company, who argues that the artificial turf is "one of the most durable mnemomorphic plastics known . . . . *mnemo*-meaning memory, and -*morphic* meaning shape or form. Literally, memory for form. This grass, you see, this grass surrogate, I should say—'remembers' its original form indefinitely and always returns to it" (149–50). The Orwellian nature of the drama is assured when the false turf begins to seem to the players like real grass.

Like the mnemomorphic plastic, the Wraiths are estranged both from nature and, despite the masks and artifice, from genuine art, a predicament that reminds us of the novel's title. Transfixed in an artificial bubble, the normally dynamic Babe Ruth is not functional except as a symbol. Hypocritical Slezak says, "destroy a *symbol* and you've destroyed an *idea*. Ideas are irreplaceable!" (250). Yet Slezak has never used the memento either as a paperweight or as a symbol. For him it was always only a "good luck charm" bought for him "casually" by his father (36). And just as the Babe is out of his natural element caught in a snowstorm so too are the Wraiths caught in their "mnemomorphic" fantasy of wanting to play like the boys they once were. Hence, Graham's resurrection of the Babe Ruth legend exploits the discrepancy between George Herman Ruth's indisputable greatness and the psychological infantilism implied by his nicknames "The Babe" and "The Bambino."

As striking as the controversy about the artificial turf is the team's discovery of books, another sign of the Wraiths' shift from innocence. After their string of successes destroys their harmony and produces conflict, the decidedly unintellectual team suddenly discovers reading. "Now, a funny thing. Around this time George Lloyd's taking up reading, too, except not photography magazines, of course. He reads books . . . . Well, come to think of it, some of the other guys are taking up reading, also—maybe it's catching" (173).

Even inarticulate, flower-loving Yak "gave up on his flowers . . . and takes up reading too. Kids' books, he reads" (174). For all of them, but especially for Yak, the first player eventually killed in the closing apocalypse, reading and the acquisition of knowledge signify a fall from innocence. As a solitary activity, reading represents the team's ultimate disintegration as a cohesive community within their baseball Eden.

Through its dystopic ending, *Babe Ruth Caught in a Snowstorm* discloses Graham's debt to Coover's *The Universal Baseball Association* (1968), a parodic black humor fantasy that also ends apocalyptically. The Wraiths' baseball Eden, now clearly exposed as an atavistic Erewhon, ends in upheaval. Having lost their innocence, their success, and the sense of community brought together by their love for baseball, the team—or rather, what's left of it—disintegrates, and the players return to their prior lives of anonymous, individual desperation, their wild dream of professional play now become as unreal, as fake, as the once heroic Ruth trapped inside a plastic snowstorm.

If Graham's use of the Ruth legend emphasizes the deterioration of uplifting human values and any possibility of purposive meaning in American society, then in Jay Neugeboren's *My Life and Death in the Negro American Baseball League* (1973) the legend functions in direct contrast as a grail myth in which "Babe Ruth" serves as both object of and means to the quest. One of the most extraordinary treatments ever written of the Babe Ruth tale, *My Life and Death* is an embedded narrative within Neugeboren's novel, *Sam's Legacy* (1973). While the narrative holds an important place within the plot of the novel, it also effectively stands alone apart from it. This discreteness is underscored by its placement in three long units within the novel, as well as by the use of an elite typescript of sixteen characters per inch instead of the larger type that is used in the rest of the book. The contrasting types set it off visually. Neugeboren also separates *My Life and Death* stylistically by switching narrators from that of Sam Berman, title figure and main character, to Mason Tidewater, Sam's friend and surrogate father.

Neugeboren portrays Mason Tidewater, narrator and protagonist of the novella, as a superstar in the Black barnstorming leagues that flourished when racial segregation was the policy of most professional sports in America (Riess, 194; see also Peterson, 1970; Brashler, 1973). An incredibly skilled pitcher who early in his career also excels as a powerful hitter, Tidewater soon becomes known as "the Black Babe," a tribute, the story proposes, that would have been reversed in a free and just society. Tidewater insists upon that "fact" at the start and throughout his "Slave Narrative": "if things had been otherwise, however, he [Babe Ruth] might have been named for me, and he often admitted as much in the privacy of our friendship" (Neugeboren, 1973; 69). That friendship centers the story, making it at once the major element of suspense in the plot and the heuristic key to the story's theme.

No opening paragraph in any work of fiction discussed in *Seeking the*

Perfect Game: Baseball in American Literature and few other works of serious contemporary fiction match the gripping power of the first sentence in *My Life and Death*.

I consider the high point of my life to have been that moment on the fifteenth day of February, 1928, in the city of Havana, Cuba, when, after I had pitched and hit my team, the Brooklyn Royal Dodgers, to a 1 to 0 triumph over a team composed of players from the New York Yankees, George Herman "Babe" Ruth mocked me again for having chosen the life that was mine, calling me a "make-believe nigger," whereupon I slammed my fist into the pasty flesh of his dark face and struck him down; it was a blow I should have struck long before that day, and one which, filling me momentarily with joy, would lead, on that same afternoon to my own death as a player in the Negro American Baseball League (69).

With that first sentence Neugeboren hooks the reader's interest and sets up the plot outline that will explain the fictive Ruth's insult, the first-person narrator's violent response, and the nature of their relationship. Neugeboren builds suspense by delaying the explanations until the final third section of the novella, but the delay is not gratuitous or sensationalist suspense. The postponement of the legendary Ruth's appearance allows the author to establish the voice and character of the narrator, Mason Tidewater, and to bind him firmly in the reader's sympathies. This bond between reader and narrator is indispensable to the believability of its crucial homosexual love scenes.

Neugeboren—who has created a fine gallery of essentially outcast first-person characters including the Black basketball player protagonist of *Big Man* (1968)—crafts Tidewater as the highly intelligent, literate youngest son of a cultured, talented Brooklyn family. In *Sam's Legacy* we meet him in his old age, the wise and loyal friend of the young and confused title character, Sam Berman. In *My Life and Death*, the manuscript of his autobiograpy that he gives to Berman, we learn of his youth and career in the early 1920s as a Negro league barnstormer. The son of a "light cocoa-colored" father who lived as a plantation slave until the death of his owner, and a West Indian mother who could have almost "passed for white," Tidewater is the only fair-skinned, European-featured child in his family, a fact that produces "a then unfashionable pride in my [Black] racial origins" because he sorely desired, as any "child would, to be like the others" in his family (71). Neugeboren's purpose thus discloses how the individual's physical reality affects the psychological condition and how both combine to pernicious effect within the societal context of racism.

What is most striking at the outset about Tidewater's characterization is the agile but decidedly formal dignity of his narrative style. Rather than create another busher voice (like Jack Keefe's and even Henry "Author" Wiggen's) or inarticulate narrative windows (like Roy Hobbs and J. Henry Waugh), Neugeboren elected to craft a narrator-as-artist figure like Melville's Ishmael or Ellison's invisible underground man. Tidewater's intelligence and sensitivity lock the reader's sympathies into his corner and contribute to the power of his experience.

We follow his entrance as a teenager into the professional baseball Negro leagues against the strenuous objections of his mother, who wished a loftier future as "a gentleman" for her talented "White Star" (75). But his extraordinary arm and bullet-hard fastball bring him the masculine praise he needs and does not receive from his father, and he enters his ball-playing career with emotional fire, intent upon being "the best of the best" (79). Only in writing his self-story—that is, *My Life and Death*—does he understand his motives. "I developed . . . a special hatred for the men who were my daily opponents, and I set myself the task of defeating them as badly as I could . . . I wanted to be the best of the blacks in order, once again, that I might be a man set apart from them . . . this is a thought which did not occur to me, as self-evident as it is, until I had begun to set down this narrative" (79–80). His "special hatred" motivates him to sharpen his native athletic talents. On the mound he discovers that "the truest thing" in the world for him is pitching: "since I commanded it, I also contained, within my small ring of earth, the entire arena. I felt, in brief, as if I were untouchable" (85).

As the heroic Black Babe, Tidewater is untouchable in his greatness, especially after he develops a power-hitting style to match his amazingly low ERA. He is not satisfied, however, by either his achievement or fame because he is prohibited from playing in the major leagues. Particularly grating to him—as it was historically to Black players before 1947 when Jackie Robinson was permitted to break the racist color barrier—is the knowledge that he and the other Negro league players were as proficient and often better than the major league White players. The Negro League teams regularly beat the teams of White major league players that they competed against in the off season. Neugeboren incorporates facts about the Negro barnstorming era that are now well-known but, when he was writing the book, were not as accessible. For example, Tidewater describes the torturous double- and occasional triple-headers played on tour in the constant discomfort of crowded cars, seedy hotels, and all-night parties used to substitute for unavailable rest. Consequently, in his first game against Ruth, he reminds himself that he is not "the first black man to outplay George Herman 'Babe' Ruth . . . [Rube] Foster himself, among others . . . had been able to handcuff the man on a given day" (192). He identifies actual players who, because of their muted African features, were able to slip into the White leagues "passing" as Cubans or Mexicans. The bitter message conveyed by these historical details explains the intensity of Tidewater's anger at the harshness of American racism.

The race issue is also a factor in Neugeboren's employment of the Babe Ruth story, and, in one of the most original strokes of imagination encountered in the entire corpus of baseball fiction, he transforms the legend. He selects certain documented facts from Ruth's life, combines them with one central rumor from the Ruth legend, and then weaves these into Tidewater's story.

The key facts Neugeboren employs are Ruth's nonpareil achievements in baseball through the 1927 season when the Yankees ruled the sports world as the "greatest team ever," a description only slightly modified by time. He also depends upon Ruth's matchless fame and heroic stature, even among non-White audiences, as a foundation on which to build his protagonist's "rabid passion" for "perfection" as a pitcher; accordingly, Tidewater secretly carries a newspaper picture of Ruth in his wallet as a talisman (183). Two other traits of the Sultan of Swat that Neugeboren includes are his highly publicized personal generosity to charities, children, and churches, and his ravenous appetites for wine, women, and food. All of these facts have been used by other writers—notably Malamud, Coover, and Roth—who depend on the public's acquaintance with the Ruth legend as a type of cultural short-hand.

It is, however, the rumor about Ruth's origins that Neugeboren exploits for the fresh reworking of the legend that gives his treatment the vigor and originality of genuine artistry. The introduction to this section describes Ruth's humble origins in a Baltimore family and his early confinement to a Catholic orphanage for wayward boys. In the yeast of popular rumor, these facts became a question and then a mystery about his racial background, one that I personally recall hearing in the 1950s as a child in much the same way that Neugeboren presents it through Tidewater's memory.

My heart stopped. It was only when I found that Johnson was staring at me . . . that I realized I had been gaping . . . . "Sure," Johnson continued. "Ruth ain't no Tarzan, but he's probably more of a nigger than our golden boy here [in other words, fairskinned Tidewater] . . . . "Everybody knows that," he added for my benefit. . . . I saw in my mind's eye photos of George Herman "Babe" Ruth . . . those features, so familiar and beloved by millions of American schoolboys, so evident in a thousand photos: the moon face, the broad flat nose and wide nostrils, the almond-shaped eyes that turned down at the outside corners, the heavy lower lip . . . . Nobody would claim that he actually had colored blood in him, but there were rumors. (181–83)

Neugeboren's point is not to use fiction as a forum to prove that Babe Ruth was part African American or even to help validate the rumor. Instead, the race issue serves as a heuristic device to encourage new insight into the meaning of race and racism in the United States. Since Tidewater is physically almost White and Ruth is physically almost Black, but both are insistently and independently their own persons, race as a fundamental essence constitutive of any real meaning is reduced to insignificance, an empty cipher. Neugeboren conveys this nothingness in the barroom scene that follows Tidewater's first strikeouts of Ruth in an exhibition game.

[Babe Ruth said]"This here's the greatest pitcher I ever faced. You jackasses can tell 'em that the Babe said so." He put his cheek next to mine then. "But who's blacker, huh? Tell me that. Who's blacker?"

"You got him beat a country mile, Babe," [Bob] Meusel said.

[Babe Ruth said]"Color ain't what counts. It's blood that does it, ain't that so kid?"

"Yes," I said. "It's blood." (303)

The "blood" they refer to is the inherited material—both genetic and cultural—that explains their greatness, that is the source of their talents, and that is as much a product of the hunger (or postmodernist "desire") that propels them, as it is of any competitive drive stemming from the game itself. Neugeboren's method also works to expose not only the terrible damage racism inflicts on society (that, after all, has long been manifest), but also to expose the utterly pointless habit of racism. Ruth's and Tidewater's racial ambiguity (like that of several key Faulknerian characters) underscores the way social custom and politics often rest on emptiness, on nothing—except the habit of hate. That is Neugeboren's point.

As powerful, but certainly more striking than his heuristic employment of race in America, is Neugeboren's astonishing treatment of the novella's love story. "There remain, then, only facts. I met George Herman 'Babe' Ruth some six weeks after the close of the 1924 World Series, and we became lovers" (298). The two characters remain passionate lovers in *My Life and Death* even while Ruth publicly maintains his high profile as a lusty womanizer and Tidewater keeps his low profile as aloof and reserved.

I [Tidewater] wanted to laugh at the sheer clownishness of the situation—the great hero of America lying in bed, locked in that most absurd of positions with a fair-skinned nigger—and yet the humor, ultimately, turned upon me, for I brought him low only by making myself lower. What might have seemed a joke to the world, had it known, held no humor for me . . . his willingness to meet me and to love me . . . made him seem to me, given who he actually was, brave . . . . [this] made me realize all the more that I was only another one of that multitude which worshipped him. (309)

Neugeboren makes it clear that his protagonist falls deeply in love with Ruth (310), but, appropriate to a famous legend, he keeps the true nature of Ruth's feeling ambiguous (310–12). The fictive Ruth clearly has a strong sexual attraction and personal affection for the younger athlete, but his carefree style and privileged lifestyle decidedly separate him from the other's world, thus making it difficult to know what truth lies beneath his everpresent inscrutable smile.

At this point what does the homosexuality in the narrative ultimately achieve? Why does Neugeboren use it? Obviously a shocking novelty, it trips the reader's fluent eye as it moves across the page. The interruption forces reflection upon the freshness and uniqueness of the work's drama, subject, characterizations, and thematic premise and context. In baseball fiction its novelty is analogous to Casey surprising Mudville by striking out instead of hitting a homerun, while in folklore it is like the unexpected fact that *even all* the king's armies can*not* put Humpty Dumpty together again despite their vast power. Accordingly, in *My Life and Death* the shock stems as much from the compelling power and beauty of the love affair as it does from the author's ingenious originality in resurrecting the Ruth legend in this way.

Another important effect produced by the homosexuality is its efficacy in capturing the characters' inner consciousness and complex feelings. To probe the emotions of baseball players, writers have usually employed romance, a love interest in women through which they refract the man's emotional states. For instance, we come to know Roy Hobbs (*The Natural*) and Dick Dixon (*The Rio Loja Ringmaster*) as sensitive, intensely feeling men because of their love relationships primarily with Iris Lemon and Conseulo, respectively. Similarly, in *My Life and Death* we come to know hidden private dimensions of both baseball superstars through their affair: "[Ruth] saw then that I was crying. 'I didn't mean nothing,' he said, lying down besides me. . . . I turned my head away . . . so that he would not know the real reason for my tears" (308).

Relatedly and as discussed in this book's introduction and chapter one, one of the chronic deficiencies of baseball fiction is its male chauvinism— that is, its uncritical chronicling of the onesided male-centeredness of the sports world and, especially, of the larger society. Baseball fiction in general presents a largely benevolent treatment of sexism. Neugeboren exploits this characteristic of the genre and subtly extends it to narcissistic extreme by drawing his romance in homoerotic terms. Thinks Tidewater in one intimate moment with his famous lover, "In the dimly lit hotel room I believed that I could see my own reflection in his eyes" (308). Like all lovers, he can, indeed, see himself in the other's eyes, and this self-reflection underscores the theme of fraternity and twinness in the tale (and, in fact, in the larger frame story of *Sam's Legacy* as in other Neugeboren creations). "In loving him, I was only pursuing that same self-love which had always consumed and driven me," muses Tidewater, "Like many before me, it was the image of my love which I desired" (314). Hence, the two pitcher-sluggers are twin figures in their awesome talent, in standing apart from other men, and, sensationally, in their romantic union.

At this point in Neugeboren's consummate baseball fiction, we understand the brilliance of the opening sentence of *My Life and Death in the Negro American League*, which opened this discussion of the narrative. Filled with every major dimension of the story to be told, the sentence serves not only the formal functions of beginning the story and hooking reader interest, but it also serves the thematic function, not apparent until the end, of presenting the conclusion within the introduction. When Tidewater writes, "it was a blow I should have struck long before that day, and one which . . . would lead, on that same afternoon to my own death as a player," Neugeboren captures the seamless interconnection between beginnings and endings, between past and prologue, between cause and effect (69). That interconnection is at work in the seamless weave between the young Tidewater and his baseball grail, the tattered newspaper photo of the Bambino in his wallet, just as it describes Tidewater the baseball star at the moment of his demise as a player, when he reacts to preserve his dignity and ends up killing his nemesis, Brick Johnson, instead.

Finally, through the homosexuality, Neugeboren emphasizes the intellectual poverty of unquestioned social conformity and rigid social categories. Through his focus on its ambiguities, Neugeboren's use of race exposes its futility and meaninglessness as a category. Likewise, his employment of homosexual attachment in a love story recounted simply and straightforwardly reveals the sham of homophobia. The story is not about gays or their persecution, but about amazing men who fall in love with one another. Only implicitly and obliquely does it address the way perverted social forms keep people apart and disrupt communities. From the hostility in Tidewater's family caused by his skin color, to the hatred and anger inflicted on his team by Brick Johnson because of his skin color and age, to the lack of trust between Tidewater and Ruth despite the genuineness of their affection because personal commitment between two men of their stature is not socially acceptable—the novel discloses how slavishly followed decayed conventions destroy effective human relations. As the subject and backdrop for this tragedy, baseball is indispensable, first, because of its centrality in U.S. history and custom and, second, because of the crucial fact of the Babe Ruth myth to American culture.

REFERENCES

Ralph Andreano, *No Joy in Mudville: The Dilemma of Major League Baseball* (Schenkman, 1965).

The Baseball Encyclopedia, 6th ed. 1985.

Peter Bjarkman, "The Glorious Diamond," unpublished manuscript, 1986.

William Brashler, *The Bingo Long Traveling All-Stars and Motor Kings* (Harper and Row, 1973).

Bruce Cook, " 'Bang the Drum': Best Sports Film Ever Made?" *National Observer* (September 15, 1973): n.p.

Arthur Cooper, "Murderers' Row," *Newsweek* (April 30, 1973): 83–84.

Robert Creamer, *Babe: The Legend Comes to Life* (Simon & Schuster, 1974).

F. Scott Fitzgerald, *The Great Gatsby* (Scribner's, 1925).

John Alexander Graham, *Babe Ruth Caught in a Snowstorm* (Houghton Mifflin, 1973).

Mark Harris, *Bang the Drum Slowly* (Knopf, 1956).

———. "Easy Does It Not" in *The Living Novel: A Symposium*, Granville Hicks, ed. (Collier Books, 1962).

———. "Home Is Where We Backed Into," *Nation* (November 1, 1958): 322–24.

———. *It Looked Like For Ever* (McGraw-Hill, 1979).

———. *The Southpaw* (Knopf, 1953).

———. *A Ticket for a Seamstitch* (Knopf, 1957).

Roger Kahn, "Intellectuals and Ballplayers," *American Scholar*, 26, 3 (Summer 1957).

Jay Neugeboren, *My Life and Death in the Negro American Baseball League* in *Sam's Legacy* (Holt, Rinehart, & Winston, 1973).

Robert Peterson, *Only the Ball Was White* (Prentice-Hall, 1970).

Steven A. Riess, *Touching Base: Professional Baseball and American Culture in the Progressive Era* (Greenwood Press, 1980).

Murray Ross, "Football Red and Baseball Green," *Chicago Review* (Jan.-Feb. 1971): 149–157.

Harold Seymour, *Baseball: The Golden Age* (Oxford, 1971).

Marshall Smelser, *The Life that Ruth Built, A Biography* (Quadrangle, 1975).

Tony Tanner, *City of Words* (Harper & Row, 1971).

David Q. Voigt, *American Baseball* (University of Oklahoma Press, 1970).

6

BASEBALL FICTION, THE PERFECT GAME

Baseball is our national game and our national passion...a timeless escape into a world of perfect balance and harmony...perhaps the most perfect game ever invented.

Peter C. Bjarkman, "The Glorious Diamond"

"I had enough one-hitters to string a rosary and enough two-hitters to send Christ back to the cross...but never the big one—never 000000000."

Lamar Herrin, *The Rio Loja Ringmaster*

Baseball, said Hall of Famer Sandy Koufax, "has a cleanness" about it: "if you do a good job, the numbers say so. You don't have to ask anyone or play politics...or wait for the reviews" (in Boswell, 52). In other words, the game's required box scores and other statistics preserve a perfect record of individual and team performances. This is what sportswriter Thomas Boswell has in mind when he observes: "In baseball, almost every act has an accountability" (246). This aspect of the sport's highly touted tradition of recordkeeping— that is, baseball's book on and of itself—accounts for a large measure of its fascination for the literary mind. Also concerned with recording experience, writers find baseball's preoccupation with its own history and story a perfect corollary to fiction. As Roger Angell notes, "a box score is more than a capsule archive. It is a precisely etched miniature of the sport itself" (12). This quality reinforces baseball's power over memory and gives it aesthetic charm.

Box scores, season records, and the meticulous sort interested in them have even spawned a new field: sabermetrics—the systematic, scientific "search for new knowledge about baseball" (James, 7). The sabermetrically inclined care deeply about registering every detail of the game in numerical code, according to carefully derived mathematical formulas. They seek to contribute to baseball new ways of keeping its permanent record, accounting for such salient, but previously undocumented, features as defensive efficiency, RBI value, park adjustment, total average, and so on. Both for historical accuracy and for the multitude of comparisons that inevitably interest followers of the game, the book on and of baseball has been vastly expanded by sabermetrics. For example, Bill James, a principal proponent of the sabermetric method, stresses the following fact in his annual *Abstract*:

The sum total of all baseball statistics is the same in every decade, the same in every season, the same in every game and in every inning and on every pitch. It is always .500. (9)

Because of the mathematical purity implied by this fact, every significant detail of the game can be translated into numbers. Some might argue that too much trivia is too often thus translated, but most authors of baseball fiction would rebut that viewpoint by questioning the premise that there are *any* trivial details in the actual playing of baseball. They would assert that the basic mathematics of its rules and its methodical pace demand that all its details, however, minute, be preserved in another form off the diamond.

Mathematics is perhaps the only form of human discourse that routinely comprehends perfection both as an abstraction and as an achievement. It has the capacity to account for observable facts, unknowns, probability, and uncertainty with consistency, logic, and proof. The language of math is thus flawlessly complete and, by its completeness, perfect. In parallel fashion the baseball's lexicon contains the idea of perfection in very specific form: the "perfect game." The term refers to a full game in which the pitcher does not allow any batter to reach first base, but retires the opposing team with twenty-seven uninterrupted consecutive outs, and with at least one run scored by his side. The difficulty of pitching a perfect game can be gauged by considering that of the thousands of major league games played since 1875, as of 1988 there were only fourteen recorded "perfect games of nine or more innings" pitched in the history of the organized professional major leagues (*Baseball Encyclopedia*, 31). A rarity of extraordinary occurrence, the perfect game requires perfection on the mound by one pitcher and by his team's defense.

The perfect game, literally and figuratively, is significant among the wealth of reasons the literary imagination has consistently found baseball the most fascinating and artistically viable of American pastimes. In its literal sense, the perfect game is a paragon of exceptional occurrence that, in its making, produces the extremes of suspense, drama, and emotion appropriate to the realization of rare excellence. On the other hand, as an ideal, the perfect game is not without controversy in baseball. Many even think it a misnomer in that by disproportionately emphasizing pitching, it does not describe the perfectly played baseball game which, ideally, should include such important aspects as hitting and stealing (Thorn, Palmer, and Reuther, *The Hidden Game*, 55). To other imaginative fans, however, this alleged minus, like negative numbers in calculus or elusive electrons in quantum mechanics, helps to prove the sport's essential completeness.

In its figurative sense, the perfect game symbolizes any paradigm of perfection in life, whether a grail-like treasure or a pinnacle of achievement sought for either personal or social reasons. The perfect game metaphor may also refer to process, particularly to the process by which the grail or achievement is sought and attained. This chapter continues an examination of baseball's efficacy as a metaphor by focusing on the idea of perfection. As it appears in the two novels discussed here and also as an indication of both novels' excellence as baseball fiction.

One measure of such excellence is the degree to which writers address the nine characteristics of the game enumerated in chapter one as important recurrent literary motifs. In summary, the nine points are: its cultural substance derived from aboriginal myth and ritual and from a rich matrix of grass-roots sociohistory, its satisfaction of basic human needs such as physical play and physical competition, the patriarchal exclusion of nonWhite men and women arising from classist social assumptions, the fallacy of the Doubleday/Cooperstown genesis legend, the collusion between the sport and commerce producing the Black Sox fall from grace, the political folklore requiring the President of the United States to toss the opening season ball, and the pure symbolism of the game's inherent, essential form. Both novels analyzed in this chapter incorporate most of these characteristics, transmuting them to highly original, intensely dramatic, and thematically complex ends.

In this vein, Bjarkman accurately points out that baseball is "the only game to sustain a serious and weighty literature of its own—an endless stream of novels, essays, poems, movies, stories" ("Glorious Diamond" n.p.). Two of the most successful novelists contributing to the canon of baseball fiction are fabulist Robert Coover and humorist Philip Roth. These writers, both iconoclastic satirists, have written masterpieces of baseball fiction that have failed to elicit their just critical acclaim largely because their subject matter is baseball, a topic consistently and unfairly, relegated to the critically debased realm of popular culture and subliterary fiction. The literary sophistication and thematic power of both *The Universal Baseball Association* and *The Great American Novel* decisively destroy that intolerant, unexamined preconception.

THE UNIVERSAL BASEBALL ASSOCIATION, INC., J. HENRY WAUGH, PROP. (ROBERT COOVER)

Several scholars have pointed out the error of Johan Huizinga's perception of play as a gratuitous, disinterested activity separate from literal multitudinous reality (see, for example, Bjarkman, Ehrmann, and Messenger). They and others have demonstrated the active, changing nature of play's intrinsic forms, and their dynamic relation to "real life" activities. On the surface, at least, everyday reality seems to be extrinsic to play, but, as shown throughout this study, play and sport are closely tied to the innermost customs and values of society. In his *Sport: A Philosophic Inquiry*, Paul Weiss makes a relevant point:

In some form sport has had a place in every culture which has a tribal memory. But apart from some awareness by officials or by players of previous exemplifications or rules, a sport exists only as a tradition sometimes does, *i.e.*, as a set of possibilities which, when realized in the present, are thereby necessarily related to all past realizations, and eventually to the first, while still allowing for further realizations in the future. (144–45)

Although his analysis relies to a significant degree on Huizinga's, Weiss stresses the vigorous dynamics connecting sport and play to ordinary life.

Importantly, the emphasis Weiss places on the dynamic interconnections between sport and society resemble the indeterminism or "uncertainty principle" articulated in the modern physics of quantum mechanics:

This thought—that there is a fundamental uncertainty about the wavelength, momentum, and position of wave-particles—is one of the more unsettling discoveries of twentieth-century science. Uncertainty now has been built into the accepted picture of natural phenomena. . . . [Q]uantum mechanics and the uncertainty principle remain cornerstones of modern physics. (Boikess and Edelson, 155)

Specifically, Weiss's reference to tradition as a "set of possibilities" that is affected by both prior and potential realizations calls to mind Werner Heisenberg's redefinition of the dynamic "reality" of subatomic particles. Heisenberg argued that "the old concepts of space and time" cannot describe "electronic orbits," which are so fluxive and variable that the very act of observation changes them. This phenomenon led him to conclude that "electronic orbits" and similar subatomic material exist "not as reality but rather as a kind of 'potentia,' " which can at best be represented as "only a tendency toward reality" (Heisenberg, 55).

Likewise, modernist perceptions of reality as fluxive multiplicity are also germane to the interaction between, and cross-fertilization of, sport and life that Weiss finds mirrored in human tradition, an interaction that has obvious cogency to this study of literature about a game. Furthermore, such perceptions underlie the postmodernist fiction of Robert Coover. A highly praised writer, Coover recognizes that the bridge between the "new scientific and the new aesthetic concepts" arose because the old concepts were "rigidified" and "impotent" with respect to the "on-goingness of the world" (Coover in Gado's *First Person*, 153). Not that he unequivocally rejects the old, instead, he seeks to throw out the bathwater of "depleted" narrative forms and approaches and to keep the baby of genuinely creative "fictional constructs" (*First Person*, 154). In this he parallels the postwar writers (for example, Borges, Barthes, Pynchon) whose fabulations mirror the flux and uncertainty of experiential reality and post-Newtonian physics.

Chief among the fictional constructs that Coover admires is the work of Cervantes. In *Pricksongs and Descants* (1969), he describes the sixteenth-century Spaniard as having combatted the omnipresent "unconscious mythic residue" of his age by recreating "the real" (79). Cervantes and other pre–industrial age innovators are seminal in Coover's view because they reconstruct the known and the ordinary in mythopoeic fictions that radically move "away from mystification to clarification, away from magic to maturity, away from mystery to revelation" (79). That kind of "fiction making," says Coover, "re-form[s] our notions of things" (*First Person*, 149–50). Cervantesque fantasy, fabulation, autotelic reflexivity, and humor are therefore crucial concepts in an analysis of Coover's *The Universal Baseball Association, Inc., J. Henry Waugh, Prop.* (1968), and, indeed, Coover's handling of the sport in this novel partakes of all these literary strategies.

Unlike other baseball fictions that deal with the sport as an offshoot of literal, played history, baseball appears in *The Universal Baseball Association* purely as a mental construct emerging from the mind of Coover's protagonist, accountant J. Henry Waugh. Baseball exists in the novel in the form of a game created by Waugh and played solitarily with dice, cards, charts, score sheets, and other familiar accessories of parlor games. Coover's absorbed knowledge and his familiarity with the historical sport and its meaning within American culture clearly derive from actual baseball, but that public, shared, communal pastime operates here only as a distant source for the book's simultaneously parodic and pastoral treatment of "THE GREAT AMERICAN GAME" (19). It is as if the author wrote this novel to reify John Steinbeck's often-quoted remark that "baseball is not a sport or a game or a contest" but "a state of mind" (in Berman, 88).

Summarized, Waugh's story is simple. Because of his preoccupation with his parlor game hobbies, and particularly because of his growing obsession with the Universal Baseball Association (UBA) game he has developed, Waugh loses his job at the accounting firm of Dunkelmann, Zauber & Zifferblatt's; his only friend, Lou Engle; and contact with the two other people ("local B-girl" Hettie and bartender Pete) even remotely involved in his "real" life. Participating in hardly any conventional plot action or dialogue, as a character Waugh fails to elicit traditional reader-protagonist identification. His solipsism effectively distances him from the reader and allows his personality to be subsumed within the more dominant plot and dynamic personalities of the fabricated players. Nor does Henry, like Roy Hobbs and Jack Keefe, travel through the usual pattern of suffering, punishments, and fresh understanding intended to produce spiritual or emotional development. Waugh's movement as a character is, finally, an inexorable withdrawal from real life and human contact, and eventually even from the foreground of the novel. Culminating in a remarkable last chapter that contains no explicit reference to the main character at all, the narrative ends with his total absorption into his fictive universe, isolated from everyone—including the reader.

As a narrative experiment, Coover's unconventional elimination of his protagonist without either killing him off or upsetting the internal logic of the fiction is, like Latin American magic realism, delightfully inventive and—especially in 1968—gloriously fresh. Moreover, Waugh's interior monologue, operating on varying levels of consciousness and issuing forth his baseball fantasy, deepens the novel to make it a complex tour de force of postmodernist writing. Inhabited by players, managers, and commissioners with fully developed histories and carefully charted careers, the Universal Baseball Association game carries the novel's thematic and symbolic burden: and imagining and playing it constitutes Henry's only real substance. By refracting his main character through multiple layers of narrative consciousness that then ultimately subsume—or, perhaps more precisely, consume—him, Coover innovatively reconstructs the known world of readerly fiction. To comprehend the fecundity of his innovative achievement requires closer apprehension of the nature and history of the fictive Association.

When J. Henry Waugh—whose initials neatly suggest a biblical acronym, JHW, for the omniscient Yahweh—first created the Association in the past outside the novel, it was only one of an extensive repertoire of games he played, "all on paper of course," usually with other people (44). But because his fascination with games was always in their intact purity *as games* quite apart from such mundane realities as "slumdwellers," "industrial giants," and "sex," the consistent "inability of the other players to detach themselves from their narrow-minded historical preconceptions depressed" him (44–45). Simply put, Henry sought an arena of play totally detached from the exigencies of real life, an arena that he perceived as the only possible enclave within which perfection could be realized.

Consequently, he "found his way back to baseball," with its essential records and statistics, and its "peculiar balances between individual and team, offense and defense, strategy and luck, accident and pattern, power and intelligence"—all requisite features appealing to his compulsion for controlled order and leading eventually to total solipsistic surrender to the UBA (45). Moreover, baseball was also far removed from the distasteful "militant clockwatch[ing]" of his boss Horace Zifferblatt and the pesky "clumsiness" of people and society—aptly represented by his chubby friend Lou (35). Both Zifferblatt and Lou represent social reality and are the targets of Waugh's distaste for mundane experience.

Waugh considers baseball the only "activity in the world" that has "so precise and comprehensive a history, so specific an ethic" that it retains much of its "ultimate mystery" (45). To capture that unambiguous sanctity of order and precision, he created the Universal Baseball Association game, with its scrupulously defined goals and rules that omnisciently anticipate every conceivable contingency. Played with three dice whose roll determines the action on the diamond, the game includes charts for regular play (pitches, hits, bases on balls, box scores, and so on), as well as the special "Stress" and "Extraordinary Occurrences" charts used only with very rare dice combinations (1–1–1, for example). Henry originally selected his teams from "baseball's early days in the Civil War and Reconstruction eras," but, like the creation of the cosmic world, that "abrupt beginning had its disadvantages," primarily because it was "too arbitrary" and "inexplicable" (45). He was thus "relieved" when "the last veteran of Year I . . . had died" and the Association's connection to the "deep past was now purely 'historic,' its ambiguity only natural" (46). Significantly, by "natural" Henry refers not to its usual meaning relating to nature but to the unimpeded outcome of the game's dice roll. Rid of ambiguity, the natural passage of time, quotidian cares, and other "nuisances" of life experience, the game is a convenient reality—predictable, flawless, and well-protected within the gray matter of his mind and the gray walls of his kitchen.

But woe is Waugh for, like *the real* world, his "universe," too, suffers the fate of human frailty. The one thing he cannot control with objective dice throws and precise charts of action is himself and his feelings. As a result, the game begins increasingly to compel its creator's interest and thought to the exclusion of everything else: the heart of the novel concerns the growing hegemony of the Association over its proprietor. Among the feelings that account for this compulsion is his unexpected love for one of the Association players, Damon Rutherford, son of the legendary Hall of Famer Brock Rutherford who, like Henry, is fifty-six years old. It is a love at once paternal and fraternal, deferent and familiar, for the Rutherfords, created in JHW's image, especially fulfill his perception of the heroism so remote from the dullness of his own real-life experience. "Brock had raised his two sons to be more than ballplayers . . . they were, in a way, the Association's first real aristocrats" and, quite unlike mild-mannered, Walter Mittyesque Henry, blessed with remarkable "poise" and "gently ironic grace" (12). Through the name Rutherford, Coover achieves two significant allusions: it alludes to the atomic physicist Ernest Rutherford and also contains an embedded reference to Babe Ruth.

The book opens on Damon's pitching of a perfect game, and Waugh is beside himself with excitement.

Rookie pitcher Damon Rutherford, son of the incomparable Brock Rutherford, was two innings—six outs—from a perfect game! Henry, licking his lips, dry from excitement, squinted at the sun high over the Pioneer Park [part of the imagined UBA world]. . . . Exceedingly rare, no-hitters; much more so, perfect games. . . . Henry paced the kitchen [part of Henry's "real" world], drinking beer, trying to calm himself, to prepare himself, but he couldn't get it out of his head: *it was on!* (4, 11)

Damon's perfect game turns out to be just the impetus needed to rekindle Waugh's interest in playing the game after months of feeling afflicted by tedious "games lived through, decisions made, averages rising or dipping, and all of it happening in a kind of fog" (13). Caught up again in the "magic of excellence" (14) possible in baseball, Waugh is not only cheered and inspired by the pitching of a perfect game, but his accountant's respect for numbers, balance, and "the beauty of [baseball's] records system" is reinvigorated by the act of logging the neat zeroes evenly in his ledger (19). "All those zeros! Zero: the absence of number, an incredible idea! Only infinity compared to it, and no batter could hit an infinite number of home runs— no, in a way, the pitchers had it better. Perfection was available to them" (57). And clean perfection is exactly what Henry, who lives amidst messy clutter and who makes numerous mistakes on his accounts at work, seeks.

Still haloed by the glow of Damon's extraordinary feat, Waugh's rekindled involvement as UBA proprietor elicits an appropriate nostalgia and leads him to a wistful retrospective of earlier teams and seasons. "Funny," thinks Henry after the perfect game, "Damon was not only creating the future, he was doing something to the past, too" (22). Through his exhaustive official archives he can call to awareness every UBA game and player that ever "existed," just as the external world's baseball record books preserve the sport and its players for all time. Coover's thematic emphasis on recordkeeping underscores the fundamental human yearning for immortality that lies behind the very concepts and origins of history and art, just as it clearly lies behind the abstraction of the perfect game. Waugh's grappling with his own (im)mortality explains why, despite his renewed interest in the Association, Damon's perfect game produces a "feeling so melancholy" in him that he wonders why the "inherently joyful moment" does not console him (49). Accordingly, the coincidence of the dice rolling a perfect game for his favorite player in the fifty-sixth season, a number corresponding both to Waugh's and the UBA's ages, "though, of course, their 'years' were reckoned differently," causes him to ask himself nervously if, somehow, "the vital moment" of dying is near (50). In this coincidence, as elsewhere, Coover taps into the well-known superstitions associated with baseball that have become a key part of its lore (Seymour, 128–29).

The analogue for Waugh's veiled worry about his mortality is Damon's astonishing death, caused by another rookie's, Jock Casey's, beaned pitch. In the words of the Extraordinary Occurrences Chart, "1–1–1: *Batter struck fatally by bean ball*" (70). Compounding the tragedy for the proprietor is father Brock's presence in the stands to celebrate Brock Rutherford Day with "all the old Pioneers from the Rutherford Era" (71). Anguished and paralyzed by the scene "before" him—that is, in his head—Henry imagines in detail the "crying and shouting" of the fans, the initial incredulity then violent anger of the players, and the mesmerizing stillness of Brock, who sits in the bleachers with his "head reared in shock and his face drawn...[looking] suddenly gray and old" (73). Unlike the noble Brock who manages to compose himself, walk to home plate where his son "gaz[es] at a sun he could no longer see," and, en route even manages to prevent Jock's lynching by Damon's angry teammates, the "Proprietor of the Universal Baseball Association...[is] brought utterly to grief." He can only bury "his face in the heap of papers on his kitchen table and cr[y] for a long bad time" (76). His intimate identification with both Damon and Brock, coupled with his anxiety about aging and death means that Henry cries partly for himself.

Moreover, although he weeps tears of genuine grief, they also represent his guilt at knowing that his own cheating led to Damon's death. He had knowingly played Damon "after only one day of rest," even though it "wasn't recommended practice," because he sought to combat the strange melancholy induced by the perfect game by seeing his hero continue his wonderful streak of scoreless innings. In addition, instead of using the Knicks' likely pitcher, "Ace southpaw, Uncle Joe Shannon," Henry, eager to assure Damon's victory, started rookie Jock Casey against him (63). Thus, like any gambler arranging a major fix, he justified his unethical departure from Association custom with defensive comments to himself, such as it really "isn't against the rules" and "the Knicks could risk losing the first" game. But in his heart he knew it was wrong and thus admonished himself: "there'd be no further concessions" (63). These rigged pregame actions, occurring in chapter two, are motivated solely by Henry's selfish interest in achieving his version of Apollonian perfection. They foreshadow the outright cheating in chapter six during actual game-play that Henry knowingly commits to cause Casey's murder (200–202). Thus, the base human motives of greed and revenge, subtly paralleling the Black Sox fix, resulted in the fateful dice roll that produced the fatal pitch, and that unfair interference helps account for the intensity of the proprietor's reaction to the "Extraordinary Occurrence" of Damon's death.

But fate is not to blame for Damon's death nor, for that matter, for the tragedy of the supposedly arbitrary dice roll that caused it. Nor ought fate be held responsible for Waugh's woes and for the "loser" personality that explains his escapist Mittyesque fantasy. Coover is responsible. He is the omniscient power controlling this imaginary universe. As the title tells us, Henry is but the "Prop.," the proprietor, yes, but also a stage prop, an object lacking any power or control. Not intended merely as an exercise in deterministic naturalism or psychological realism or existential black humor— although thematically it looks like all three—this novel about a game is itself a game. A game called fictionmaking.

As Neil David Berman rightly observes, "Henry is as much a part of his own game as his players...and [he] and the Association are parts of a larger game being played by still another god figure—Robert Coover" (93), whom we know to be the "impersonal force of fate" that is "physically represented by the dice" (96). While all writers are omniscient vis à vis their literary creations, not all writers emphasize that omniscience through stylistic foregrounding as Coover does. Subtly achieved so that on one level the reader is caught up in the story of Damon Rutherford, Jock Casey, and the UBA, and in Henry's obsession, on another level the reader must consider the narrative structure. The special emphasis on narrative consciousness becomes strikingly manifest in chapter eight, when we encounter the proprietor's total disappearance. That fact forces the reader to wonder why he is gone, and what has happened. In other words, we must confront the author's underlying premises and ultimate purpose.

Coover's purpose, seen in his fiction and disclosed in interviews, is re-mythification—that is, re-forming in the manner of Cervantes the depleted myths and values of the past in order to *reform* them for the present (see, for example, Coover, *Pricksongs and Descants* [1969] and *First Person* [1973]). As Arlen J. Hansen explains in "The Dice of God: Einstein, Heisenberg, and Robert Coover," an especially perceptive analysis of the novel, by "offering a re-consideration, a re-conceptualization, the act of remythifying acknowledges the hypothetical and fictive nature of the values and order that [humans] perceive. Re-mythification, in short, reminds us that our knowledge is tied intrinsically to the concepts we use to think with" (57–58). The novel thus challenges our preconceptions about literature as well as our expectations of a work of fiction. We have already seen, for example, how Coover remythifies the convention of novelistic plot, which ordinarily consists of characters within fictive realities that engage reader interest through characterization and suspense. Coover re-forms that convention by focusing on the fantasy "reality" of the UBA, making it the primary plot focus, even to the extent of replacing Waugh's fictive reality as a character. When we realize that the protagonist, the traditional narrative threshold into the center of the story, has vanished without explanation but, importantly, without sacrifice of the work's thematic integrity, the prior "concepts we use to think with," at least about plot and character, have been decidedly reshaped.

The novel's central remythification, however, concerns the transformation of the Association from a parlor baseball game and fantasy controlled by Waugh to a dystopian ritual detached from him and occurring years later. What is more, we neither question the validity of the change nor the thematic significance of the UBA—existing somewhere in the far distant future, but now completely disembodied from the mind and imagination of its proprietor. What explains the daydream's independence from the daydreamer?

One answer is that the emotions that assault Henry after Damon's death are so overpowering that the same revulsion defining his interaction with others begins to creep into his attitude toward the UBA. Following his elaborate imagining of Damon's funeral and the subsequent drunken "gathering of the grieving" at "Jake's Bar," for instance, the thoughts of all the key UBA figures appear colored by the same gray brush of gloom we have already come to associate with Henry's characteristic mood (75–123). From Bancroft's lofty rationalizations about the "[p]syche, up against the wall" to Rooney's "staggering flop-limbed" drunkenness masking his being "afraid to die," and to Chancellor Fenn's cold-blooded worry about how "officially" to handle the killing, the Association has been infected by the same "melancholy" that grips its proprietor.

Sealing that negativity in terms comparable to the tragedy of Damon's killing, Waugh concludes Damon's imagined funeral and "raucous" wake with the gang rape of a nameless girl in the "back room" of Jake's (really "Pete's" in Henry's non-UBA world). Through Pappy Rooney's mind, he thinks "[t]hey'd had enough of the putrefaction phase, they'd passed through the dissolutions and descensions and coagulations: what they wanted now was union. . . . [I]t was a goddamn inundation" (115). Coover shows the mob psychology lurking beneath the surface of sports in a patriarchal society through their total masculine disregard for the pathetic rape victim. This scene makes clear that the same psychosocial dysfunction that has made Henry's life a ledger of deficits has inevitably entered his fantasy. Likewise, the extended UBA scenes where the proprietor is a nonentity submerged by his fantasy foreshadow his eventual disappearance from the novel.

Relatedly, Henry's carelessness in letting the Association spill over into his non-UBA life signals a loss of control that worsens his conduct in, and hold on, both arenas. By using Damon's accident (the proverbial absentee alibi of a "death in the family") as an excuse to Zifferblatt for his shoddy work performance and absenteeism, he sullies the sacred meaning that the game's precision and beauty formerly held for him, and he demonstrates as well his abject finiteness as a "Prop." Another indication of the severity of his psychic disintegration is his attempt to include first Hettie then Lou in the actual playing of the game, for, as noted earlier, he had years before already seen the futility of playing with other people, who were bound by "history" and the "dull" desire to win. Thus, when he enlists blundering Lou to join him in an effort to "save" either the game or "him from the game" (170), we are not surprised that, as Berman points out, "Lou's presence 'proves to be truly disruptive' because he is incapable of conceiving the Association in any but profane terms" (Berman, 97).

Fittingly absurd to reflect the chasm that divides their basically opposite natures, the scene in which the two men play is hilarious for the reader but solemnly traumatic for Henry and can only end in chaos.

"Lou! Listen, I just rolled a triple six. That moves the game to the special Stress Chart!"

"Well, that's nice, Henry, but—"

Lou yawned . . . and slapped his hat to the table—*bop!* the beer can somersaulted and rolled bubbling out over charts and scoresheets and open logbooks and rosters and records—

"*Lou!*" screamed Henry. He leaped for the towel . . . but everywhere he looked ink was swimming on soaked paper. . . . When he'd got up the worst of it, he sank into his chair, stared at the mess that was his Association. It's all over, he realized miserably, finished. (198–99)

What is over is his control. His hold on reality—even the imaginary reality of his fantasy. Despite Waugh's occasional awareness that "total one-sided participation in the league would soon grow even more oppressive than his job" (141), such clear-eyed knowledge is not enough to actually alter his modus vivendi. He even asks:

So what were his possible strategies? He could quit the game. Burn it. But what would that do to him? Odd thing about an operation like this league: once you set it in motion you were yourself somehow launched into the same orbit; there was growth in the making of it, development, but there was also a defining of the outer edges . . . the urge to annihilate [it] . . . seemed somehow alien to him, and he didn't trust it. (141–42)

His response to these questions—that is, his succumbing entirely to the game's "orbit"—is evident as soon as he realizes there are no other alternatives for him, and his decision to "finish out the season" leads inevitably to the ultimate submission of his own consciousness to the Association's momentum.

To "finish out the season," at the end of chapter six, Henry puts into action the vengeful feelings against Jock Casey that had stalked him since Damon's death. Even though he as proprietor had fudged the rules enough to cause the killing, his identity is so fused with that of the UBA that it is as if Jock had actually killed a real person in Henry's life, not a fantasy character in a fiction. Thus, deliberately setting the dice ("One by one. Six. Six. Six" [202]), he arranges Jock's death. The effect of his cold premeditated murder is overwhelming guilt, which causes him to suddenly convulse with spasmodic vomiting (compare Roy's post-binge catharsis in *The Natural*), after which he falls into a "deep deep sleep" (compare Joe's sleep at the end of *The Year the Yankees Lost the Pennant*). After the accidental homicide of Damon's death, the raping of the girl during the wake, Jock's murder follows the overarching pattern of destructive events with totally consistent logic. Consequently, Henry's own self-destruction (that is, disappearance) in the last chapter meshes perfectly with the plot's growing violence, and with the takeover of "reality" by the protagonist's interior monologue.

Yet another explanation of Henry's disappearance while his daydream lives on lies in an exploration of the dystopia into which his game and imagined universe has been transformed in chapter eight. The chapter depicts the evolved Association in the future of its CLVII[TH] year. The Association is in a dreadful state, drastically changed from its original form into a rigid religious rite meaningful solely in terms of its rituals. No longer resembling baseball as sport or play, the UBA has become pure mythology, centered around the "annual rookie initiation ceremony, the Damonsday reenactment of the Parable of the Duel" (220). Presented as theater of the sacred with appropriate mystery and secrecy, the parable recreates the deaths of Damon and Casey, now transformed by time, as a rookie initiation ceremony. Three chosen rookies perform the roles of the two pitchers and that of Royce Ingram, the player whose linedrive killed Jock in the fateful LVITH season. Ironically, however, like ritual sacrifices of old, the young initiates are actually killed during the Damonsday (that is, doomsday) Parable of the Duel.

Coover presents the parable through the mind of current baseball star, Hardy Ingram, the rookie tapped to play the role of Damon. Significantly, his name aurally evokes the "hardiness" in "grammar" (that is, in words) that constitutes the essence of literature. Moreover, the etymology of "ingram," meaning a "perverted form of *ignorant*" (*Oxford English Dictionary*), underscores the character's role as innocent. Hardy's ruminative questioning of the significance and exact nature of the Duel and its origins precisely parallels the attitude expressed earlier by Waugh; Hardy "wants to quit—but what does he mean 'quit'? The game? Life? Could you separate them?" (238) Facing imminent death for a cause that at best bewilders him, Hardy-as-Damon "doesn't know any more whether he's a Damonite or a Caseyite or something else again . . . it's all irrelevant . . . even . . . that he's going to die" (242). In the end, "all that counts" is his ultimate acceptance of being because, simply, *it is* (242). Nor is any of it a genuine "trial" or even a "lesson," observes Hardy-Damon, "It's just what it is" (242). At the end, then, he remythifies the ritual for himself and returns it to its presacerdotal form as a game. He "holds the baseball up . . . hard and white and alive in the sun. . . . 'It's beautiful, that ball' " (242).

The transformation of the Association into an unquestioned societal ritual far removed from its origin thus replicates within the novel's plot Coover's idea of remythification. From the mythos of self-gratification within Waugh's private utopia, the UBA is remythified into a decadent dystopia in which human sacrifice provides a debased but accepted form of public gratification.

Furthermore, re-creation also occurs thematically if we read chapter eight as a postmodernist Christian allegory, as Arlen J. Hansen has suggested. The name of the original hero, Damon Rutherford, conjures up four important images—Damon the biblical martyr, demonism, Babe Ruth, and the physicist Rutherford—that, singly and in combination, convey the heroic power and dynamic talents of the original "son" of Jahweh-Henry. Oppositely, Jock Casey, the original antagonist, brings forth two important images—J. C. for Jesus Christ (also this JHW's "son") and the tragicomic "Casey at the Bat." Coover's version thus reverses orthodox values and makes, as Hansen notes, the "daemon" a true "life-force" who injects the UBA with the "vital anarchy" needed to revivify it, while J. C. is depicted as "a rather bland, un-*Christ*like figure" (Hansen, 57).

Remythification continues after the two players' deaths when their followers form two polarized camps, the Damonites and the Caseyites; but far from reflecting the complexity that produced the originals, the evolved "religions" hold the "simplistic and pious view of Damon as Good and Jock as Evil" and vice versa (Coover, 1968, 221). Moreover, the Damonites evolve from their early radicalism into tradition-bound conformists, an evolution which mirrors the historical shift from Christ's original radicalism and the unorthodoxy of early Christianity to the ossified fragments that form today's Christian institution. In this way Coover remythifies the "traditional Christian 'reading' of the legendary material" (Hansen, 57).

In sum, chapter eight's Association mythos, now distended beyond recognition from its solipsistic source (hence, the incomplete Jahweh "*JHW*") illustrates how myth grows and develops its own momentum out of banal experience, a process which eventually destroys the energy and dynamics that catalyzed its genesis. In Hansen's words,

any given myth should be considered a temporary expedient, constructed strategically and consciously so as to generate the meaning currently most rich and functional. When a particular myth becomes calcified, pernicious, or unsatisfactory for one reason or another, a new fiction has to be created. . . . Myth and language, then, are matters of epistemological convention. No given myth or name can capture all possible qualities, a fact which accounts for the rejuvenation that occurs with subsequent remythification. (57)

Accordingly, *The Universal Baseball Association* demonstrates the primal human need for mythic frameworks. The primal need requires that fluid, patternless life, even within the perfect order of a game, be governed by an implicit teleology, a schema of infinity and omniscience that emerges from the very finite and limited human mind and temperament. The dystopian fantasy that concludes the novel makes Waugh's UBA truly "universal" as it transcends the limits of its particularity—the proprietor himself—to parallel the function of primitive myth. In this light the fact of the book's eight-chapter format obtains new meaning in the recognition that the ninth "inning" was programmed by the extra-textual "Jahweh," Coover, to play in the reader's mind where it tugs at racial memory.

We must bear in mind, of course, that part of the fun of this novel is the author's playing with the reader, especially with the literary critic. In the same methodical way he toys with his "Prop.," Coover toys with the critic who cannot avoid discussing the UBA's plot and characters—their baseball performances, personal lives, murders, future form, and metaphysics—as if they were real fictional characters, not just the figments of a pathetic fictional bookkeeper's escapist daydreams. But this insight links unavoidably to another postmodernist idea concerning the very nature of fiction: it is itself unreal, obviously so. Yet Coover reminds us here that the important thing is that fiction is not untrue: according to the novel's epigraph from Kant it is not "requisite to prove that such an *intellectus archetypus* is possible, but only that we are led to the Idea of it."

As playwright Edward Albee once observed, "fiction is fact distorted into truth," a wisdom artfully proven in *The Universal Baseball Association* and one that underlies the controlling idea of this chapter, seeking the perfect game. The novel captures, through Waugh's obsession with his game's statistics and recordkeeping of its history, baseball's mathematical purity and its required "book" on itself, twin essentials of its perfection. The proprietor's fixation on the achievement of perfect performances by Damon Rutherford and Jock Casey further illustrates this study's title metaphor. Likewise, if "a box score is...a precisely etched miniature of the sport" (Angell, 12), then *The Universal Baseball Association* is a perfectly etched miniature of the transformation process that turns actual experience into stylized ritual and ephemeral myth, a process that parallels art. Baseball fiction writers, for example, turn the actual experience of baseball in American culture into stories that communicate both the reality of baseball and the culture while simultaneously transcending them into the encompassing truth of art.

THE GREAT AMERICAN NOVEL (PHILIP ROTH)

In *City of Words* Tony Tanner affirms that Ralph Ellison's *Invisible Man* is a germinal work in modern American fiction. It is, he says, a novel that ties up the various ends bequeathed to the next century by nineteenth-century prose fictionists and that also supplies fresh metaphors for the interpretation of our postwar experience (50–52). Philip Roth also notes and acknowledges the importance of Ellison's contribution by closing his often-cited essay, "Writing American Fiction," with a sobering existential allusion to *Invisible Man*:

Consequently, it is not with this image [Styron's hero, Henderson] that I should like to conclude, but instead with the image that Ralph Ellison gives to us of his hero at the end of *Invisible Man*. For here too the hero is left with the simple stark fact of himself. He is as alone as a man can be. Not that he hasn't gone out into the world; he has gone out into it, and out into it, and out into it—but at the end he chooses to go underground, to live there and to wait. And it does not seem to him a cause for celebration either. (158)

Smitty, *The Great American Novel*'s hero-narrator, trapped in what he perceives as the "tyranny" of a nursing home, begins and ends similarly with only "the simple stark fact of himself" (*Great American Novel*, 18). For followers of Roth's fiction, Smitty's alienation exactly resembles that of Alexander Portnoy of *Portnoy's Complaint* and Peter Tarnopol of *My Life as a Man*. And although Smitty has written what he believes to be "the great American novel," his accomplishment gives him no "cause for celebration" because no one celebrates alone in the anonymity of the world's neglect (18–19).

Developed around the idea that baseball is so ingrained within America's culture that it serves as a microcosm of it, *The Great American Novel* reminds us of folklorist Tristram Coffin's comment that "somewhere, somehow, out of all this an elusive dream has begun to take shape—a dream of using this national pastime to write the great national novel" (192). Roth cleverly merges two basic Yankee myths, that of baseball as emblem of the nation's rugged frontier innocence, and that of the great national novel as the culture's literary grail. In keeping with the subjective, even sentimental, nature of both grand myths, Roth employs mock heroic satire—he calls it "satyre" to stress the Dionysian "pure pleasure of exploring the anarchic." *The Great American Novel*, then, is his monument to the national pastime (*Reading Myself*, 76).

Roth's characteristic humor and reductive irony pervade the novel, however, making the investigation of themes and literary signs a hazardous undertaking. For this reason, one reviewer described this novel as "ideally suited to Roth's lesser, more eclectic skills; his talent for parody and caricature" (Prescott, 125). Another disagrees, calling *The Great American Novel* "tour de force," and the author's "funniest, most purely comic novel" (Rodgers, 109). Chroniclers of baseball fiction in general rate it highly.

The Great American Novel presents the satirical baseball "history of the Patriot League [and]...the Ruppert Mundys" (53) through the point of view of protagonist Word "Smitty" Smith. Smitty's twofold purpose in telling the history to his "fans" is, first, to correct the falsified record about a League which he stresses has been *"willfully erased from the national memory"* (18) and, second, to sketch "the plight of the artist" (414). The imaginary Patriot League once rivaled and nearly surpassed the greatness of the American and National Leagues, according to Smitty; the Ruppert Mundys, the team name obliquely alluding to Jacob Ruppert, who built Yankee Stadium, were a part of that glory. Roth summarizes his purpose at the start of chapter one, a device he uses to open each chapter in the manner of eighteenth-century novels.

Containing as much of the history of Patriot League as is necessary to acquaint the reader with its precarious condition at the beginning of the Second World War. The character of General Oakhart—soldier, patriot, and President of the League. His great love for the rules of the game.... By way of a contrast, the character of Gil Gamesh, the most sensational rookie pitcher of all time....A brief history of the Ruppert Mundys, in which their decline from greatness is traced. (53)

During the course of the novel, Smitty explains why he thinks that the Patriot League's extraordinary talents made it a natural target for the corruptions of money-grubbing opportunists. Clearly, he should know, for he was once a renowned sportswriter privy to the sport, its behind the scenes interactions, and all strata of its players, those on the diamond and in the boardrooms.

Critic Bernard F. Rodgers points out that Roth's narrator and narrative are direct descendants of the tall tale of Southwestern oral humor (categorized earlier by Walter Blair) which require "extensive use of the vernacular... emphasized masculine pastimes, employed exaggeration and popular myth, and dealt chiefly with the lower classes of society" (1978, 110). Accordingly, reduced to the nursing home indignities associated with age in America, Smitty does not go gently into the night of institutional geriatrics. Instead, combining all the local colorisms of the Southwestern yarn-spinners, his grand monologue constitutes his fight against both nature (his age and infirmities) and civilization (the nurses and staff of the "home"). With the focused intensity of the very old, he has considered death and, consequently, chosen to pursue immortality through writing about baseball.

The relationship between art and immortality appears explicitly in the novel's prologue and implicitly in the epilogue, where Smitty's abundant literary allusions, including a conversation with Ernest Hemingway, and self-conscious fictionmaking take on a life of their own quite apart from the Patriot League story. Presented as parody by Roth, Smitty's literary hyperbole places him in the august literary tradition of Chaucer, Shakespeare, and Melville, whom he cites with princely casualness (15). He compares himself to these canonical writers and artists to underscore his single-minded effort to present the truth, presumably about the Patriot League but also about himself as a misunderstood "wordsmith." "There is nothing so wearing in all of human life," he moans, "as burning with a truth that everyone else denies. You don't know suffering, fans, until you know that" (24). Because Smitty and his suffering belong more appropriately with the Canterbury pilgrims, Lear, and Ishmael, rather than with their creators, his exaggerated sense of himself compounds the novel's fundamental irony.

Smitty's literary allusions are of three main types. First, there is the slightly veiled reference to a writer and his work without explicit use of the writer's name. This type is typified in the narrative's *Moby Dick*-like opening: "Call me Smitty." Although similar references appear frequently throughout the novel, and pointedly at the end ("The drama's done"), Roth clutters chapters six and seven with them. For example, in Smitty's digression on Ulysses Fairsmith, exemplar baseball loyalist and manager of the Mundys, the scene shifts from America to the African jungle. We are told, "Mistah Baseball—he dead," by a native whose people are given to shouts of "Omoo!" and "Typee!" (328–334), references to Conrad's *Heart of Darkness* and to two early Melville novels.

Similarly, in describing the woeful peregrinations of Gil Gamesh, extraordinary athlete expelled from the game for life, Roth alludes to Sherwood Anderson's minor classic: "one day down in Winesburg, Ohio, unable to bear any longer his life as a lonely grotesque," Gamesh decides to rejoin America's mainstream (372). The apparent disconnectedness of these references aid Smitty's self-serving purpose, to demonstrate the genuine ties between his simple tale about a homely pastime and the supposedly loftier literary works. The writerly subtext, however, in its allusive anarchy, is pure parody—Roth's ironic undermining of accepted literary conventions.

More fully developed, the second type of allusion includes names of writers and, usually, the titles of specific works. These references begin with Smitty's recollection of a 1936 sailfishing expedition with "my old friend (and enemy) Ernest Hemingway." Besides being a realistically acerbic, occasionally sympathetic but totally plausible portrait of Hemingway, Smitty's memory also introduces the novel's title subject.

> Then he took up the subject of the Great American Novel again, joking that it would probably be me of all the sons of bitches in the world who could spell cat who was going to write it. "isn't that what you sportswriters think. . . . That some day you're going to get off into a little cabin somewhere and write the G.A.N.? Could do it now . . . if only you had the Time." . . . "Stand close to me . . . let me look into a human eye. The Great American Novel. Why should Hemingway give chase to the Great American Novel?" (28, 34)

Despite the black humor, the Hemingway section portrays the chilling compulsion felt by certain artists to create a definitive masterpiece that will assure immortality. In sympathy with the pain of Hemingway's obsession, Smitty appears in this section less cynical, even tender. The chapter's humor, cynicism, and tenderness are all captured nicely in Smitty's reverie at the end of the Papa Hemingway section: "Every once in a while I would get a Christmas card from Hem, sometimes from Switzerland or Idaho, written in his cups obviously, saying more or less the same thing each time: use my style one more time . . . and I'll kill you. But of course in the end the guy Hem killed for using his style was himself" (39).

Besides the macho Nobel laureate, Smitty also discusses Hawthorne, Twain, and Melville, whom he calls "my precursors, my kinsmen." Examining the most acclaimed works written by these authors, he offers critiques of *The Scarlet Letter, The Adventures of Huckleberry Finn* and *Moby Dick* within the context of "the story I have to tell" (41). The strained and ironic nature of both his critiques and the contrived relationships he draws between his book and his "precursors" are not in themselves as significant as the need he feels to make them. His literary insecurity compels him to try to legitimize both baseball as a fit subject, and himself as a worthy crafter of authentic belles lettres. The American writer's inferiority complex is not new, of course, but, at least since the Second World War, it has been viewed as moot.

Patronized by the nurses and fellow patients that surround him in the Valhalla Home for the aged, Smitty needs the approval of others to eat, sleep, or write. Without official approbation, he cannot publish nor even mail off his exposé of the Ruppert Mundys. Moreover, by opening the novel with Smitty among actual historical authors—as well as presidents, popes, and baseball commissioners—Roth gives weight to the parodic seriousness of his narrator. The gallery of famous names also prepares the reader for Smitty's later association with more immediate contemporary figures like Walter Cronkite and Ralph Nader, as well as for the unlikely letter to Mao Tse-Tung that concludes the novel.

The novel's third type of allusion is that of parallel structure. As Smitty admits when he discusses the introduction to his great American novel, the form of his narrative recalls that of the symmetrical *Scarlet Letter*; it consists of an even nine chapters, counting the prologue and epilogue. His role as narrator, and the fact that he alone lives to tell the tale of the Patriot League, also recalls Poe's *The Narrative of Arthur Gordon Pym*. The lone survivor motif suggested here is at least as old as the Book of Job: "And I only am escaped alone to tell thee" (Job1:14–19). Melville, whom Smitty addresses as his "blood brother," uses the Job messenger's refrain prominently in *Moby Dick*, the masterpiece that provides the primary model for *The Great American Novel*'s form. Not only do the prologue and epilogue refer directly to *Moby Dick*, but Melville's solemn frustration at the unchanged nature of human imperfection rings thunderously throughout Roth's work.

One significant similarity between Melville's and Roth's novels concerns the omniscience of the first-person narrator. Like Ishmael, who alone lives to record a full account of Ahab's story, so too Smitty relates the Patriot League's rise and fall with total omniscience. The fact of fiction—that is, that both Ishmael and Smitty must fictionalize, even to hyperbole, to present a completely truthful record is conspicuous in both narratives. Smitty's kinship to Melville can be traced to the latter's lifelong struggle with both the metaphysics and the ideology of literature—just as Roth has struggled as a teacher and writer with the meaning of literature. The possibility that art is perhaps the only vehicle to immortality binds Smitty to his most illustrious and revered precursor.

Another technique in Roth's novel to reinforce the theme of immortality through literature is the excessive foregrounding of language—of words, grammar, and signs. Through Smitty's name, revealing his former trade, "wordsmithing," and through his admitted greatest vice, uncontrolled alliteration, the novel emphasizes the linguistic basis of literature with the unrestrained bombast and ribaldry of Roth's style in *Portnoy's Complaint*. "The alphabet! That dear old friend," effuses Smitty, "Is there a one of the Big Twenty-Six that does not carry with it a thousand keen memories for an archaic and humorous, outmoded and outdated and oblivionbound sportsscribe like me?" (8).

Smitty responds to the threat of oblivion with his monument of words, the novel in our hands: an exhaustive anatomy of the world he once knew, a heap composed of senile impressions, baseball lore, belles lettres, and numerology, conveyed through his profligate use of the number seven. As a symbol, seven is important in the Western tradition because of the mythical creation of the world in seven days. It also has special meaning in baseball through the superstition of the lucky seventh inning (Coffin, 181–82). With incredible persistence Smitty milks dry his numerical emblem, invoking it whenever a number or a symbol is called for, just as he does the alphabet.

According to him, the measure of a writer's life and immortality arises from the ceaseless flow of words he manages every day (10). The joke is, of course, that as he awaits death in the nursing home, Smitty's indulgence in puns and free association of words literally constitutes the very extent of his eighty-seven-year-old life. Writing is no longer his means to something else; it is life itself. Similarly, his interest in immortality is hardly an academic or artistic concern; his mortality is nearly done. For each word or sentence that he can stretch endlessly in his manuscript, Smitty extends his existence. He thus writes pages full of alliterated sentences, unharnessed ad nauseam to suggest the infinite possibilities of language. Through his tireless alliterations, ubiquitous puns, and prevalent synonyms, Smitty seems intent upon proving the linguistic axiom that the expressive capacity of language is infinite.

Roth's calculated purpose behind the endless language games and unrestrained "lexicomedy" is literary analogy. He aims to parallel through his narrator's style and the novel's central themes two of baseball's singular features: its inherent timelessness and its implied immortality through recordkeeping. According to baseball rules, any given game could theoretically last to infinity, because the rules do not regulate time limits, only innings played. Similarly, the rules require exact recordkeeping of all significant action of every regulation game played. Baseball's record book on itself, therefore, eternalizes in writing each game played. These inherent qualities of the game help explain its unique and persistent attraction to writers and other literati.

Through its emphasis on literary allusion and language, *The Great American Novel* resembles the metafictional quality of Coover's *The Universal Baseball Association*. Like Coover, Roth foregrounds the mechanics of literature through a narcissistic protagonist acting in a plot self-contained within his consciousness. Roth draws attention to the act and craft of fiction in a number of ways. Through the hyperbole of his narrator's insistence on the unadulterated authenticity of his story, as well as through his forcing of connections between it and literary classics, the novel asserts its place within the tradition of American romance. Roth also emphasizes the process of fiction creating itself through Smitty's repeated references both to his career as a writer and to the self-conscious act of writing the book we are reading. Smitty's repeated allusions to his manuscript as "the great American novel"—which one critic calls "this country's elusive Loch Ness monster" (Prescott, 125)—also suggest fiction in the act of becoming and defining itself. Finally, the novel's framing structure (that is, the Prologue and Epilogue) and framework story (that is, the history of the Patriot League and one of its teams, the Ruppert Mundys) further underscores its quintessential fictiveness.

If Roth's major theme in the novel concerns seeking immortality, the ultimate perfect game, through art, then the framework story about the lost Patriot League and the Ruppert Mundys demonstrates his theme. The novel opens with Smitty's "sad fact" that the League and the Mundys, once visible and viable parts of mundane but essentially artless reality, have totally vanished from public memory in less than a generation's time. To Smitty, the public amnesia about an integral part of its social identity demonstrates the vulnerable mortality of everything—unless, he and the novel argue, they are redeemed by art, which confers immortality.

The demise of the Patriot League begins with the Mundys' patriotic sacrifice of their baseball park to the United States War Department for use as a GI embarkation camp in World War II. In reality, the team is exiled from its ballpark because of a lucrative financial deal arranged by the owner. "To help save the world from democracy," as leftist dissident Smitty describes it, the homeless Mundys must henceforth play all their games on the road. Their fanatical fundamentalist Christian manager, Ulysses Fairsmith, likens his team's fate to the expulsion and forced diaspora of the Old Testament Israelites. The vicissitudes of their forced exile provide the main plot action of *The Great American Novel*, producing a Mundys recordbreaking season of catastrophic proportions. For example, their 120 losses were the "most games lost in a season," and their 302 errors were the "most" ever committed by one team (184).

But the fall of Port Ruppert's team represents only one particularly unpleasant aspect of the entire league's dissolution. Although the fate of the Mundys, with their perpetual homelessness and relentless losing, is the worst of all, other teams in the Patriot League suffer during the war years too. The Kakoola Reapers, for instance, are destroyed by their gimcrack financial success, won through carnival-like gimmicks. Roth parodies the circus ostentation of the Chicago White Sox, owned by Bill Veeck. According to Smitty, the show business excesses of the Kakoola Reapers—including the use of midgets, cripples, and disguised women to play on the team—helped wreck ("as in Veeck") them.

In relating the demise of the Patriot League, Smitty stresses the connection between the deteriorating league and the decaying state of the nation in general. He reiterates several times his belief that "baseball is this country's religion" (101) and that "the fate of the Mundys and of the republic were inextricably bound together" (146). Roth devises this social context to lampoon the hysteria known as the "red scare," that right-wing affliction that simplistically blames all world ills and problems on "*COMMUNISM.*" Nurtured by money-grubbers, politicians, and militarists since before the Cold War, the red scare is America's twentieth-century ideological bogeyman, just as the "bloodthirsty savage Indians" served that function in the seventeenth and eighteenth centuries.

Roth lampoons this infantilism that masks as democratic ideology by sketching a communist conspiracy against baseball. Angela Whittling Trust, another reactionary league owner, and sexual partner of legions of ballplayers, sees the conspiracy in these terms:

Do you remember what it is that links in brotherhood millions upon millions of American men, makes kin of competitors, makes neighbors of strangers, makes friends of enemies, if only while the game is going on? *Baseball!* And that is how they propose to destroy America . . . that is their evil and ingenious plan—to *destroy our national game!* . . . In order to destroy America, the Communists in Russia and their agents around the world are going to attempt to destroy the major leagues. They have selected as their target the weakest link in the majors—our league. And the weakest link within our league—the Mundys. (288)

Mrs. Trust's influence eventually helps generate an investigation of the Patriot League by the House Un-American Activities Committee (HUAC), a direct allusion to the illegal persecution of American citizens known as the "McCarthy witch hunts." In the novel, the government's criminal invasion of privacy and usurpation of constitutional rights eventually leads to the suspension of over seventy-two league members. The ultimate result of the government persecution is "temporary" suspension of league operations. To Smitty's undying regret, the unfair suspension ends up lasting forever.

This brief summary of *The Great American Novel* hardly touches on the novel's multiple strands of plot and theme. "When a novelist writes about baseball," comments reviewer Sheldon Frank, "he is automatically writing about America—the lust for victory at any price, the greed of the owners, the zombified aggression of the fans. Baseball is the Nineteenth-Century sport" (1973, 23). Roth's use of the baseball metaphor conforms to Frank's perception, and his mordant characterizations offer a grim portrayal of the nation and its values. Each of the scores of characters, and certainly each major character, serves as the center of a developed allegory that often deviates from the novel's main storyline. Bernard Rodgers describes Roth's cast: "Midgets, dwarfs, halfwits, cripples, one-armed and one-legged characters swarm through the pages . . . and typically tasteless—and hilarious—fun is made of their physical maladies" (111). Out of this satiric abundance, the tales about General Oakhart, Ulysses S. Fairsmith, Mrs. Trust, Gil Gamesh, and Roland Agni warrant close examination.

Representing unswerving loyalty to patriarchal rules and regulations is General Oakhart, coach of the Mundys. His military background clearly recalls that of Abner Doubleday, the spurious founder of baseball. Oakhart's staunch upholding of baseball's rules ("take away the Rules and Regulations, and you don't have civilized life as we know and revere it" [57]) conflicts directly with his "official" decision to force the Mundys to play all their games on the road, which is, in fact, against the sport's rules. The General's pragmatic nature—his oak-like self—permits him to see what the rest of the country fails to see: "that it is not an overflow of patriotic emotion that had drawn [the Mundy brothers] into the deal to lease their ball park to the government," but, rather, it was greedy capitalism, to the "tune of fifty thousand dollars a month, twelve months a year" (89).

On the other hand, Oakhart's overpowering sentiment—his "heart"—about the rules of the game, prevents him from recognizing that their intransigent application will inevitably shorten the life of baseball, which must "keep pace" with the country's explosion of fashion and trends if it is to survive (87). Despite his reactionary zeal, his pristine vision provides the novel's greatest source of pastoral rhapsody about baseball.

You could not begin to communicate through *words*, either printed or spoken, what the game was all about—not even words as poetical and inspirational as those Mister Fairsmith was so good at. As the General said, the beauty and meaning of baseball resided in the fixed geometry of the diamond and the test it provided of agility, strength, and timing. Baseball was a game that looked different from every single seat in the ball park. . . . You might as well put an announcer up in the woods in October and have him do a "live" broadcast of the fall, as describe a baseball game on the radio. (96–97)

However pristine, such narrowness of vision also accounts for the General's life record of near misses.

General Oakhart's smashing failures emphasize that he is not a reasonable alternative to the nation's progressive decadence. He had once "almost" been selected Commissioner of Baseball, we learn, but the owners decided instead on the "less principled" Judge Kenesaw Mountain Landis—the actual commissioner appointed ostensibly to rid the sport of the dishonor brought to it by the Black Sox fix. Likewise, Oakhart was supposedly mentioned as a possible vice-presidential running mate for Warren Harding, who instead chose a candidate from among his Teapot Dome cronies—another historical crime scandal resulting from the money-grubbing collusion of business and politics.

Oakhart finally decides to nominate himself for President on the Patriot Party ticket. He chooses for his running mate a similarly right-wing former baseball player, a midget, whose size symbolizes the shrunken nature of the General's politics. Exposed at last as undeniably ludicrous in his fascist regard for rules (that is, ruling) and nationalism (that is, exclusionism), the General loses: "alas, the American people didn't seem to care any more than the politicians for a man who lived by and for the Rules and Regulations" (59). Roth's point is that the people get the lowest common denominator of corruption disguised as *ruler* that they deserve.

Comparable to the General in loyalty to baseball is Ulysses Fairsmith, also known as Mister Baseball, manager of the Ruppert Mundys. Twenty years before the hapless 1943 season, Mister Fairsmith, a Christian zealot, journeyed far from home like his illustrious Greek namesake. As a missionary to Africa he "organized the first game of baseball ever played on the continent of Africa between all-native teams" (320). Describing Fairsmith's mission in meticulous, hilarious detail, Smitty tells us that the African villagers displayed natural mastery of the sport—until "the trouble erupted over sliding" (321). The trouble stems from the African love for sliding into base—even when unnecessary as, for example, in a slide to first on a base-on-balls. Because of his prohibition of "extraordinary" sliding (note the embedded pun to religious backsliding), Fairsmith and his young American assistant narrowly escape being served à la carte to the entire tribe.

Mister Baseball's challenging African experience and brush with martyrdom prepare him well for the shocking collapse of the Mundys: he countenances their "fall" with ecstatic faith in the "will of the Lord" (99). Nevertheless, despite his moral fitness, the "unprofessional, undignified, *immoral*" behavior of the 1943 Mundys eventually brings him to the "very edge of an abyss even more terrifying than the one he had glimpsed twenty years earlier" (319–20). Appropriately, the undiluted Calvinist, incapable of tolerating (back)sliding of any kind, winds up a "martyr" after all, for he dies immediately after hearing a Mundy player employ gambling as a figure of speech: " 'You say you *gambled*? My God,' and rolling off his rocker, [he] died on the floor of the visitors' dugout" (344).

Unlike Fairsmith, Angela Whittling Trust was originally a woman with "the morals of a high school harlot" who routinely made lovers of baseball players. Her conquests included legends Babe Ruth and Ty Cobb. What saves Mrs. Trust from further iniquity is her dying husband's final request: "Learn to be a responsible human being, Angela" (285). Now the widow of the Tri-City Tycoons' founder, she accedes to her husband's deathbed wish with a fervor that soon rivals the ultra-conservative zeal of Oakhart and Fairsmith. Opposed to any changes in the game's original form, she is the only owner who speaks out publicly against the sport's increasingly "liberal" style of entertainment. Typical of her right-wing fanaticism, to her the unacceptable liberalism includes allowing Blacks to play as equals in the major leagues.

Like Oakhart and Fairsmith, whose baseball fundamentalism matches their obsession with law and order and with Christ, respectively, Trust's purism also matches an obsession: fear of communism.

They are going to turn the people, not only against the national game, but simultaneously against the profit system itself. Midgets! Horse races! And [Reaper owner Mazuma will] have colored on that team soon enough, just wait and see. I've had him under surveillance now since the day he came into the league. I know every move he makes before he makes it. Colored . . . colored major league players! And that is only the beginning. Only wait until Hitler is defeated. Only wait until the international Communist conspiracy can invade every nook and cranny of our national life. (289)

In her mind, Smitty tells us, as in the minds of countless average millions, White is Right—both morally and politically speaking. Ironically, it is her action that "signal[s] the demise of [the] Patriot League" when, after the HUAC investigation, she "releases 'loyal' Tycoon players from contracts for business reasons" (408). Her final words in the novel constitute a fitting epitaph: "Better dead than Red" (408).

Although Angela Trust's reactionary views are transparent, those of super-star player Gil Gamesh remain obscure and contradictory throughout the narrative. Overbearing and boastful, Gamesh enters the league when Oakhart becomes its president. One of Oakhart's first official decisions is to assign Mike the Mouth Masterson, "the toughest, fairest official who ever wore blue," to Gamesh's team to make the cocky pitcher "obedient to the Rules and Regulations" (77). Their expected confrontation occurs in the first chapter, when Masterson ruins a third consecutive no-hitter for Gamesh, this one twenty-six strikeouts toward the perfect game most pitchers dream about. With the umpire's back turned, justifiably we're told, Gamesh pitches his 27th strikeout, but Masterson cannot call it because he did not see it. To retaliate for Masterson's interference with the immortality that the perfect game would have assured him, Gamesh tries to kill Masterson with a fastball to the umpire's throat. The near-lethal pitch not only costs Mike the Mouth his voice but also leads to Oakhart's expulsion of Gamesh from organized baseball for life.

At this early stage in the novel, and in an oblique allusion to Waugh's disappearance from the main action of *The Universal Baseball Association*, Gamesh disappears temporarily from the novel's pages. He returns to the center of Smitty's story in the seventh chapter, still cloaked in enigma, compounded by ideological ambiguity. His return to the novel's foreground brings a bizarre tale of being "an agent of the Communist Party, just returned from six years in Moscow, four of them at the International Lenin School.... My mission is to complete the destruction of the Patriot League" (356). The skeptical narrator, Smitty, makes it clear that he does not believe Gamesh's Russian chronicle. In time the former star pitcher disavows communism, because it lacks baseball or a satisfying equivalent for it, and instead becomes a double agent bent on "rooting out" the other Commie infiltrators of the organized sport.

Through these labyrinthine twists and cul-de-sacs, he comes to embody the full allegorical significance of his name. A man of strong will—that is, "Gil"—who once excelled at a professional game, Gamesh now gamely assumes a double—"G.G."—identity. Moreover, with the use of his Russian cover name, Marshal Gilgamesh the novel presents one of its most original contributions to baseball fiction: Roth's reconstruction of the "Epic of Gilgamesh."

One of the grandest of humankind's ancient legends, the "Epic of Gilgamesh" is acknowledged as the oldest epic preserved in writing and, thus, is considered an ur-Odysseus myth. Gilgamesh, whose name means "hero," was a Sumerian king whose heroism emerged from his reputed extraordinary nobility and his amazing deeds of valor. Like mythic heroes before and after him, however, the force that drove him was his obsessive desire to escape death and enjoy eternal immortality. The epic, traditionally divided into five parts, gives the Gil Gamesh thread in Roth's baseball narrative a structural framework like that provided in Joyce's *Ulysses* by its Homerian antecedent and that discussed in chapter four concerning the mythic structure of Mala-mud's *The Natural*.

In summary, the epic tells of Gilgamesh's search for immortality through a treacherous journey to the sun god, Shamash, whom he hopes will grant his wish. He encounters awesome perils with each league he travels, including impenetrable darkness, cyclonic winds, the wrath of the goddess Ishtar (whose revenge is meted out by her father in the form of an attacking bull), and similar epic-scale adversities. Finally, after traveling distant leagues the light of dawn shines for Gilgamesh when he meets Shamash on the twelfth league of his journey. With the god's sunlight streaming brightly upon him, the king who does not want to die, Gilgamesh presents his case, arguing that his godlike attributes and amazing mortal conquests entitle him to everlasting life.

And to Gilgamesh [Shamash] said, "You will never find the life for which you are searching." Gilgamesh said … "Now that I have toiled and strayed so far in the wil-derness, am I to sleep, and let the earth cover my head for ever? Let my eyes see the sun until they are dazzled with looking. Although I am no better than a dead man, still let me see the light of the sun!" (Eliot, *Myths*, 246).

The king learns what all heroes must, what even his subjects know, that despite their greatness, brilliance, and (con)quests, even heroes cannot, in the end, shed their humanness.

From the variant renditions of the Sumerian myth and the variety of trans-lations and interpretations, Roth selects key elements as prototypes for use in *The Great American Novel*'s characterization of the Patriot League's leg-endary pitcher. For example, Gil Gamesh's early experiences parallel the epic's "Gilgamesh and the Land of the Living" section in which the great king distinguishes himself as a remarkable but despotic ruler. He is feared for his rapaciousness toward female subjects, whether maiden or married, willing or not. To neutralize the king's power, the gods send Enkidu, a wildman identical to the animals of the forest that raised him. Although Gilgamesh outwits Enkidu, the two form a deep fraternal bond that leads the king to seek even greater conquests outside his kingdom. This parallels the pitcher's reliance on his catcher in the novel.

The pitcher's bitter confrontation with Mike Masterson in chapter five of the novel are like the legend's "Gilgamesh and the Bull of Heaven" sequence. In it the legendary king rejects Ishtar's erotic advances, thus eliciting her wrathful revenge that comes in the form of a raging bull's ferocious attack on the king. Saved by Enkidu, the bull is destroyed and the king mocks the goddess to the din of the people's chanting: "Gilgamesh is dazzling among men!" (*New Larousse*, 71). If umpire Masterson resembles the raging bull, then Angela Whittling Trust echoes the goddess Ishtar. Both Ishtar's arbitrary abuse of power and her great libido coincide in Roth's portrayal of owner Trust, the harlot turned patriot.

Gamesh's expulsion from professional baseball and his subsequent sojourn in Russia resemble the myth's tales of the original king's departure from his kingdom as he journeys far away and undergoes awesome trials, including descent into the Underworld, and astonishing successes in his quest for immortality. Although Gamesh's Russian exploits are not presented, Smitty provides some evidence that the former star pitcher underwent comparable epic tribulations, and, indeed, that his return to America in full strength is itself a heroic feat. The ancient myth's "Underworld" tales appear in contemporary narrative with the pitcher's return to the United States in a subversive new role as baseball's double agent. Sent to expose and, by so doing, undermine the money-grubbing corruptions of the American baseball industry, Gamesh enters a metaphoric Underworld of sports quite unlike the mainstream realm he formerly inhabited as one of its superstars.

The traditional function of such a "descent" is to instruct the hero, and in turn the audience, in the meaning of suffering, powerlessness, and death— that is, in the essential nature of human mortality. Like his ancient namesake, double agent Gamesh learns the lesson poignantly well, which may explain why he, virtually alone in the cast of characters, is depicted seriously, without the broad humor and irony found in the work's other portrayals. Accordingly, Roth's use of double agency in this characterization recalls the legend's emphasis on twins and doubling as a means of conveying the complexity and mystery of life.

To conclude his version of the "Epic of Gilgamesh" Roth employs the same veil of mystery and ambiguity found in the original epic regarding the ancient king's end. The original mystery is intensified by the numerous variants of the legend, the absence of large sections of material, and the fragmentary nature of the extant text. Analogously, Roth ignores his baseball hero for many pages until, toward the conclusion, he slips in a brief mention of Gamesh's death almost as an afterthought.

MAY 1949. PRAVDA photograph of Moscow May Day celebration reprinted in American papers; hatted figure between Premier Stalin and Minister of State Security Beria identified as Marshal Gilgamesh. . . . *MARCH 1953. PRAVDA* photograph of Stalin funeral mourners reprinted in American press; hatted figure between Minister of State Beria and First Secretary Malenkov identified as Marshal Gilgamesh. . . . *MARCH 1954.* Marshal Gilgamesh sentenced to death as "double agent" and executed (409).

Just as King Gilgamesh is buried by centuries of temporal and spatial distance, Roth buries his protagonist into a hidden corner of the narrative under pages of prose and multiple twists of plot.

Following the pattern of the novel's other allegorical characters, Gamesh's life ironically inverts in mock-heroic fashion such conventional heroic virtues as modesty, humility, and loyalty by showing American despots corrupting them, both politically and corporately. In addition, like heroes of old and, pointedly, like contemporary instant celebrities, Gamesh's life captures the contradictions facing adventurous heroes whose far-flung travels and exotic experiences transform their self-image and worldview but do not affect the adulation of their fans. Andreano's analysis of the Folk Hero Factor (see chapter two, "Emergence of a Metaphor") applies here, particularly in his observation that hero worship blinds the idolators to the imperfections of their sports idols. The public's fascination with celebrity, with fame of any sort and for any reason, seems to celebrate without distinction both talent and mediocrity, leaders and led, theologian and thief (for example, Abraham Lincoln and Warren Harding, Charles Lindburgh and Oliver North, Eleanor Roosevelt and Tammy Faye Bakker). Roth's parody suggests that media fame in the United States constitutes mass culture's acceptance of a new royalty, an aristocracy not necessarily based on birth or merit, but on fame and wealth, however achieved.

Another Patriot Leaguer whose baseball heroics emerge from traditional legend and then become corrupted by instant fame is eighteen-year-old Roland Agni, the only authentic baseball player on the Mundy roster. An exceptional athlete in the Apollonian mode, Agni's performance compares with that of the finest historical stars at their zenith. In typical reductive irony, Roth satirizes the public worship of sports figures by emphasizing, like Lardner and Malamud before him, the fact that most great athletes do not deserve to be idolized outside the playing field. In Agni's case, his arrogance makes him unbearable even to his parents. To give him a taste of ordinariness—an experience foreign to one who had "hurled a perfect sandlot game at age six" (134)—his parents force him to play for the disastrous Mundys. Frustrated beyond his level of toleration by a team that loses as if instinctively, Roland arranges to be bought by another League team, the Greenbacks. He collaborates with Isaac Ellis, a child prodigy and the son of the Greenbacks' owners, and their fix results in a major miracle, a Mundy winning streak. The introduction to chapter seven summarizes the situation:

Isaac Ellis makes another appearance; a conversation on "the Breakfast of Champions," wherein the desperate hero [Roland] of this great history learns the difference between the Wheaties that are made in Minneapolis by General Mills and those that are manufactured in an underground laboratory by a Jewish genius, and something too about Appearance and Reality. Roland succumbs. . . . A short account of the Mundy miracle . . . the bewilderment of Roland. (303)

Roland's bewilderment arises from his belief, drilled into him by his scrupulous parents, that his every act is selfishly motivated (134). He has therefore learned to distrust himself, his judgment, and his perceptual powers, making him an easy target for exploitation. A perfect embodiment of his errant namesake, he suffers manifold assaults on his heroism in the novel's own "Chanson de Roland," a parody of the medieval chivalric romance (387–89).

In the end he gains "fame throughout the land" when he becomes cor-rupted baseball's martyr as a result of the swindle by Isaac Ellis. The genius Ellis concocts a moneymaking scheme based on a phony product (like the mnemomorphic grass in *Babe Ruth Caught in a Snowstorm* and the use of steroids today), and he needs Roland to make it work.

> "Throwing a game, Roland, is *losing* when you're supposed to be *winning*. Winning instead of losing is what you're supposed to do."
> "But not by eatin' Wheaties!"
> "*Precisely* by eating Wheaties! That's the whole idea of Wheaties!"
> "But that's *real* Wheaties! And they don't make you do it anyway!"
> "Then how can they be 'real' Wheaties, if they don't do what they're supposed to do?"
> "That's what *makes* them 'real'!"
> "No, that's what makes them *unreal*. *Their* Wheaties say they're supposed to make you win—and they don't! *My* Wheaties say they're supposed to make you win—and they do!" (314)

Like the Black Sox Scandal and the destruction of Shoeless Joe Jackson's career, the doctored Wheaties swindle ruins naïve Roland forever. This se-quence underscores the Orwellian effects of capitalism, with its zeal for profit at any cost that results in the promotion of an inferior product as superior.

Subsequently, Roland is mistakenly shot and killed by the aged umpire Masterson, who was actually gunning for another baseball star, the ubiquitous Gamesh. Roland (again like Shoeless Joe), innocent to the end, becomes a posthumous media hero, and because of his reputed conversion to the cause of ultra-anticommunism, he finally even gains the unqualified acceptance of his parents. In this ludic and ludicrous context, the idea that he died for his country is exposed as the absurdity it is: an unnecessary death caused by greed and political corruption.

Each of the comic allegories in *The Great American Novel* concerns itself with crucial ideological and moral constituents of the U.S. national character. In presenting his view of those constituents, Roth employs baseball for di-dactic purposes, in what has been described as his characteristic "moral artistic intentions" realized through the "method" of "realism" (McDaniel, 202). Relatedly, Rodgers finds illuminating Roth's "dedication" to "finding subjects and techniques which will reveal the effect of the interpenetration of reality and fantasy in the lives of his representative Americans" (Rodgers, iii). Roth also uses the baseball metaphor as propaganda to poke ironic fun at the eighteenth-century morality handbooks that sought to convey Christian, and right-wing nationalist demogoguery, through limericks and jingles about the game.

Roth's critique of the dominant right-wing national consensus forms the core of his message. Each of the novel's allegories exposes the essentially commercial greed underlying the nation's so-called progress (345–97). Oakhart and Fairsmith, for instance, represent the earliest interplay of religion and mercantilism at the core of Yankee nationalism. In the same vein, hardly the "angel," Angela stands for the "Whittling" away at grassroots mutual "Trust" in democracy by the greed-inspired, greed-promoted "fear" of communism. Similarly, by signifying only tainted vestiges of the genuine heroism associated with their mythic namesakes, Gil Gamesh and Roland Agni prove that Patriot League history—like that of the nation—is debased myth, and represents humanity corrupted by vanity, venality, and avarice.

Although Smitty mocks the beliefs motivating Oakhart, Fairsmith, and Trust's desire to preserve the fantasy of baseball's "purity," his pervasive irony undermines their antitheses as well. This is evident in his farcical approach to the only alternatives offered in the story, hedonism and communism. Especially in chapters five and six, Smitty vilifies baseball's entrepreneurial roots and development, and suggests that it should seek to overturn the baseball-as-business ethos in favor of values encouraging baseball as play and diversion (366). All simple explanations of the corruption of baseball and America elicit only satire from Smitty. Humor and ironic detachment seem to be the narrator's only possible manner of coping with what he considers an unalterably decayed society.

The novel's two major themes, immortality through art presented in the prologue and epilogue, and the corruption of Yankee democracy, presented in chapters one through seven, unite in the single narrative voice of Word Smith, who serves as the diffusive, episodic work's cohering principle. As firsthand observer of the Patriot League's life, as recorder of its past, and as insistent prophet of the culture's future, Smitty's several roles converge into one: the persecuted artist who must present truth as he sees it.

CHAIRMAN [of HUAC Hearings]: But surely as a writer, Mr. Smith, you know the old saying that truth is stranger than fiction.

SMITTY: So are falsehoods, Mr. Chairman. Truth is stranger than fiction, but stranger still are lies. (407)

In its pithy summary of the ceaseless hypocrisies committed by the powerful elite against the commonweal, this exchange applies equally well to such other official government lies as those of Teapot Dome, Watergate, the deceptions that led to the U.S. debacle in Vietnam, and the more recent Iran/ Contra crimes. By shedding light on the "lost" Patriot League, Smitty is motivated by a belief in the need to maintain a free press by exposing facts that cause discomfort to politicians and their colluders. And, like any artist, he also aches for immortality.

Despite his valiant and acerbic efforts to fulfill his artistic function, old Smitty is dismissed as a senile blatherer. Far removed from his former life as a sportswriter, his nursing home confinement punishes him pitilessly. Like John Alexander Graham's Babe Ruth, transfixed in the alien setting of the paperweight, Smitty is trapped inside a geriatrics ward totally inimical to his energetic spirit and still potent imagination.

> "Smitty," said the slit [nurse], still smiling, "why don't you act your age?"
> "And what the hell does that mean?"
> "You know what it means. That you can't always have what you want."
> "Suppose what I want is for them [the official sports community] to admit THE TRUTH!"
> "Well, what may seem like the truth to you," said the seventeen-year-old bus driver and part-time philosopher, "may not, of course, seem like the truth to the other fella, you know." (21)

Fettered not only by his circumstances, Smitty is also imprisoned by his "factual" knowledge and enlightened opinions which, in the context of over-whelming mass ignorance and slavish devotion to false heroes, make him a man lonely in thought and deed.

He was once considered an authority through his sports column, "One Man's Opinion." Age, presumption, and outrage at the unfairness have painted the lonely storyteller into the corner of truth. His unharnassed irony and bitterness alienate the few others who enter his life, thus isolating him from society. In the end, he joins a lonely company of American characters: the vanished Waugh in *The Universal Baseball Association*, the "invisible man" gone underground in Ellison's novel, Lily Bart chloroformed to death by Wharton's *House of Mirth*, Edna Pointellier wandering out into the sea in Chopin's *Awakening*, and Huck Finn about "to light out for the territory," and many others. That is, he joins a remarkable cast of American existential figures who are unable, or unwilling, to live content within the pulsing hypocrisy of the false American Dream.

Smitty, *The Great American Novel*'s true protagonist, meets the penultimate challenge of confinement with his lofty pose as alienated writer. (His ultimate challenge is, of course, death.) A clue to his survival is found in the name of his cursed "rest" home, Valhalla, which derives from the mythic gathering place for the returned souls of dead Norse heroes. Valhalla is a fit home for those who survive only as disembodied voices; word smiths as a class and Smitty individually, are finally that. The artist as observer, chronicler, and seer is the ultimate hero. Despite the humor of the novel's conclusion, which consists of Smitty's personal letter to Chairman Mao of the People's Republic of China, Roth conveys the solemnity of the closing message more forcefully than its comic aspects, for his point is utterly serious. Smitty concludes:

[My belief is in] art—an art, sir, not for its own sake, or the sake of national pride or personal renown, but art for the sake of the record, an art that reclaims what is and was from those whose every word is a falsification and a betrayal of the truth. "In battle with the lie," said Alexander I. Solzhenitsyn, "art has always been victorious, always wins out, visibly, incontrovertibly for all! The lie can stand against much in the world—but not against art." (416)

The passage shows Smitty, as both imaginary character and surrogate author, seeking the honesty of historical facts as the sine qua non of moral truth in society. Like most of the protagonists of baseball fiction, and their creators, he seeks the perfect game of art to record both the fulfillment and the betrayal of the American Dream.

In addition, the impassioned, painstakingly argued letter to Mao returns us to "the plight of the artist, fans," a subject that reminds us again of the *Invisible Man* symbol. What is there in this contaminated, fallen world for the seer of truth, Smitty asks throughout his work in progress, *the novel in our hands*. His existential answer, like that of Ellison's protagonist is: nothing. "But I will wait. I will wait, and I will wait, and I will wait. And need I tell you what that's like for a man without the time for waiting, or the temperament?" (414). After all the humor, satire, and irony, we are left with "the simple, stark fact" of Smitty alone—like J. Henry Waugh and Oscar Petashne at the end of their respective narratives—awaiting his ineluctable end, "and it does not seem...a cause for celebration either" (Roth, *Writing American Fiction*, 158).

CONCLUSION

What we are left with, to paraphrase Philip Roth, at the end of this first book-length critical survey of baseball fiction, leaves ample cause for celebration. The works discussed here (and an assortment of others which, for reasons of space or theme, were not included) testify to the enduring importance of baseball in the dynamic reality of American life. They also, in their respective insights, approaches, frequent eloquence, and occasional brilliance, attest to the persistent genius of the national literature in serving as conscience of the nation, in daring to cast an omniscient eye over the forest whole to show the blight of the trees as well as their beauty, our destruction of flora and fauna as well as our respect for their pristine nature.

Defined as poetry and prose literature that derives its primary subject and controlling metaphors from the sport of baseball, its history, or its particular ambience, baseball fiction occupies a valuable place in our literary and cultural history. *Seeking the Perfect Game: Baseball in American Literature* demonstrates the genre's importance through a survey of the baseball imagery and symbol in primarily American novelistic fiction, and through its thesis that just as the game has evolved from primitive ritual and folk play, to sport and business, and finally to art and literature, the genre of baseball fiction has itself evolved in parallel fashion. Its development has been from subliterary treatments sentimentalizing the game, such as the work of Gilbert Patten; to greater realism, as in Ring Lardner, and greater literary sophistication as in Bernard Malamud; to multidimensional metafictions encapsulating baseball as fiction within fiction, for example the work of Coover and Roth.

Moreover, the nine characteristics of baseball that recur as definitive motifs shape the contours of the genre, as the material analyzed in this study illustrates. *The Rio Loja Ringmaster* and *The Bingo Long Traveling All-Stars and Motor Kings*, for example, exemplify the first three characteristics concerning the sport's aboriginal roots, folk development, and basic restorative function. The fourth and fifth traits, regarding the game's fulfillment of agonistic human impulses and its early American history as gentlemanly activity, appear in *The Year the Yankees Lost the Pennant* and *The Celebrant*, among others. The false genesis legend of baseball's origin, the sixth characteristic, is milked by every major and most minor authors writing baseball fiction (for example, *The Great American Novel*, *Babe Ruth Caught in a Snowstorm*, and *You Know Me Al*). *The Natural* and *The Great American Novel* join several important titles capturing traits seven and eight concerning baseball's ingrained links to American commerce and politics. Likewise, *The Universal Baseball Association, Inc., J. Henry Waugh, Prop.* and the Henry "Author" Wiggen books present especially well the ninth trait, that baseball *as baseball* is uniquely conducive to extensive literary treatment.

Ultimately, I hope that this study explains its title by demonstrating that baseball's ideal of the perfect game connotes with peculiar precision the literary and moral search for truth about America's history and society that occupies most of the writers discussed in these pages. That search, or more precisely, that *seeking*, reflects the inspiration underlying much of our country's artistic and intellectual life for over three centuries. The success of their search depends upon our comprehension of it.

REFERENCES

Roger Angell, *The Summer Game* (Popular Library, 1962).

Baseball Encyclopedia, 6th ed. (Macmillan, 1985).

Neil David Berman, *Playful Fictions and Fictional Players: Game, Sport, and Survival in Contemporary American Fiction* (Kennikat Press, 1980).

Peter Bjarkman, "The Glorious Diamond," manuscript, 1986.

Robert S. Boikess and Edward Edelson, *Chemical Principles*, 2nd ed. (Harper and Row, 1981).

Thomas Boswell, *How Life Imitates the World Series*, 1983.

Tristram Coffin, *The Old Ball Game*, 1971.

Robert Coover, *Pricksongs and Descants* (Dutton, 1966).

———. *The Universal Baseball Association, Inc., J. Henry Waugh, Prop.* (Random House, 1968).

———. *In First Person* (Union College Press, 1973).

Jacques Ehrmann, "Game, Play, Literature," *Yale French Studies*, 41 (1968).

Alexander Eliot, *Myths* (McGraw-Hill, 1976).

Ralph Ellison, *Invisible Man* (Random House, 1952).

Arlen J. Hansen, "The Dice of God: Einstein, Heisenberg, and Robert Coover," *Novel* (Fall 1976): 49–58.

Ihab Hassan, *Radical Innocence: Studies in the Contemporary American Novel* (Princeton, 1961).

Werner Heisenberg, *Physics and Philosophy* (Harper and Row, 1958).

Robert Henderson, *Ball, Bat and Bishop*, 1947.

Lamar Herrin, *The Río Loja Ringmaster*, 1977.

Johan Huizinga, *Homo Ludens*, 1955.

Bill James, *The Bill James Baseball Abstract*, 1986.

Patrick A. Knisley, "The Interior Diamond," 1978.

Sandy Koufax quoted in Thomas Boswell, *How Life Imitates the World Series*, 1983.

John Nobel McDaniel, *The Fiction of Philip Roth* (Haddonfield House, 1974).

Christian K. Messenger, *Sport and the Spirit of Play in American Fiction*, 1981.

The New Larousse Encyclopedia of Mythology (Prometheus Press, 1972).

Michael Novak, *The Joy of Sports* (Basic Books, 1976).

Peter S. Prescott, "At the Old Ball Game," *Newsweek*, May 14, 1973.

Bernard F. Rodgers, Jr., *Philip Roth*, (Twayne, 1978).

Philip Roth, *The Great American Novel* (Bantam, 1973).

———. "Reading Myself," *Partisan Review*, 40 (1973): 404–417.

———. "Writing American Fiction" in *The American Novel Since World War II* (Fawcett, 1969): 142–58.

John Steinbeck, "And Then My Arm Glassed Up," *Sports Illustrated*, 23 (December 20, 1965).

Tony Tanner, *City of Words*, 1971.

John Thorn and Pete Palmer with David Reuther, *The Hidden Game of Baseball: A Revolutionary Approach to Baseball and Its Statistic* (Doubleday, 1984).

University Physics, Sixth Edition, F. W. Sears, M. W. Zemansky, and H. D. Young (Addison-Wesley, 1984).

Paul Weiss, *Sport: A Philosophic Inquiry*, (Southern Illinois University, 1969).

BIBLIOGRAPHY

PRIMARY SOURCES

Fiction

Alger, Horatio. *Ragged Dick*. 1867ff. New York: Collier, 1962.

Algren, Nelson. *Never Come Morning*. 1941. New York: Harper & Row, 1963.

Asinof, Eliot. *Man on Spikes*. New York: McGraw-Hill, 1955.

Beaumont, Gerald. *Hearts and the Diamond*. New York, 1921.

Brady, Charles. *Seven Games in October*. Boston: Little, Brown, 1979.

Brashler, William. *The Bingo Long Traveling All-Stars and Motor Kings*. New York: Harper and Row, 1973.

Broun, Heywood. *The Sun Field*. New York: G. P. Putnam's Sons, 1923.

Carkeet, David. *The Greatest Slump of All Time*. New York: Harper & Row, 1984.

Charyn, Jerome. *The Seventh Babe*. New York: Arbor House, 1979.

Coover, Robert. "McDuff on the Mound." *Iowa Review* 2 (Fall 1971): 111–20.

———. *The Origin of the Brunists*. New York: Putnam, 1966.

———. *Pricksongs and Descants*. New York: Dutton, 1969.

———. *The Universal Baseball Association, Inc., J. Henry Waugh, Prop*. New York: Random House, 1968.

Davies, Valentine. *It Happens Every Spring*. New York: Farrar, Straus, 1949.

Farrell, James T. *Father and Son*. Cleveland: World, 1940.

———. *Invisible Swords*. New York: Doubleday, 1971.

———. *My Days of Anger*. Cleveland: World, 1943.

———. *No Star Is Lost*. New York: Vanguard, 1938.

———. *Studs Lonigan Trilogy*. New York: Random House, 1938.

———. *A World I Never Made*. New York: Vanguard, 1936.

Fitzgerald, F. Scott. *The Great Gatsby*. New York: Scribner's, 1925.

Graber, Ralph S., ed. *The Baseball Reader*. New York: A. S. Barnes, 1951.

Graham, John Alexander. *Babe Ruth Caught in a Snowstorm*. New York: Houghton Mifflin, 1973.

Greenberg, Eric Rolfe. *The Celebrant*. 1983. New York: Penguin, 1986.

Harris, Mark. *Bang the Drum Slowly*. New York: Knopf, 1956.

———. *It Looked Like For Ever*. New York: McGraw-Hill, 1979.

———. *The Southpaw*. New York: Knopf, 1953.

———. *A Ticket for a Seamstitch*. New York: Knopf, 1957.

Hawthorne, Nathaniel. *Selected Tales and Sketches*. 3rd ed. Hyatt Waggoner, ed. New York: Holt, Rinehart & Winston, 1970.

Hemingway, Ernest. *The Old Man and the Sea*. New York: Scribner's, 1952.

Herrin, Lamar. *The Río Loja Ringmaster*. New York: Viking, 1977.

Higdon, Hal. *The Horse That Played Center Field*. New York: Holt, Rinehart & Winston, 1969.

Kinsella, W. P. *The Iowa Baseball Confederacy*. New York: Houghton Mifflin, 1986.

———. *Shoeless Joe*. Boston: Houghton Miflin, 1982.

Lardner, Ring[gold Wilmerding]. *First and Last*. New York: Scribner's, 1938.

———. *The Portable Ring Lardner*. Gilbert Seldes, ed. New York: Viking, 1946.

———. *The Ring Lardner Reader*. Maxwell Geismar, ed. New York: Scribner's, 1963.

———. *Round Up*. 1924. New York: Scribner's, 1929.

———. *You Know Me Al*. 1914. New York: Scribner's, 1960.

Lewis, Sinclair. *Babbitt*. New York: Harcourt, Brace & World, 1922.

Malamud, Bernard. *The Natural*. New York: Farrar, Straus and Giroux, 1952.

Malory, Sir Thomas. *The Morte Darthur*. Charles R. Sanders and Charles E. Ward, eds. New York: Holt, Rinehart & Winston, 1970.

Maxwell, James. "Shine Ball." *The New Yorker*, Oct. 7, 1950: 39–60.

Mayer, Robert. *The Grace of Shortstops*. Garden City, NY: Doubleday, 1984.

Molloy, Paul. *A Pennant for the Kremlin*. New York: Doubleday, 1964.

Morgenstein, Gary. *Take Me Out to the Ballgame*. New York: St. Martin's, 1980.

———. *The Man Who Wanted to Play Centerfield for the New York Yankees*. New York: Atheneum, 1983.

Morris, Wright. *The Huge Season*. Lincoln: University of Nebraska Press, 1954.

Neugeboren, Jay. *Before My Life Began*. New York: Simon and Schuster, 1985.

———. *Corky's Brother*. New York: Farrar, Straus and Giroux, 1964.

———. *Listen Ruben Fontanez*. Boston: Houghton Mifflin, 1968.

———. *An Orphan's Tale*. New York: Holt, Rinehart & Winston, 1976.

———. *Parentheses: An Autobiographical Journey*. New York: Holt, Rinehart & Winston, 1972.

———. *Sam's Legacy*. New York: Holt, Rinehart & Winston, 1973.

Norris, Frank. "This Animal of a Buldy Jones." 1897. In *Collected Stories*. New York: Doran, 1928.

Nye, Bud. *Stay Loose*. New York: Doubleday, 1959.

Proust, Marcel. *Remembrance of Things Past*. 2 vols. C. K. Scott Moncrieff and Terence Kilmartin, trans. New York: Random House, 1982.

Quigley, Martin. *Today's Game*. New York: Viking, 1965.

Rheinheimer, Kurt. "Umpire" in *New Stories from the South*, S. Ravenal, ed. Chapel Hill: Algonquin Books, 1986.

Richler, Mordecai. *St. Urbain's Horseman*. New York: Knopf, 1971.

Roth, Philip. *The Great American Novel*. New York: Holt, Rinehart, & Winston, 1973.

———. *Portnoy's Complaint*. New York: Random House, 1969.

Runyon, Damon. "Baseball Hattie." In *Take it Easy*. New York, 1931.

Shaw, Irwin. *Voices of a Summer Day*. New York: Delacorte Press, 1965.

Smith, H. Allen. *Rhubarb*. New York: Doubleday, 1946.

Stein, Harry. *Hoopla*. New York: Knopf, 1983.

Thurber, James. "You Could Look It Up." *Saturday Evening Post*, Sept. 14, 1942: 19–25.

Twain, Mark. *A Connecticut Yankee in King Arthur's Court*. 1889. New York: Hill and Wang, 1960.

Updike, John. *Rabbit Redux*. New York: Knopf, 1971.

———. *Rabbit Run*. New York: Knopf, 1960.

Van Loan, Charles E. *The Lucky Seventh*. Boston: Small, Maynard, 1913.

Wallop, Douglass. *The Year the Yankees Lost the Pennant*. New York: Norton, 1954.

Warren, Robert Penn. "Goodwood Comes Back." In *The Circus in the Attic and Other Short Stories*. New York: Harcourt, Brace, 1947.

Willard, Nancy. *Things Invisible to See*. New York: Knopf, 1985.

Wolfe, Thomas. *The Letters of Thomas Wolfe*. Elizabeth Nowell, ed. New York: Scribner's, 1956.

———. *Look Homeward Angel*. New York: Scribner's, 1929.

———. *Of Time and the River*. New York: Scribner's, 1935.

———. *The Web and the Rock*. New York: Harper & Brothers, 1940.

———. *You Can't Go Home Again*. New York: Harper & Brothers, 1940.

Poetry

Adams, Franklin P. "The Ball Game." In Edward Lyman, *Baseball Fanthology: Hits and Skits of the Game*. New York: Scribner's, 1924.

———. "Baseball Note." *The New Yorker* (Feb. 19, 1938): 699.

———. "Baseball's Sad Lexicon." In *The Melancholy Lute*. 1936. New York: Dover Publications, 1962.

Anonymous. "Base-ball" and "Poisoned Ball (Baseball)" in Robert W. Henderson's *Ball, Bat and Bishop*. 1744. New York, 1947.

Anonymous. "Fan Letter" in Lawrence Ritter's *The Glory of Their Times*. New York, 1966.

Corso, Gregory. "Dream of a Baseball Star." In *The Happy Birthday of Death*. New York: New Directions, 1960.

Creeley, Robert. "The Ball Game." In *For Love Poems 1950–1960*. New York, 1962.

Deutsch, Jordan A. "The Great Mississippi." In *The Sports Encyclopedia: Baseball*, David S. Neff et al., eds. New York, 1974.

Ebert, P. K. and R. R. Knudson, eds. *Sports Poems*. New York: Dell, 1956.

Fitzgerald, Robert. "Cobb Would Have Caught It." In *In The Rose of Time*. New York: New Directions, 1943.

Fogel, Aaron. "Lou Gherig [sic]." *Columbia Forum* 13.4 (Winter 1970): 22.

Francis, Robert. "The Base Stealer" and "Pitcher." In *The Orb Weaver*. Middletown, CT: Wesleyan University Press, 1953.

Holden, Jonathan. "A Personal History of the Curveball." *Harper's Magazine* (April 1986): 32–33.

Humphries, Rolfe. "Junior" and "Polo Grounds." In *Collected Poems of Rolfe Humphries*. Bloomington: University of Indiana Press, 1965.

Johnson, Allen. Untitled article. In Harold Seymour, *Baseball: The Golden Age*. New York, Oxford University Press, 1971.

Kirk, William. "Sunday Baseball." *Baseball Magazine* (May 1980): 29.

Koch, Kenneth. *Ko or A Season on Earth*. New York: Grove, 1959.

Metzger, Deena. "Little League Women." 1969. In *Sports in Literature*, Henry B. Chapin, ed. New York: David McKay, 1976.

Moore, Marianne. "Hometown Pieces for Messrs. Alston and Reese." In *The Complete Poems of Marianne Moore*. New York: Viking, 1959.

Patchen, Kenneth. "The Origin of Baseball." In *Selected Poems of Kenneth Patchen*. New York: New Directions, 1942.

Rogin, Gilbert. "Spring." In *The Spectacle of Sport*. Englewood Cliffs, NJ: Sports Illustrated, 1957.

Sanchez, Sonia. "on watching a world series game." In *We A BadDDD People*. New York: Broadside Press, 1970.

Schwartz, Delmore. "Genesis." *American Poetry Review* (Jan.-Feb. 1974).

Swenson, May. "Analysis of Baseball." 1971. In *Sports in Literature*, Henry B. Chapin, ed. New York: David McKay, 1976.

Thayer, Ernest Lawrence. *The Annotated "Casey at the Bat."* 1888. Martin Gardner, ed. New York: Clarkson N. Potter, Inc., 1967.

Updike, John. "Tao in the Yankee Stadium Bleachers." In *The Carpentered Hen and Other Tame Creatures*. New York: Harper & Row, 1956.

Wallace, Robert. "The Double Play." 1961. In *Sports in Literature*, Henry B. Chapin, ed. New York: David McKay, 1976.

Williams, William Carlos. "At the Ball Game." In *Selected Poems*. New York: New Directions, 1949.

Juvenile

(Readers are advised to consult the excellent "Checklist of American Sports Fiction" in Michael Oriard's *Dreaming of Heroes* [See below Secondary Sources] for thorough listings of juvenile baseball fiction through 1980.)

Archibald, Joe. *Bonus Kid*. Philadelphia: Macrae Smith, 1959.

———. *Centerfield Rival*. Philadelphia: Macrae Smith, 1974.

———. *Outfield Orphan*. Philadelphia: Macrae Smith, 1961.

———. *Payoff Pitch*. Philadelphia: Macrae Smith, 1971.

———. *Right Field Rookie*. Philadelphia: Macrae Smith, 1967.

———. *Southpaw Speed*. Philadelphia: Macrae Smith, 1966.

Barbour, Ralph Henry. *For the Honor of the School*. New York: Appleton, 1900.

———. *Weatherby's Inning*. New York: Appleton, 1903.

Bishop, Curtis. *Lank of the Little League*. New York: Lippincott, 1958.

———. *Larry Comes Home*. Austin, Tex.: Steck, 1955.

———. *Larry Leads Off*. Austin, Tex.: Steck, 1954.

———. *Larry of the Little League*. Austin, Tex.: Steck, 1953.

———. *Little League Amigo*. New York: Lippincott, 1964.

———. *Little League Little Brother*. New York: Lippincott, 1968.

———. *Little League Visitor*. New York: Lippincott, 1966.

Carol, Bill J. *Clutch Single*. Austin, Tex.: Steck, 1963.

Chadwick, Lester. See Stratemeyer, Edward.

Christopher, Matthew. *Baseball Flyhawk*. New York: Little, Brown, 1963.

———. *Baseball Pals*. New York: Little, Brown, 1956.

———. *Catcher With a Glass Arm*. New York: Little, Brown, 1964.

———. *Mystery Coach*. New York: Little, Brown, 1964.

———. *Shortstop from Tokyo*. New York: Little, Brown, 1970.

———. *The Year Mom Won The Pennant*. New York: Little, Brown, 1968.

Corbett, Scott. *The Baseball Bargain*. New York: Little, Brown, 1970.

———. *Baseball Trick*. New York: Little, Brown, 1965.

———. *The Home Run Trick*. New York: Little, Brown, 1973.

Cox, William R. *Big League Sandlotters*. New York: Dodd, 1971.

———. *Chicano Cruz*. New York: Bantam, 1972.

———. *Trouble at Second Base*. New York: Dodd, 1966.

Creighton, Don. *Secret Little Leaguer*. New York: Hale, 1966.

Grey, Zane. *The Redheaded Outfield and Other Baseball Stories*. New York: Grosset and Dunlap, 1920.

———. *The Short Stop*. Chicago: A. C. McClurg, 1909.

Jackson, C. Paul. *Big League in The Small League*. New York: Hastings, 1968.

———. *Little Leaguer's First Uniform*. New York: Crowell, 1952.

———. *Little Major Leaguer*. New York: Hastings, 1963.

———. *Tom Mosely, Midget Leaguer*. New York: Hastings, 1970.

Johnson, Owen. *The Hummingbird*. New York: Baker and Taylor, 1910.

Knott, Bill. *Junk Pitcher*. Chicago: Follett, 1963.

Lord, Beman. *Bats and Balls*. New York: Walck, 1962.

———. *Perfect Pitch*. New York: Walck, 1965.

———. *The Trouble With Francis, Shortstop*. New York: Walck, 1958.

McCormick, Wilfred. The Bronc Burnett Series of Baseball Books.

———. The Rocky McCune Series of Baseball Books.

Olson, Gene. *Bonus Boy*. New York: Dodd, 1963.

Pallas, Norvin. *The Baseball Mystery*. New York: D. Washburn, 1963.

Patten, Gilbert. Big League Series of Baseball Stories.

———. College Life Series of Baseball Books.

———. The Lefty Locke Series of Baseball Books.

———. The Frank Merriwell Series of Baseball Books.

Russell, Patrick. *Going, Going, Gone*. New York: Doubleday, 1967.

Scholz, Jackson. *Batter Up*. New York: Morrow, 1946.

———. *Big Mitt*. New York: Morrow, 1968.

———. *Dugout Tycoon*. New York: Morrow, 1963.

———. *Fielder from Nowhere*. New York: Morrow, 1948.

———. *The Perfect Game*. New York: Morrow, 1959.

———. *Spark Plug at Short*. New York: Morrow, 1966.

Standish, Burt L. See Patten, Gilbert.

Stratemeyer, Edward. The Baseball Joe Series of Baseball Books.

Tunis, John R. *Buddy and the Old Pro*. New York: Morrow, 1955.
————. *The Kid from Tomkinsville*. New York: Harcourt, Brace & World, 1940.

SECONDARY SOURCES

Literary Criticism

Bachner, Saul. "Baseball as Literature: *Bang the Drum Slowly*." *English Record* 25.2 (1974).
Bank, Stanley. *American Romanticism: A Shape for Fiction*. New York: Putnam, 1969.
Baumbach, Jonathan. *The Landscape of Nightmare: Studies in the Contemporary American Novel*. New York: New York University Press, 1965.
Berkow, Ira. "Farrell and Sports." *Twentieth Century Literature* 22.1 (February 1976): 105–110.
Berman, Neil David. *Playful Fictions and Fictional Players: Game, Sport, and Survival in Contemporary American Fiction*. Port Washington, NY: Kennikat Press, 1980.
Berryman, John. "The Case of Ring Larnder: Art and Entertainment." *Commentary* 22 (July-Dec. 1956): 416–23.
Bjarkman, Peter. "The Glorious Diamond: Baseball in American Literature and American Culture." Manuscript.
Boon, James A. *From Symbolism to Structuralism: Levi-Strauss in a Literary Tradition*. New York: Harper & Row, 1972.
Brosnan, Jim. "The Fantasy World of Baseball." *Atlantic Monthly* (April 1964): 69–72.
Candelaria, Cordelia (Chávez). "Baseball in American Literature: From Ritual to Fiction." Ph.D. Diss. University of Notre Dame, 1976.
————. *Chicano Poetry, A Critical Introduction*. Westport, CT: Greenwood Press, 1986.
————. "A Decade of Ethnic Fiction by Jay Neugeboren." *MELUS* 5.4 (Winter 1978): 71–82.
————. "Jay Neugeboren." *Dictionary of Literary Biography*, vol. 28. Detroit: Gale Research, 1984:181–88.
————. "Literary Fungoes: Allusions to Baseball in American Fiction." *Midwest Quarterly* 23.4 (Summer 1982): 411–25.
————. "Seeking the Perfect Game in the Playfields of Fiction." *American Book Review* 9.2 (March-April 1987): 1–9.
Carothers, James B. "The Literature of Baseball," *American Studies*, 27 (Fall 1976): 111–114.
Chase, Richard. *The American Novel and Its Tradition*. Garden City, NY: Doubleday, 1957.
Coffin, Tristram Potter. *The Old Ball Game: Baseball in Folklore and Fiction*. New York: Herder and Herder, 1971.
Cook, Bruce. " 'Bang the Drum': Best Sports Film Ever Made?" *National Observer* (September 15, 1973), n.p.
Cooper, Arthur. "Murderers' Row." *Newsweek* (April 30, 1973): 83.
Cutler, John Levi. "Gilbert Patten and His Frank Merriwell Saga: A Study in Sub-Literary Fiction, 1896–1913." *Maine University Studies* 2.31 (1934).
Dahlberg, Edward. *Can These Bones Live*. New York: Harcourt, Brace & World, 1941.

DuCharme, Robert. "Art and Idea in the Novels of Bernard Malamud." Ph.D. diss., University of Notre Dame, 1970.

Ehrmann, Jacques. *Game, Play, Literature*. New Haven, CT: *Yale French Studies*, 41, 1985.

Elder, Donald. *Ring Lardner, A Biography*. New York: Doubleday, 1956.

Ellison, Ralph. *Invisible Man*. New York: Vintage Books, 1947.

———. *Shadow and Act*. New York: Random House, 1964.

Ellmann, Mary. *Thinking About Women*. New York: Harcourt Brace Jovanovich, 1968.

Evans, Walter. "The All-American Boys: A Study of Boys' Sports Fiction." *Journal of Popular Culture* 6 (Summer 1972): 104–21.

Farrell, James T. "Baseball As It's Played in Books—Some Jeers, Cheers and Hopes." *New York Times Review* (August 10, 1958): 5.

Fitzgerald, F. Scott. "Ring" (1933). In *The Crack-Up*. Edmund Wilson, ed. New York: New Directions, 1956.

Gado, Frank, ed. *First Person* [Interviews]. Schenectady, NY: Union College Press, 1973.

Gardner, Martin: See Thayer, Ernest Lawrence.

Garland, Hamlin. *Crumbling Idols*. New York: Macmillan, 1894.

Geismar, Maxwell. *Ring Lardner and the Portrait of Folly*. New York: Crowell, 1972.

Gilman, Richard. "Ball Five: A Review of *The Great American Novel*." *Partisan Review* 40 (1973): 467–71.

Graber, Ralph S. "Baseball in American Fiction." *English Journal* 56.8 (November 1967): 1107–1114.

Grobani, Anton, ed. *Guide to Baseball Literature*. Detroit: Gale Research, 1975.

Hansen, Arlen J. "The Dice of God: Einstein, Heisenberg, and Robert Coover," *Novel* (Fall 1976): 49–58.

Hardwicke, Elizabeth. *Seduction and Betrayal: Women and Literature*. 1970. New York: Random House, 1974.

Harris, Mark. "Easy Does It Not." In *The Living Novel: A Symposium*, Granville Hicks, ed. New York: Collier Books, 1962.

———. "Home Is Where We Backed Into," *Nation* (November 1, 1958): 322–24.

Hassan, Ihab. *Radical Innocence: Studies in the Contemporary American Novel*. Princeton, NJ: Princeton University Press, 1961.

Heckard, Margaret. "Robert Coover, Metafiction, and Freedom." *Twentieth Century Literature* 11.1 (May 1976): 210–26.

Johannsen, Albert. *The House of Beadle and Adams*. 2 vols. Norman: University of Oklahoma Press, 1950.

Knisley, Patrick Allen. "The Interior Diamond: Baseball in Twentieth-Century American Poetry and Fiction." Ph.D. diss. University of Colorado at Boulder, 1978.

Lavers, Norman. *Mark Harris*, Boston: G. K. Hall, 1978.

Leonard, John. "Cheever to Roth to Malamud." *Atlantic* 231.6 (June 1973): 112–116.

Lewis, R.W.B. *The American Adam: Innocence, Tragedy and Tradition in the Nineteenth Century*. Chicago: University of Chicago Press, 1955.

McDaniel, John Nobel. *The Fiction of Philip Roth*. Haddonfield, NJ: Haddonfield House, 1974.

Mellard, James M. "Malamud's Novels: Four Versions of the Pastoral." *Critique* 9.2 (1967): 43–61.

Mencken, H. L. *The American Language*. 1919. 4th ed. New York: Knopf, 1955.

Messenger, Christian K. *Sport and the Spirit of Play in American Fiction: Hawthorne to Faulkner*. New York: Columbia University Press, 1981.

Moore, Marianne. "Baseball and Writing." In *Collected Poems*. New York: Macmillan, 1968.

———. "Some Questions on Sports." In *Ten Answers: Letters from an October Afternoon, Part II*. New York: Lawrence E. Brinn and Louise Crane, 1965.

Murdock, Eugene C. *Mighty Casey: All-American*. Westport, CT: Greenwood Press, 1984.

The New York Times Film Review 1913–1968. 6 vols. New York: Times and Arno Press, 1970.

Oriard, Michael. *Dreaming of Heroes: American Sports Fiction, 1868–1980*. Chicago: Nelson-Hall, 1982.

Patrick, Walton R. *Ring Lardner*. New York: Twayne, 1963.

Patten, Gilbert. *Frank Merriwell's Father: An Autobiography by Gilbert Patten*. Norman: University of Oklahoma Press, 1964.

Podhoretz, Norman. "Achilles in Left Field." *Commentary* 15 (March 1953): 321–26.

Prescott, Peter S. "At the Old Ball Game," *Newsweek* (May 14, 1973).

Rahv, Philip, ed. *A Malamud Reader*. New York: Farrar, Straus and Giroux, 1967.

Richman, Sidney. *Bernard Malamud*. New York: Twayne, 1966.

Rodgers, Bernard F., Jr. *Philip Roth*. Boston: Twayne, 1978.

Rogers, Katharine M. *The Troublesome Helpmate: A History of Misogyny in Literature*. Seattle: University of Washington Press, 1966.

Ross, Murray. "Football Red and Baseball Green," *Chicago Review* (Jan.-Feb. 1971): 149–157.

Roth, Philip. "Reading Myself," *Partisan Review* 40 (1973): 404–17.

———. "Writing American Fiction." In *The American Novel Since World War II*, Marcus Klein, ed. Westport, CT: Greenwood Press, 1969: 142–158.

Rourke, Constance. *American Humor*. New York: Harcourt, Brace & World, 1931.

Rupp, Richard H. *Celebration in Postwar American Fiction 1945–1967*. Coral Gables, FL: University of Miami Press, 1970.

Schultz, Max F. *Black Humor Fiction of the Sixties*. Athens: Ohio University Press, 1973.

Seldes, Gilbert, ed. *The Portable Ring Lardner*. New York: Viking, 1946.

Shelton, Frank W. "Humor and Balance in Coover's *The Universal Baseball Association, Inc.*" *Critique* 17 (Sept. 1975): 78–90.

Simpson, Claude M., ed. *The Local Colorists: American Short Stories 1857–1900*. New York: Harper & Brothers, 1960.

Smith, Leverett. "Ty Cobb, Babe Ruth and the Changing Image of the Athlete Hero." In *Heroes of Popular Culture*; Ray B. Browne et al., eds. Bowling Green, OH: Bowling Green University Press, 1972.

Stein, Allen F. "This Unsporting Life: The Baseball Fiction of Ring Lardner," *Markham Review*, 3 (Oct. 1971): 27–33.

Tanner, Tony. *City of Words: American Fiction*. New York: Harper & Row, 1971.

Taylor, Mark. "Baseball as Myth." *Commonweal* 96.10 (May 12, 1972): 237–39.

Turner, Frederick W. "Myth Inside and Out: Malamud's *The Natural*." *Novel* 1.2 (Winter 1968): 133–39.

Wakefield, Dan. "The Purple and the Gold." *Atlantic Monthly*, 234 (August, 1974): 84.

Wasserman, Earl R. "*The Natural*: Malamud's World Ceres." *Centennial Review* 9.4 (Fall 1965): 438–60.

Weinberg, Helen. *The New Novel in America: The Kafkan Mode in Contemporary Fiction*. Ithaca, NY: Cornell University Press, 1970.

Wise, Suzanne. *Sports Fiction for Adults: An Annotated Bibliography of Novels, Plays, Short Stories, and Poetry with Sporting Settings*. New York: Garland, 1986.

Wolff, Geoffrey. "Eye on the Ball: A Review of *The Río Loja Ringmaster*." *New York Times Book Review* (Feb. 27, 1977): 26.

Woolf, Virginia. "American Fiction." In *Collected Essays*, vol. 2. New York: Harcourt, Brace & World, 1967.

Yardley, Jonathan. *Ring*. New York: Random House, 1977.

Baseball Material

Andreano, Ralph. *No Joy in Mudville: The Dilemma of Major League Baseball*. Cambridge, MA: Schenkman, 1965.

Angell, Roger. *The Summer Game*. New York: Popular Library, 1962.

Asinof, Eliot. *Eight Men Out: The Black Sox and the 1919 World Series*. New York: Holt, Rinehart, & Winston, 1963.

The Baseball Encyclopedia. 6th ed. Joseph L. Reichler, ed. New York: Macmillan, 1985.

Boswell, Thomas. *How Life Imitates the World Series*. New York: Penguin, 1983.

———. *Why Time Begins on Opening Day*. New York: Penguin, 1984.

Bouton, Jim. *Ball Four*. New York: World Publishing, 1970.

Boyd, B. C. *The Great American Baseball Card Flipping, Trading and Bubble Gum Book*. New York: Little, Brown, 1973.

Bueter, Robert J. "Sports, Values and Society." *Christian Century* 86.14 (April 5, 1972): 389–92.

Cohen, Marvin. *Baseball the Beautiful*. New York: Link Books, 1974.

Cook, Earnshaw. *Percentage Baseball*. Cambridge, MA: MIT Press, 1966.

Creamer, Robert W. *Babe: The Legend Comes to Life*. New York: Simon and Schuster, 1974.

Crepeau, Richard C. *Baseball: America's Diamond Mind 1919–1949*. Orlando: University of Central Florida Press, 1980.

Cummings, Parke. *The Dictionary of Baseball*. New York: A. S. Barnes, 1950.

Davis, John P., ed. *The American Negro Reference Book*. Englewood Cliffs, NJ: Prentice-Hall, 1966.

Drury, Joseph F., Sr. *The Fireside Book of Baseball*. New York: Simon and Schuster, 1956.

Farrell, James T. *My Baseball Diary*. New York: A. S. Barnes, 1957.

Fowles, John. "A Pitch for Cricket." *Sports Illustrated* 38.30 (May 21, 1973).

Geyer, Alan. "On the Sociological Dodgers." *Christian Century* 86.14 (April 5, 1972): 383.

Gmelch, Gordon. "Baseball Magic." *TransAction* 8 (June 1971): 128–37.

Grella George. "Baseball and the American Dream." *The Massachusetts Review* (April 1977).

Henderson, Robert W. *Ball, Bat and Bishop*. New York: Rockport Press, 1947; Detroit: Gale Research, 1974.

Jackson, Reggie. "We Have a Serious Problem that Isn't Going Away," *Sports Illustrated*, (May 11, 1987), 40–48.

James, Bill. *The Bill James Baseball Abstract*. New York: Ballantine, 1986.

Kahn, Roger. *The Boys of Summer*. New York: New American Library, 1971.

———. "Intellectuals and Ballplayers." *American Scholar* 26.3 (Summer 1957).

Lane, F. C. "A Fallen Idol," *Baseball Magazine* (November 1925): 556–66.

Lieb, Fred. *Baseball As I Have Known It*. New York: Grosset and Dunlap, 1977.

Mead, William. *Even the Browns: the Zany, True Story of Baseball in the Early Forties*. New York: Contemporary Books, 1978.

Meany, Tom. *Baseball's Greatest Teams*. New York: A. S. Barnes, 1949.

Musick, Phil. *Who Was Roberto? A Biography of Roberto Clemente*. Garden City, NY: Doubleday, 1974.

Neft, David S., et al. *The Sports Encyclopedia: Baseball*. New York: Sports Illustrated, 1974.

Peterson, Robert. *Only the Ball Was White*. Englewood Cliffs, NJ: Prentice-Hall, 1970.

Riess, Steven A. *Touching Base: Professional Baseball and American Culture in the Progressive Era*. Westport, CT: Greenwood Press, 1980.

Ritter, Lawrence. *The Glory of Their Times*. New York: Macmillan, 1966.

Rogosin, Donn. *Invisible Men: Life in Baseball's Negro Leagues*. New York: Atheneum, 1983.

Rosenberg, John M. *The Story of Baseball*. New York: Random House, 1962.

Senzel, Howard. *Baseball and the Cold War*. New York: Harcourt Brace Jovanovich, 1977.

Seymour, Harold. *Baseball: The Early Years*. New York: Oxford University Press, 1960.

———. *Baseball: The Golden Age*. New York: Oxford University Press, 1971.

Smelser, Marshall. "The Babe on Balance." *American Scholar* 44.2 (Spring 1975): 299–304.

———. *The Life that Ruth Built*. New York: Quadrangle, 1975.

Smith, Leverett T., Jr. *The American Dream and the National Game*. Bowling Green, KY: Popular Press, 1975.

Steinbeck, John. "And Then My Arm Glassed Up," *Sports Illustrated*, 23 (Dec. 20, 1965).

Suehsdorf, A. D. *The Great American Baseball Scrapbook*. New York: Random House, 1978.

Thorn, John, Pete Palmer and David Reuther. *The Hidden Game of Baseball: A Revolutionary Approach to Baseball and Its Statistics*. Garden City, NY: Doubleday, 1984.

Umphlett, Wiley L. *The Sporting Myth and the American Experience*. Cranbury, NJ: Associated University Press, 1975.

United States. Cong. House. Subcommittee on the Study of Monopoly Power. Hearings. 82nd Cong., 1st sess. Part 6, *Organized Baseball*. Washington: Government Printing Office, 1952.

Voigt, David Quentin. *American Baseball: From the Commissioners to Continental Expansion*. Norman: University of Oklahoma Press, 1970.

———. *American Baseball: From Gentleman's Sport to the Commissioner System*. Norman: University of Oklahoma Press, 1966.

Wagenheim, Kal. *Clemente!* Maplewood, NJ: Waterfront Press, 1984.

Wallop, Douglass. *Baseball: An Informal History*. New York: Norton, 1969.

Other

Barzun, Jacques. *God's Country and Mine*. New York: Vintage, 1954.

Boikess, Robert S. and Edward Edelson. *Chemical Principles*, 2nd ed. New York: Harper and Row, 1981.

Caillois, Roger. *Man, Play, and Games*. 1958. Meyer Barash, trans. New York: Free Press of Glencoe, 1961.

Cantwell, Robert. "Sport Was Box Office Poison." *Sports Illustrated* 15 (September 1969): 108–16.

Cozens, Frederick W. and Florence Scovil Stumpf. *Sports in American Life*. Chicago: University of Chicago Press, 1953.

Eliot, Alexander. *Myths*. NY and Chicago: McGraw-Hill, 1976.

Frazer, Sir James George. *The Golden Bough*. 1922. New York: Collier, 1963.

Harris, Mark. *Killing Everybody*. New York: Doubleday, 1973.

Heisenberg, Werner. *Physics and Philosophy: The Revolution in Modern Science*. New York: Harper and Row, 1958.

Higgs, Robert J. *Laurel & Thorn: The Athlete in American Literature*. Lexington: University of Kentucky Press, 1981.

Huizinga, Johan. *Homo Ludens: A Study of the Play Element in Culture*. 1950. Boston: Beacon Hill Press, 1955.

Klotman, Phyllis Rauch. *Frame by Frame—A Black Filmography*. Bloomington, IN: University of Indiana Press, 1979.

Levi-Strauss, Claude. *The Raw and the Cooked: Introduction to a Science of Mythology*. John and Doreen Weightman, transl. New York: Harper Torchbooks, 1969.

———. *The Savage Mind*. Chicago: University of Chicago Press, 1966.

———. *Structural Anthropology*. New York: Basic Books, 1963.

Massingham, H. J. "Origins of Ball Games." In *The Heritage of Man*. London, 1929.

McWilliams, Carey. *The Idea of Fraternity in America*. Berkeley: University of California Press, 1973.

New Larousse Encyclopedia of Mythology. 1959. London: Hamlyn, 1968.

Novak, Michael. *The Joy of Sports*. New York: Basic Books, 1976.

Reese, W. L. *Dictionary of Philosophy and Religion: Eastern and Western Thought*. Sussex, England: Harvester Press, 1980.

Spivey, Donald, ed. *Sport in America: New Historical Perspectives*. Westport CT: Greenwood Press, 1985.

Weiss, Paul. *Sport: A Philosophic Inquiry*. Carbondale: Southern Illinois University Press, 1969.

INDEX

I gratefully acknowledge the fine indexing assistance provided by my niece, Kirsten Eloida Kelly.

About the Author

CORDELIA CANDELARIA is Associate Professor of English and Chicano Studies at the University of Colorado, Boulder. Her publications include *Chicano Poetry: A Critical Introduction* (Greenwood Press, 1986), *Multiethnic Literature of the United States: Critical Essays and Classroom Resource*, a volume of poetry, and numerous articles on American and Chicano literature.